Reviews of Physiology, Biochemistry and Pharmacology 134

Springer-Verlag Berlin Heidelberg GmbH

Reviews of

134 Physiology Biochemistry and Pharmacology

Special Issue on Signal Transduction
in Smooth Muscle
Guest Editor: R. A. Murphy

Editors
M.P. Blaustein, Baltimore   R. Greger, Freiburg
H. Grunicke, Innsbruck   R. Jahn, Göttingen
W.J. Lederer, Baltimore   L.M. Mendell, Stony Brook
A.Miyajima, Tokyo   D. Pette, Konstanz   G. Schultz,
Berlin   M. Schweiger, Berlin

With 50 Figures and 1 Table

Springer

Guest Editor
Dr. Richard A. Murphy
University of Virginia Health Sciences Center
Department of Molecular Physiology and Biological Physics
P.O. Box 10011
229060011 Charlottesville, Virginia
USA

ISSN 0303-4240

ISBN 978-3-662-31200-1   ISBN 978-3-540-68932-4 (eBook)

DOI 10.1007/978-3-540-68932-4

Library of Congress-Catalog-Card Number 74-3674

© Springer -Verlag Berlin Heidelberg 1999

Originally published by Springer-Verlag Berlin Heidelberg New York in 1999.
Softcover reprint of the hardcover 1st edition 1999

The use of general descriptive names, registered names, trademarks, etc. in this publication does not imply, even in the absence of a specific statement, that such names are exempt from the relevant protective laws and regulations and therefore free for general use.

Product liability: The publishers cannot guarantee the accuracy of any information about dosage and application contained in this book. In every individual case the user must check such information by consulting the relevant literature.

Production: PRO EDIT GmbH, D-69126 Heidelberg
SPIN: 10551736   27/3136-5 4 3 2 1 0 – Printed on acid-free paper

Contents

Indexed in Current Contents

Signal Transduction and Regulation in Smooth Muscle: Problems and Progress

Richard A. Murphy

Department of Molecular Physiology and Biological Physics, Health Sciences Center, University of Virginia, Charlottesville, Virginia 22906, USA

Inappropriate responses of smooth muscle contribute to most morbidity and mortality in developed countries including hypertension, atherosclerosis, coronary and cerebral vasospasm, asthma, and a variety of gastrointestinal and urogenital diseases. Thus, the medical significance of smooth muscle pathophysiology is unquestioned. Nevertheless, smooth muscle is often regarded as unsuitable for serious studies of muscle function. The reviews that constitute this issue show this view is no longer true. The fact remains, however, that in many respects our understanding of smooth muscle is at a point attained some thirty years ago for skeletal muscle. *It is generally accepted that the smooth muscle myosin motor isoforms and chemo-mechanical transduction are described by the sliding-filament/cross-bridge paradigm. However, signal transduction and regulation of cross-bridge cycling in smooth muscle differ from that in striated muscle. This confers unique properties on smooth muscle that are not fully understood, such as the capacity to reduce cross-bridge cycling rates and ATP consumption in sustained contractions (Latch).* The reviews by Gunst and by Pfitzer and Arner provide an extensive discussion of the unique mechanical and energetic properties of smooth muscle. The reasons for the slow progress in studies of smooth muscle function merit consideration as a background for evaluating developments in the field. Arguably there are three major reasons for the slow pace. One of these categories is well recognized, but the others remain impediments to progress.

The recognized difficulties include the diversity of smooth muscle which has an important role in the function of most organ systems. The field is split into airway, urogenital, vascular, visceral and other specialties with a corresponding fragmentation of publications, meetings, and investigator interactions. Rather than a single final extracellular input regulating contraction (a motor nerve), there is a vast and still imperfectly identified array of extracellular signals. These include many different excitatory and inhibitory neurocrine, endocrine, paracrine and autocrine signals; as well as mechanical stimuli, electrical and chemical coupling of cells, and a notable fraction of all drugs in the pharmacopoeia. A single signal transduction event at the cell membrane (neuromuscular transmission at the motor end plate) is replaced by a profusion of receptors, channels, active and passive transport mechanisms and signal transduction cascades that provide great functional diversity. This complex area is reviewed by Kotlikoff et al. Almost all smooth muscle is found in inconvenient packages for experimental analysis: tissues typically include multiple layers of smooth muscle cells and connective tissue elements with varying orientations, and a variety of other cell types. Furthermore, two dimensional imaging techniques have failed to provide the structural insights in the absence of a uniform sarcomeric structure that underlie the sliding filament/cross-bridge paradigm for striated muscle. Finally, smooth muscle cells are not terminally differentiated, and exhibit considerable phenotypic variability within a tissue; a topic reviewed by Sartore et al. This variability can be extreme *in vitro* or pathologically.

A second category of factors impacting progress in understanding smooth muscle contractile function involves assumptions on the applicability of paradigms derived from skeletal muscle. Both muscle types are specialized for the expression of high external forces generated by myosin II motors coupled through a myofilament/cytoskeletal structure. Nevertheless, the function of the two cell types basically differs even though contraction can be characterized by similar mechanical (force-length and velocity-load) and energetic parameters. The functional unit of a skeletal muscle is a cell, or more properly, the group of cells comprising a motor unit. The cells are normally relaxed with most gravitational loads opposed by the skeleton. Activation is all or none, with gradation of force provided by recruitment and summation (tetanization). The output is typically characterized by a high power output matched by high ATP consumption rates and metabolic specialization between fiber types. Function is very different for smooth muscle in the walls of hollow organs. *In vivo*, the cells are normally active and play a skeletal role in maintaining organ dimensions against imposed loads. ATP consumption is minimized in the face of continuous contraction by very slow cross-bridge detachment rates as low power outputs are consistent with organ system function. Cells are not the basic contractile unit in smooth muscle and the cells are typically electrically, chemically, and mechanically coupled to form an integrated mechanical unit. Contractile output is varied by uniformly regulating activation in the linked cells: *in smooth muscle, cross-bridge recruitment and cycling rates are varied by regulation of the myoplasmic [Ca⁺⁺] and other mechanisms.* These are reviewed by Somlyo et al. Regulation of cross-bridge kinetics (and thereby correlated parameters such as shortening velocities, power output, and ATP consumption) confer unique mechanical and energetic properties, and are attributable to covalent regulatory mechanisms. A corollary is that the role of smooth muscle contractile protein isoforms in contributing to functional diversity is minimal compared with striated muscle.

A basic concept in striated muscle is activation: a measure of the output of the contractile system. Activation is simply assessed as steady state active stress or stiffness if a steady-state is not assured. In molecular terms activation estimates the number of mechanically coupled cross-bridges that can attach to the thin filaments and cycle as determined by the myoplasmic [Ca⁺⁺]. Since activation is physiologically all-or-none in skeletal muscle cells, activation is of little functional importance and now receives scant attention. *By contrast, activation of smooth muscles is a critical functional variable. The assessment of activation is fundamental to interpretation of virtually all experiments directed at elucidating the signal transduction processes and activation-contraction coupling.* Unfortunately there is no simple, unambiguous estimate of activation in smooth muscle. Stress and stiffness measurements do not discriminate between the contributions of more cycling cross-bridges and an enhanced duty cycle (slower detachment rates). Stress values are little changed when cross-bridge phosphorylation levels exceed some 25% even though shortening velocities, power output, or ATP consumption continue to increase in proportion to phosphorylation. Clearly a force transducer does not allow the accurate assessment of activation in the sense that it cannot fully reflect the biological response elicited by a stimulus. Cross-bridge phosphorylation appears to estimate activation under many circumstances, as the dependence of the mechanical output is a function of steady-state phosphorylation levels. However, phosphorylation determinations are destructive, time consuming, and subject to a long list of artifacts. Furthermore, while phosphorylation is generally a valid estimate of activation for intact smooth muscles exposed to neurotransmitters, hormones, and depolarization, there are situations in which the Ca⁺⁺-dependence of phosphorylation can be modified. Such changes in the Ca⁺⁺-sensitivity are reflections of changes in the ratio of myosin light chain kinase activity to phosphatase activity for a given [Ca^{++}], and this necessarily alters the mechanical response for a given level of phosphorylation. Both Pfitzer and Arner and Somlyo et al. address aspects of this subject in their reviews. No universally valid estimate of activation is available, and it should be recognized that force measurements (dose-response curves, etc.) in smooth muscles do not provide the same direct insights into cross-bridge function as in skeletal muscle.

The third factor negatively impacting progress in smooth muscle research is the fact that cross-bridge regulation and cross-bridge cycling are not experimentally separate issues. This differs from striated muscle where troponin acts as a thin filament Ca^{++} switch that enables cycling without affecting the kinetics of cycling. In smooth muscle, cross-bridge phosphorylation normally appears to be a requirement for attachment to thin filaments (i.e. recruitment, see Pfitzer and Arner). However, the kinetics of cross-bridge phosphorylation and dephosphorylation are comparable to the kinetics of cross-bridge cycling. Changes in the ratio of myosin light chain kinase activity to phosphatase activity modulate cycling rates, and unloaded shortening velocities are directly proportional to phosphorylation. *The classic reductionist techniques that were fundamental in the elucidation of the sliding filament/cross-bridge mechanism in striated muscle reduce or abolish regulation of cross-bridge kinetics. Yet this is the distinguishing characteristic of smooth muscle.* The latter property (Latch) depends on the presence of *in vivo* myosin light chain kinase and phosphatase activities. These enzymes are removed during purification of myosin and actin for biochemical studies or for use in motility assays including the new single myosin motor measurements. Enzyme activity loss also occurs in varying degrees in permeabilized tissues. The effect is to make phosphorylation a simple switch for cross-bridge recruitment with abolition of changes in cross-bridge cycling rates that depend on a dynamic flux of phosphate at the myosin regulatory light chain. ATPase activity becomes more or less directly proportional to phosphorylation as does force in some permeabilized tissues.

Another reductionist technique has undoubtedly distorted our understanding of *in vivo* function in smooth muscle. This is the far greater modulation of behavior associated with the study of intact smooth muscle *in vitro* compared with striated muscle. Provided that ATP production rates are not limited, isolated skeletal muscle cells exhibit unaltered contraction and relaxation in response to generation of action potentials. With few exceptions, isolation of smooth muscle greatly modifies the *in vivo* level of activation associated with *in vivo* extracellular signals. Most *in vitro* experimental studies are of the contraction elicited by a neurotransmitter, hormone or drug and the relaxation following its removal or receptor blockade. An explosion of recent research shows the importance of the nitric oxide (NO)-cGMP inhibitory pathway in virtually all smooth muscles, whether NO is generated as a neurotransmitter, released from endothelial or other cells, or generated by smooth muscle in response to receptor occmpancy. The absence of this inhibitory signal *in vitro* would be of limited significance if the effects of NO on muscle activation were limited to the well characterized reductions in the myoplasmic [Ca^{++}], reversing or limiting the normal excitatory pathways that determine cross-bridge phosphorylation rates. However, evidence is starting to accumulate that this pathway has other actions that alter the relationship between force and cross-bridge phosphorylation. *The implication is that some in vivo relaxation mechanisms are not simply the reversal of excitatory pathways.*

Single isolated skeletal muscle cells are important preparations for providing information about the basic contractile unit. Single smooth muscle cells are not physiological contractile units in the same sense, and isolation changes their properties. This includes proteolytic cleavage of the various junctions that serve both to couple the force-transmitting cytoskeleton from cell to cell or to connective tissue as well as the electrical and chemical coupling between cells. Isolation virtually abolishes the force-length behavior. Cells may shorten to a length where compressive forces are significant and produce the relaxation after a stimulus that super-contracts isolated, unloaded cells. Variable damage to integral membrane proteins including receptors and channels also occurs. Attempts to find cell or organ culture conditions that maintain the *in vivo* phenotype in a cell type that is not terminally differentiated have had very limited success. While isolated or cultured cells are important for certain electrophysiological studies (see Kotlikoff et al.), as well as determination of factors that may play a role in differentiation or proliferation (Sartore et al.), they do not provide a contractile model equivalent to the isolated skeletal muscle cell.

The objectives of the chapters in this issue are (*i*) to critically review current information on the mechanisms coupling extracellular regulatory signals to regulation of cross-bridge cycling and proliferation in smooth muscle, and (*ii*) identify significant gaps or unresolved issues that are important topics for future research. The experimental and analytical difficulties discussed above are increasingly recognized and surmounted.

Elucidation of the molecular and cellular events underlying the biological properties of smooth muscle is in the midst of a period of rapid progress. While the reviews reveal many gaps to be filled and illustrate areas of contention, they also capture the excitement of new discoveries.

Applicability of the Sliding Filament/Crossbridge Paradigm to Smooth Muscle

Susan J. Gunst, Ph.D.

Indiana University School of Medicine

Contents

1
Introduction

An evaluation of the mechanisms that regulate the mechanical behavior of smooth muscle necessarily must be based on interpretation of the physical interactions of the various structural components within the cells, and of their integrated role in mediating the transduction of chemical energy into mechanical function. The objective of this review is to describe our current state of knowledge with respect to the structure of the contractile apparatus of smooth muscle cells and the molecular and cellular basis of chemo-mechanical transduction. Great strides were made recently in our understanding of the molecular organization of the contractile filaments and of the mechanism for the transduction of chemical energy into movement by the myosin molecule. There have also been fundamental advances in the state of our knowledge of the molecular interactions that regulate the organization of contractile and cytoskeletal filaments within the smooth muscle cell. Despite these advances, we still do not have an integrated model of the molecular interactions of contractile and structural proteins that can account for the distinctive functional properties of smooth muscle cells and tissues.

The first section of this review will describe the general organization of the contractile apparatus of smooth muscle cells, and the basis for the transmission of force generated by the contractile apparatus across the smooth muscle cell membrane. In succeeding sections, the molecular structure of both the thick and thin filaments and of their organization in smooth muscle cells will be considered. The final section will evaluate the contractile behavior of smooth muscle in relation to our present knowledge of smooth muscle ultrastructure.

2
General organization of smooth muscle cells and tissues

Smooth muscle tissues occur ubiquitously throughout the body and are present in almost every organ, where they are subjected to varied mechanical, hormonal and neural stimuli. Although generalizations can be made regarding many features of smooth muscle, it should be recognized that great diversity exists in the structure and properties of smooth muscle cells and tissues. This diversity enables the functional properties of smooth muscle cells to be highly specialized to the physiologic needs of different organs and conditions. A survey of the known differences in the ultrastructural features of smooth muscle from different tissues is outside the scope of this review. However, some attempt will be made to point out important distinctions in the cellular structure of different smooth muscle tissues. This section will review the general organization of the filamentous structures that constitute the contractile apparatus of smooth muscle cells, as well as the structural basis for the transmission of force from the contractile apparatus to the exterior of the cell.

2.1
Structure of smooth muscle tissues

Most individual vertebrate smooth muscle cells are very small in comparison to skeletal muscle cells, approximately 100–200 μm in length by five μm in diameter. The volume of a smooth muscle cell is comparable to that of a monocyte, which is a sphere of approximately 19 μm in diameter (Gabella, 1990). (An exception to this is amphibian stomach smooth muscle where the cells may be 10 times larger). In the relaxed state, smooth muscle cells are usually long and spindle-shaped in form, with a high surface to volume ratio; however, they can undergo large changes in length and width during contraction.

Individual smooth muscle cells are usually densely packed within a tissue. The amount of extracellular space ranges from as little as 15% in some visceral smooth muscles to as much as 50 to 60% in some large arteries, such as the rat aortic media (Gabella 1990). Adjacent smooth muscle cells within a tissue exhibit junctions along the plasma membrane that enable them to be functionally coupled. Gap junctions, which are much more common in some muscle types than in others, provide hydrophilic channels that connect the cytoplasm of adjoining muscle cells, and allow for the exchange of ions and small molecules. Intermediate junctions, also called attachment plaques, are junctions of the plasmalemma of adjacent cells at the sites of attachment of the myofilaments. These junctions provide mechanical coupling between adjacent smooth muscle cells (see Section 2.3).

The overall structural configuration of groups of smooth muscle cells varies widely among different organs. In some tissues, such as the taenia coli, smooth muscle cells are grouped into cords or bands that run approximately parallel to the long axis of the tissue. The walls of many hollow organs, as well as of parts of the vascular system, are made up of multiple sheets or layers of smooth muscle cells, with adjacent layers sometimes running at different angles. The three-dimensional organization of smooth muscle tissue may also vary significantly between different regions of the same tissue type. For example, the orientation of the smooth muscle cells is primarily circumferential in large bronchi; whereas muscle cells may be oriented obliquely or helically in small bronchi and distal regions of the bronchial tree (Macklin 1929).

2.2
Organization of cytoplasmic filaments in smooth muscle cells

Smooth muscle cells contain a single centrally located nucleus with its associated organelles. The peripheral and distal regions of the smooth muscle cell are packed with filaments (Figure 1). Three species of filaments are observed: the thin filaments, approximately seven nm in diameter, that are composed primarily of actin; the relatively less abundant thick filaments, 12–15 nm in diameter, composed primarily of myosin; and the intermediate filaments, approximately 10 nm in diameter, composed primarily of desmin in visceral muscles (Bennett et al 1978) or vimentin in vascular muscles (Frank and Warren 1981, Gabella 1984, Bagby 1990, Somlyo et al 1973, Somlyo 1980). The actin and myosin containing thick and thin filaments, the myofilaments, are generally considered to constitute the contractile apparatus; whereas the intermediate filaments, which are much less numerous than the myofilaments (Bagby 1990), are generally believed to play a largely structural role. Intermediate filaments are grouped into bundles that run the length of the cell and exhibit ramifications to the cell membranes (Tsukita and Ishikawa 1983, Draeger et al 1990). In transverse sections of vertebrate smooth muscle, actin filaments can be seen packed in hexagonal arrays that form cable-like bundles. The spaces around the actin filament bundles are occupied by myosin filaments (Figure 1) (Gabella 1984, Cooke et al 1987, Ashton et al 1975, Somlyo 1980).

The length of myosin filaments in smooth muscle has been measured in the range of 1.5–3 microns (Ashton et al 1975, Cooke et al 1989, Small 1977, Small et al 1990, Somlyo 1980). Measurements of actin filament lengths from different laboratories vary significantly. Small et al (1990) reported actin filament lengths in chicken gizzard smooth muscle cells to range from 3 to 6 μm. These thin filament lengths are significantly longer than those of striated muscles. In contrast, Drew and Murphy (1997) reported

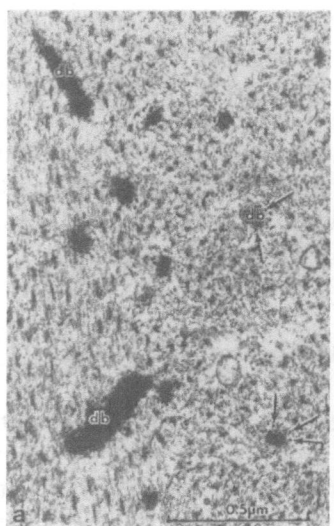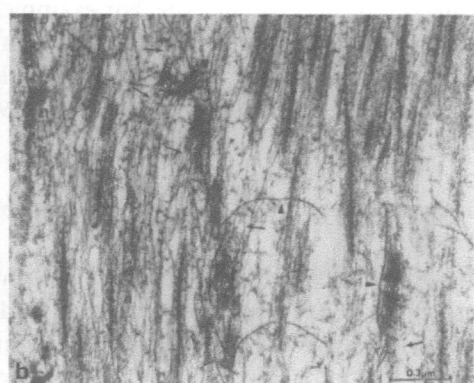

Fig. 1. Organization of filaments in smooth muscle cells. *A.* Vas deferens smooth muscle cell shown in transverse section. Thick filaments are surrounded by multiple thin filaments throughout the cytoplasm. Dense bodies ("db") are round or oval in shape in transverse section but elongated in shape in oblique sections. Arrows point to intermediate filaments (10 nm) that surround the dense bodies. *B.* Longitudinal section of portal anterior mesenteric vein smooth muscle cell. Thin filaments (arrows) can be seen inserting on both sides of the dense bodies (db). Intermediate filaments (arrowheads) extend laterally from dense bodies and sometimes connect a series of dense bodies to form a chain. Magnification: x 70,000. Reproduced from Bond and Somlyo (1982) The Journal of Cell Biology 95:403–413 with permission of Rockefeller University Press

the average length of individual actin filaments in swine stomach smooth muscle to be 1.35 μm, a length which is similar to that of actin filaments in striated muscles. The ratio of actin to myosin filaments varies among different smooth muscle tissues, ranging from as low as 8:1 in chicken gizzard (Nonomura 1976) to approximately 15:1 in vascular muscle (Somlyo et al 1973), to as high as 50:1 in isolated amphibian visceral muscle (Cooke et al 1987).

There is no apparent lateral register between myofilaments in smooth muscle cells. No striations or regular repeats are visible by electron or light microscopy comparable to those observed in striated muscle tissues. In vertebrate visceral smooth muscles, bundles of parallel actin filaments are oriented along the long axis of the cell (Gabella 1984, Cooke et al 1987). As the long bundles of actin filaments split and merge with one another, they do not run perfectly parallel for the entire cell length, but insert acutely into the plasmalemma at points all along the cell (Gabella 1984). In vascular smooth muscle cells, which are shorter than visceral smooth muscle cells, bundles of myofilaments have been observed to extend obliquely from the luminal to the adventitial side of the cell (Gabella 1990).

2.3
Membrane associated dense bands

In electron microscopic sections of mammalian vascular and visceral smooth muscle cells, dense areas of plasma membrane specialization are observed. The plasma membrane dense areas appear to be longitudinal bands oriented along the long axis of the cell membrane (sometimes described as ribs) which alternate with membrane areas containing longitudinal rows of invaginated vesicles called surface caveolae (Gabella 1984, Gabella 1990, Inoue 1990, Small 1985). The longitudinal dense bands range from 30–40 nm to 1 micron wide depending on the muscle type and may occupy 30–50% of the cell surface in the nuclear region of the cell and as much as 100% of the surface toward the ends of the cell (Gabella 1984, Gabella 1990, Bagby 1990, Small 1985). However, this rib-like geometry of the plasma membrane dense areas may differ in different types of smooth muscle. Membrane dense areas visualized by immunofluorescence staining of vinculin, talin or α-actinin in chicken gizzard smooth muscle appeared as regularly spaced transverse streaks, plaques or chevrons (Draeger et al 1989, Fay et al 1983). Because of the differences in appearance in different studies, membrane associated dense areas have been variously referred to as membrane-associated dense bands," dense bodies" or dense-plaques" (Bagby 1990).

There is considerable evidence that the membrane-associated dense bands serve as the attachment points of actin filaments to the plasma membrane and that they mediate the transmission of force across the membrane. Bundles of actin filaments and intermediate filaments can be seen penetrating the inner surface of the membrane-associated dense bands in electron micrographs (Small 1985, Gabella 1984, Ashton et al 1975, Pease and Molinari 1960). Dense bands are sometimes coupled to each other in adjacent cells with an intercellular gap of 40–60 nm which is occupied by electron dense material, forming attachment plaques. As attachment plaques are coupled to bundles of myofilaments within each of the cells, they provide mechanical coupling between the cells (Gabella 1990). Because the actin filaments of smooth muscle cells terminate at points along the entire length of the cell membrane, the transmission of force in smooth muscle cells is diffused along the entire length and width of the cell surface, rather than being concentrated at a few loci within the cell.

The plasma membrane dense areas of smooth muscle cells are similar in their molecular composition to the focal adhesion sites of cultured cells (Turner and Burridge 1991). The primary transmembrane components of these sites are transmembrane integrins, which attach to matrix proteins in the extracellular space at one end and to cytoplasmic proteins within the cell at the other. More than a dozen known cytoplasmic proteins localize to these plasma membrane sites. The most abundant are vinculin and talin. Although there is considerable biochemical evidence that these proteins function to link actin filaments to transmembrane integrins, the exact nature of the molecular interactions that serve to link actin filaments to integrins remains poorly understood (reviewed in Burridge and Chrzanowska-Wodnicka 1996). Recent evidence indicates that the binding of talin to integrins is critical for the attachment of actin to occur (Lewis and Schwartz 1995).

Non-muscle cells transmit mechanical tension generated by the actin cytoskeleton to the extracellular matrix via focal adhesion sites (Ingber 1991). Conversely, when external mechanical stress is applied to trans-membrane integrin receptors, it is transmitted to the actin cytoskeleton (Wang et al 1993). In non-muscle cells, the integrin receptors at focal adhesion cites can initiate an intracellular signaling cascade in response to mechanical stimulation that leads to cytoskeletal rearrangements and other cellular events. There is currently little information regarding the role of the membrane-associated dense plaque proteins in mediating mechanotransduction processes in smooth muscle. However, both talin and paxillin, which are constituents of the dense plaques in smooth muscle cells, undergo phosphorylation during contractile stimulation in tracheal muscle (Pavalko et al 1995, Wang et al 1996), and the level of paxillin phosphorylation is sensitive to changes in muscle length (Tang et al 1998). The similarities in the biochemical composition of focal adhesion sites in non-muscle cells and smooth muscle dense plaques suggest that they are likely to share many functional properties.

2.4
Cytoplasmic dense bodies

Cytosolic electron dense areas are also observed in smooth muscle cells that are referred to as dense bodies." Dense bodies are distributed relatively uniformly throughout the cytoplasm (Tsukita and Ishikawa 1983, Bond and Somlyo 1982, Fay et al 1983). They are obliquely oriented elongated structures that can appear circular or oval in shape under electron microscopy depending on the plane of section (See Figure 1). Cytoplasmic dense bodies have been measured up to 1.5 µm in length (Bond and Somlyo 1982, Tsukita and Ishikawa 1983, Ashton et al 1975, Fay et al 1983).

There is extensive microscopic evidence that documents the association of cytoplasmic dense bodies with actin filaments as well as with intermediate filaments (Ashton et al 1975, Bond and Somlyo 1982, Tsukita and Ishikawa 1983, Bagby 1990). Bond and Somlyo (1982) describe chains of elongated dense bodies visible in longitudinal sections that appear to have actin filaments inserting at both ends as well as along the entire length of the dense body (Figure 1). Others have described the insertions of actin filaments to be mainly at the poles of the dense bodies (Tsukita and Ishikawa 1983). Actin filaments appear in association with virtually all of the cytoplasmic dense bodies. The decoration of actin filaments with myosin subfragments (S-1) indicates that the filaments are attached with the right polarity for force generation (arrowheads pointing away from the dense bodies)(Bond and Somlyo 1982, Tsukita and Ishikawa 1983). Opposite polarities are observed on either side of a given dense body, that has led to the suggestion that the cytoplasmic dense bodies are functionally equivalent to the Z-bands of striated muscle.

Intermediate filaments can be seen on the lateral sides of the dense bodies, often forming loops at their lateral surface (Tsukita and Ishikawa 1983, Bond and Somlyo 1982) (Figure 1). Chains of dense bodies appear to be connected by intermediate filaments. However, the 10 nm filaments that surround a given dense body do not run parallel to the contractile unit and associate with the next dense body in series; but are oriented obliquely toward another dense body of a different contractile unit. This suggests that the network of 10 nm filaments connects the force generating units though their attachment to the dense bodies (Bond and Somlyo 1982) . Kargacin et al (1989) and Draeger et al (1990) analyzed the movement of dense bodies in contracting isolated smooth muscle cells, and suggested that dense bodies provide mechanical coupling between the contractile apparatus, the cytoskeleton, and the cell surface. Kargacin et al (1989) found that groups of dense bodies that are laterally aligned with respect to the vertical axis of the cell remain at fixed distances from each other during contraction. This result suggests that they are linked or constrained by semi-rigid non-contractile elements. Laterally aligned dense bodies move rapidly toward one another during contraction, indicating that they are attached to contractile filaments. Similarly, Draeger et al (1990) observed dense bodies to be organized in a regular geometric arrangement that is retained during contraction, although the intermediate filaments that link the dense bodies become crumpled and disordered. These observations are consistent with a structural role for the dense bodies as sites which anchor actin filaments and thereby serve to organize the arrangement of contractile units within the cell.

The molecular mechanisms that couple the contractile filaments to the dense bodies are not established. Dense bodies contain the actin-cross-linking protein, α-actinin, as do the Z-lines of skeletal muscle. Dense bodies also contain actin and probably calponin, although it is currently unclear whether the α and γ ("smooth muscle") isoforms of actin (See Section 4.2) are both present in dense bodies or only the β "non-muscle" isoform (Small 1995, North et al 1994a, 1994b, Mabuchi et al 1996). Ultrastructural and biochemical data obtained by Mabuchi et al (1997) suggests that one function of calponin may be to couple actin filaments and intermediate filaments at dense bodies.

3
Molecular structure of the thick filaments

Smooth muscle myosin has distinctive characteristics that may form the basis for many of the unique functional properties exhibited by smooth muscle tissues. The following section will first review the molecular structure of smooth muscle myosin and the functional implications of its distinctive characteristics. This will be followed by a discussion of the regulation of the assembly of myosin into thick filaments, and the molecular organization of the thick filaments of smooth muscle tissues.

3.1
Molecular structure of the smooth muscle myosin molecule

The thick filaments of smooth muscle are made up of monomeric myosin molecules that polymerize to form filaments. The smooth muscle myosin molecule is grossly similar to that of skeletal muscle myosin. It is a large asymmetric protein (MW approximately 520 kDa) which is made up of six polypeptide chains: two ~205 kDa heavy chains that form a dimer, and two pairs of light chains, the 20 kDa "regulatory" light chains and the 17 kDa "essential" light chains (Reviewed by Adelstein and Sellers 1996) (Figure 2). The myosin heavy chain dimer makes up the main body of the molecule, with each heavy chain containing a slightly elongated globular head at the amino terminus. The myosin globular heads are connected to a long α-helical coiled tail of approximately 120 kDa that aggregates to form the rod-like backbone of the thick filament. Each myosin globular head contains the functional motor domains of the molecule which include the nucleotide and actin-binding regions. A single essential and regulatory light chain are associated with each head. The heavy chain tails end in a short sequence at the carboxy terminus that is not predictive of a coiled structure. The entire molecule is about 165 nm long, the rod portion being about 150 nm in length.

The myosin heads can be cleaved enzymatically from the rest of the heavy chain by mild proteolysis to yield "subfragment-1" (S-1), which contains a single globular myosin head and its associated essential and regulatory light chains (Ikebe and Hartshorne 1985, Adelstein and Sellers, 1996). The S-1 fragment is very soluble, and contains all the necessary elements to generate movement of actin during ATP hydrolysis (Toyoshima et al 1987, Rayment et al 1993a and 1993b, Itakura et al 1993, Lowey and Trybus 1995).

Three-dimensional structural analysis of chicken skeletal muscle myosin subfragment-1 has been performed by Rayment et al (1993b) (Figure 3). As smooth and skeletal muscle myosins show considerable homology in the primary sequences of the head region, particularly in the regions which are thought to be involved in nucleotide or phosphate binding, it is probable that the three-dimensional structure of the smooth muscle myosin S-1 subfragment is similar to that of skeletal muscle (Warrick and Spudich 1987, Adelstein and Sellers 1996).

The asymmetric S-1 fragment of myosin can be divided into a globular component containing the motor domain that is formed exclusively from

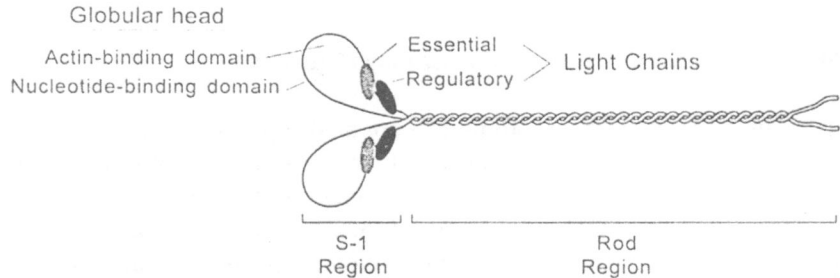

Fig. 2. Structure of the myosin molecule. Schematic representation (components are not drawn to scale)

the myosin heavy chain, and an extended α-helical motif that binds the essential and regulatory light chain subunits. The myosin motor domain has two proteolytically sensitive regions. Digestion at these sites gives rise to three fragments, referred to as the 25 kDa, the 20 kDa, and the 50 kDa fragments, that constitute a large six-stranded β-sheet of unknown function (Balint et al 1975, Mornet et al 1979, Rayment et al 1996). The nucleotide-binding pocket is located at one end of the central strand of this sheet. The 50 kDa fragment is split by a large cleft that separates it into an upper and a lower domain. Several potential actin binding regions have been identified, that generally lie on both sides of the large cleft on the opposite side of the head from the nucleotide-binding pocket.

The essential and regulatory light chains are localized along an α-helical segment of the heavy chain at the junction of the globular head and the coiled-coil rod portion of myosin subfragment-1. The regulatory light chain is attached to the C-terminal region of the S-1 heavy chain while the essential light chain is localized closer to globular head (Rayment et al 1993b). The portion of the α-helical rod at which the light chains are attached is frequently referred to as the "neck" region. The α-helical heavy chain or "neck" has a gradual bend in the region between the two light chains and a sharp bend near the junction of the globular head to the rod. The organization of the structures suggest that the conformation of the α-helical rod portion of the myosin heavy chain molecule may be stabilized by the binding of the essential and regulatory light chains.

The non-helical tailpiece present at the carboxy-terminus of smooth and non-muscle myosin rods is distinct from the short tailpiece present in

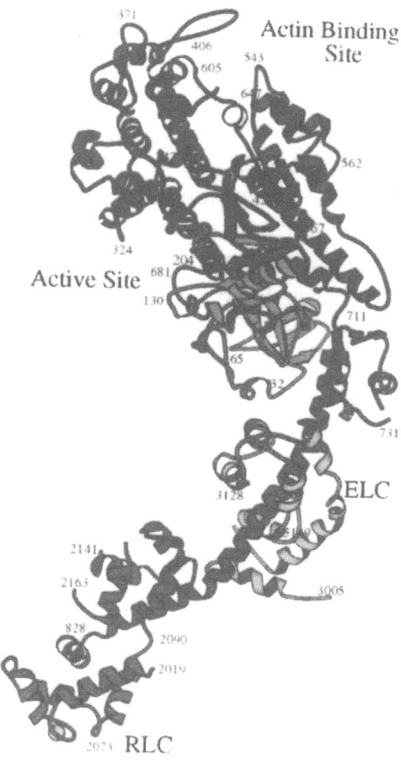

Fig. 3. Structure of myosin S-1. Ribbon representation of model of myosin S-1 showing the active (nucleotide-binding) and the actin-binding sites. Cleft can be seen at the actin-binding site. Residues of the essential and regulatory light chains are distinguished by the addition of 2000 and 3000 respectively. Reproduced with permission from Rayment et al (1993) Science 261:50–58. Copyright 1993 American Association for the Advancement of Science

striated muscle myosins. In smooth and non-muscle cells, two different length isoforms of the tailpiece have been identified, that result in 200 kDa (SM2) and 204 kDa (SM1) heavy chain isoforms (Nagai et al 1989, Babij and Periasamy 1989). These isoforms are expressed to the same extent at the protein level in some smooth muscle cell tissues (Kelley et al 1991, 1993), but are expressed unequally in other smooth muscle tissues (e.g., Morano et al 1993, Mohammad and Sparrow 1989, Cavaille et al 1986, Rovner et al 1986, Sparrow et al 1988). The expression of these isoforms may also vary during development (Eddinger and Murphy 1991, Eddinger and Wolf 1993, Kuro-o et al 1989). A number of studies have attempted to correlate functional properties of smooth muscle tissues with the differences in the expression of these heavy chain isoforms. However, a consistent correlation between the myosin heavy chain isoform expression and the functional properties of intact smooth muscle tissues has not emerged (e.g., Sparrow et al 1988, Hewett et al 1993). Studies of functional correlations of isoform expression in tissues are complicated by the fact that the expression of smooth muscle heavy chain composition may vary significantly between individual cells within a tissue, and that heavy chain isoforms can form either SM-1 or SM-2 homodimers or SM-1/SM-2 heterodimers (Tsao and Eddinger 1993, Meer and Eddinger 1996). However, no differences are detected in the rates at which the 200 and 204 kDa isoforms of myosin propel actin filaments in vitro motility assays (Kelley et al 1992).

Another set of smooth muscle heavy chain isoforms that have either the insertion or omission of 7 amino acids in the head domain near the ATP-binding region have been identified (Babij 1993, Kelly et al 1993, White et al 1993). The insertion appears to be present in visceral smooth muscle tissues but is absent in tonic vascular smooth muscle, and appears to confer functional differences in the kinetic properties of smooth muscle myosins from visceral smooth muscle. There is evidence from studies in the motility assay that the visceral smooth muscle isoform has higher ATPase activity and moves actin filaments faster than the vascular isoform (Kelly et al 1993, Rovner et al 1997). However, because the head domain isoforms may be associated with different tail domain isoforms in different tissues, as well as with different light chain isoforms, it has been difficult to establish causal relationships between myosin heavy chain isoforms and myosin kinetics with certainty. (Reviewed by Somlyo 1993, Murphy et al 1997).

Two isoforms of the 17 kDa isoform of smooth muscle myosin light chains have been identified (LC17a and LC17b). These isoforms have a difference in five of the nine carboxy terminal amino acids (Lash et al 1990, Nabeshima et al 1987). The LC17a/LC17b ratio tends to be higher in phasic fast than in tonic slow muscles (Malmqvist and Arner 1991, Fuglsang et al 1993) although high LC17a/LC17b ratios have also been reported in some tonic slow muscles (Helper et al 1988, Malmqvist and Arner 1991). There are no reported isoform variants of the 20 kDa smooth muscle and non-muscle regulatory light chains that have functional correlations (Taubman et al 1987, Kumar et al 1989)

3.2
The myosin motor

In smooth muscle cells, the development of force results from the MgATP-dependent cyclical interaction of myosin crossbridges in thick filaments with actin in the thin filaments. Myosin is believed to generate force and /or motion by mechanical cycles during which the myosin head repetitively attaches to actin, undergoes a conformational change that results in a power stroke and then detaches (Reviewed by Warrick and Spudich 1987). The energy required for the mechanical power is generated by the enzymatic hydrolysis of ATP by the globular myosin head. The ATPase cycle underlying the transduction of chemical energy into mechanical force by crossbridges is believed to be similar in striated and smooth muscles (Taylor 1987).

Based on structural information obtained on the myosin S-1 fragment and on actin and the actomyosin complex (Holmes et al 1990, Kabsch et al 1990, Rayment et al 1993), Rayment et al proposed a structural hypothesis to explain the conversion of chemical energy into directed movement. This model, sometimes referred to as the "lever-arm hypothesis," has subsequently been modified based on additional structural information (Fisher et al 1995a and 1995b). (A recent review of this model has been published by Rayment et al (1996). An important tenet of this hypothesis is that myosin interacts with actin in a stereospecific manner which requires a unique orientation of the myosin molecule for strong bindingto actin and for the power stroke to be initiated. The binding of ATP to myosin while it is bound to actin is proposed to open the cleft that splits the central 50-kDa region of the heavy chain and disrupts the orientationally-specific strong actin-myosin interaction. This results in a series of conformational changes in the myosin globular head that reduces its affinity for actin and allows it to hydrolyze ATP to ADP and Pi. As the myosin rebinds to actin and releases the nucleotide, the conformational changes are reversed causing movement of the myosin relative to actin. In this model, the light-chain binding domain (the "neck" region) pivots about a fulcrum near where the globular head and light chain binding domains abut one another, resulting in the displacement of actin relative to myosin. Support for this hypothesis was provided by studies using myosin mutants that had different "neck" lengths (Uyeda et al 1996, Kurzawa et al 1997). As predicted, the length of the neck region is correlated with filament sliding velocity in a motility assay. The lever arm hypothesis has been recently reviewed by Block (1996). Studies using smooth muscle myosin mutants in which the length of the neck region is modified have confirmed that this mechanism is essentially the same for smooth muscle and skeletal muscle myosins (Guilford et al 1997b).

Smooth and skeletal muscle myosins have important functional differences with respect to their motor activities and their regulation. The differences in motor properties are evident in the behavior of smooth and skeletal muscle myosins in an *in vitro* motility assay. Purified smooth muscle myosin propels actin filaments at one tenth the velocity of skeletal muscle myosin and produces an average of 3–4 times more force per unit time period than skeletal muscle myosin, as measured by a micro-needle assay (Warshaw et al 1990, Van Buren et al 1994) . These differences in the functional properties of smooth and skeletal muscle myosins at the molecular level parallel differences in the functional properties of smooth and skeletal muscle tissues. Smooth muscle tissues produce the same isometric force per cross-sectional area as skeletal muscle, but contain only one fifth as much myosin (Murphy et al 1974). In addition, the maximal shortening velocities of smooth muscle tissues are 1–2 orders of magnitude slower than those of skeletal muscles (Murphy et al 1997).

Guilford et al (1997a) evaluated the molecular basis for the differences in the motor properties of smooth and skeletal muscle myosin, using a laser-trap to measure the force and displacement generated by individual smooth and skeletal muscle myosin molecules in vitro. Although individual smooth and skeletal muscle myosin molecules produced similar levels of unitary force and displacement, smooth muscle myosin remained attached to actin for longer average time periods than skeletal muscle myosin, suggesting that it has longer "duty cycle." Thus, the greater average force obtained with smooth muscle myosin in the motility assay can be attributed to differences in the average attachment time of smooth and skeletal muscle myosins to actin. These differences are consistent with previous biochemical measurements using actomyosin solutions that demonstrated slower rate constants for many of the steps in the actomyosin ATPase cycle for smooth muscle myosin relative to skeletal muscle myosin (Marston and Taylor 1980, Siemankowski et al. 1985). Sata et al

(1997) demonstrated that the difference in the maximum actin-activated ATPase activity of smooth and skeletal muscle myosin is completely determined by the motor domain of the myosin head by using chimeric myosin molecules composed of combinations of skeletal and smooth muscle myosin.

The difference in the regulatory properties of vertebrate smooth and skeletal muscle myosins lies in the role of the 20 kDa regulatory light chain. In smooth muscle tissues the phosphorylation of a single serine residue (serine-19) in the N-terminus of the regulatory light chain is the switch for turning on actin-activated ATPase activity and contraction. In vitro, smooth muscle regulatory light chain phosphorylation increases actin-activated ATPase activity by 50 to 100-fold. In contrast, in skeletal muscle phosphorylation of the 20 kDa myosin regulatory light chain is not necessary for the activation of actin-activated myosin ATPase activity, and the effects of regulatory light chain phosphorylation or even the presence of this light chain on myosin ATPase activity are small (2-fold) (Hartshorne 1987, Wagner and Giniger 1981, Sivaramakrishnan and Burke 1982).

Considerable progress has been made recently in evaluating the molecular basis for the regulation of actin-activated ATPase activity of smooth muscle myosin by the regulatory light chain. The COOH-terminal portion of the regulatory light chain appears to be essential for the full regulation of motor activity in smooth muscle myosin both with respect to the complete inhibition and the complete activation of the myosin molecule (Trybus 1994a, 1994b). Sata et al (1997) reported that the sensitivity of myosin to regulation by light chain phosphorylation appears to be conferred by the light chain-associated regulatory region of the myosin head at the C-terminal and not the N-terminal globular region, that determines the rate of ATP hydrolysis. The two-headed structure of myosin is critical for phosphorylation-mediated regulation (Matsu-ura and Ikebe, 1995, Sata et al 1996, Cremo et al 1995). The data of Sata et al (1996) suggests that the interaction between the myosin heads at the C-terminal portion of S-1 (the regulatory domain) is important for the regulation of myosin motor activity. Recent work by Trybus et al (1997) suggests that the myosin rod mediates specific interactions with the globular head that are required to obtain the complete inactivation of myosin. Myosin cannot be fully inactivated unless a length of rod approximately equal to the myosin head is present (Trybus et al 1997).

3.3
Myosin filament assembly.

Smooth muscle myosin monomers assemble to form myosin filaments both in vitro and in vivo (For review see Trybus 1996). The region of the myosin molecule that regulates the assembly of monomeric myosin into filaments appears to be localized to a 150-residue fragment at the C-terminal of the rod (Cross and Vandekerckhove 1986). Smooth and non-muscle myosins share a common feature in that the phosphorylation of the 20 kDa regulatory light chain can regulate the assembly of monomeric myosin into filaments *in vitro*, a feature not shared by skeletal muscle myosin (Scholey et al 1980 and 1981, Suzuki et al 1978, Trybus et al 1982, Trybus and Lowey 1984, Craig et al. 1983, Smith et al 1983, Kendrick-Jones et al 1987, Trybus 1989, Trybus and Lowey 1987). Studies of myosin in vitro have demonstrated that if the regulatory light chain of myosin filaments is not phosphorylated, the addition of stoichiometric amounts of MgATP results in the disassembly of myosin filaments into a soluble monomeric conformation of myosin in which the heavy chain rod is bent into thirds (Trybus and Lowey 1984, Trybus et al 1982). Phosphorylation of the regulatory light chain results in the unfolding of the rod and the reassembly of the monomers into myosin filaments.

Despite several decades of investigation, the role of phosphorylation of the myosin regulatory light chain in regulating the process of myosin filament assembly and disassembly in vivo is not completely settled. A number of early ultrastructural studies suggested that the myosin filaments of vertebrate smooth muscle exist in a labile state of organization in vivo. Myosin filaments were initially observed by electron microscopy only in contracted smooth muscles (Kelly and Rice 1969, Shoenberg 1969). The discovery by Watanabe and colleagues (Suzuki et al 1978) that smooth muscle light chain phosphorylation promoted myosin filament assembly in vitro fueled interest in the possibility that the state of myosin light chain phosphorylation might regulate myosin filament assembly in vivo. However subsequent electron micrographic studies clearly demonstrated myosin filaments in both relaxed and contracted smooth muscle tissues, and it was concluded that the earlier findings in which myosin filaments were not observed resulted from problems with fixation techniques (Somlyo 1980, Devine and Somlyo 1971, Cooke and Fay 1972, Small 1977). Subsequent electron microscopic images of frozen relaxed smooth muscle clearly showed numerous thick filaments in muscles in which more than

95% of the myosin filaments were determined to be dephosphorylated (Somlyo et al 1981, Tsukita et al 1982, Xu et al 1996).

Although there is now general agreement that dephosphorylated myosin remains in filamentous form in uncontracted smooth muscle, the possibility remains that there is a pool of folded monomeric myosin molecules in vivo which is recruited to myosin filaments when the muscle is activated and phosphorylation of the regulatory myosin light chains occurs (Xu et al 1997). This could regulate an activation-dependent modulation of myosin filament number or length. There are several studies which provide support for this idea. Cande et al (1983) reported that the thick filaments of isolated glycerinated chicken gizzard smooth muscle tissues are labile under conditions resembling the relaxed state, mimicking the labile properties of myosin filaments in vitro. These investigators suggested that in vivo dephosphorylated thick filaments may be in equilibrium with a small pool of monomeric myosin that diffuses out of glycerinated tissues driving the disassembly of thick filaments. Other evidence comes from light and electron microscopic studies, in which higher level of birefringence and a greater density of thick filaments has been observed in contracted than in relaxed rat anococcygeus smooth muscle (Gillis et al 1988, Godfraind-De Becker and Gillis 1988). Recently, Xu et al (1997) reported a 23% increase in myosin filament density in the rat anococcygeus smooth muscle using low temperature electron microscopic techniques, although they found no change in myosin filament density with contraction of the guinea pig taenia coli. However, using monoclonal antibodies to specifically label monomeric myosin, Horowitz et al (1994) detected only trace amounts of monomeric myosin in both relaxed and contracted gizzard smooth muscle and concluded that the assembly /disassembly of myosin is unlikely to play a significant role in the contraction/relaxation cycle in gizzard smooth muscle. Overall, the accumulated evidence does not provide support for an extensive process of assembly and disassembly of myosin filaments during the contraction/relaxation cycle in smooth muscle tissues. However, it is possible that increases in the number or length of existing myosin filaments may play a role in regulating the length-adaptive properties of some smooth muscle tissues (Pratusevich et al 1995) (See Section 5.2).

The question of why dephosphorylated smooth muscle myosin filaments exhibit greater stability in vivo than in vitro remains unanswered. One possibility is that the intracellular concentration of myosin in vivo exceeds the critical concentration necessary for filament assembly even for dephosphorylated myosin (Kendrick-Jones et al 1987). Alternatively, myosin filaments may be stabilized by caldesmon or telokin (Katayama et al 1995, Shirinsky et al 1993), a protein identical to the C-terminal portion of myosin light chain kinase which is expressed only in smooth muscle tissues (Ito et al 1989). The assembly of smooth muscle myosin into thick filaments is promoted in the presence of F-actin, which could favor the formation of filamentous myosin in vivo (Applegate and Pardee 1992). The ability of myosin to assemble may also be regulated by the structure of the nonhelical tailpiece, which occurs in two isoforms (Ikebe et al 1991, Hodge et al 1992). Thus myosin heavy chain isoforms might play a role in determining *in vivo* myosin filament lability in different tissues.

On the basis of in vitro studies with whole myosin, Wagner and Vu (1986,1987) hypothesized that the state of assembly of myosin might be more important in determining myosin ATPase activity than the state of phosphorylation of the regulatory light chain. They found that when unphosphorylated and phosphorylated gizzard myosins were monomeric, their MgATPase activities were not activated by actin. In contrast, when myosin was in filamentous form, MgATPase activity could be stimulated by actin even when the filaments were not phosphorylated, although the MgATPase activity was lower than for phosphorylated filamentous myosin (Wagner and Vu 1987). However, this idea is not supported by the results of Trybus (1989). Using an antibody to stabilize dephosphorylated filamentous myosin, she found that regulatory light chain phosphorylation alone is sufficient to regulate the ATPase activity of filamentous myosin in the absence of changes in the state of myosin assembly. However, the effect of regulatory light chain phosphorylation on ATPase activity was further increased if myosin was allowed to disassemble to the folded monomeric state when it was dephosphorylated.

3.4
Structure of myosin filaments in vivo

Largely because of the lability of filamentous myosin (Suzuki et al 1978), controversy has surrounded the question of the three-dimensional arrangement of myosin molecules within myosin filaments in vivo. Some early ultrastructural data suggested a bipolar symmetry and a helical arrangement of myosin molecules along a rod-shaped filament, similar to the organization of myosin filaments found in skeletal muscle (Ashton et al 1975, Shoenberg and Stewart 1980). In a bipolar filament, the myosin heads

Bipolar

Side-polar

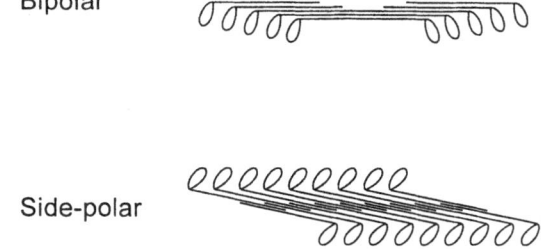

Fig. 4. Myosin filament structure. Diagrams of myosin monomer packing in non-helical side-polar and helical bipolar filaments. For simplicity, only one myosin head per monomer is shown. A bare zone is observed at the center of the bi-polar filament, and at each end of the side-polar filament

must reverse polarity at the center of the filament, resulting in a central bare zone. In vitro, the assembly of smooth muscle myosin monomers results preferentially in side-polar filaments, in which all myosin heads have the same polarity along one edge of the filament, and the opposite polarity on the other edge, with bare zones at each filament end (Craig and Megerman 1977, Trybus and Lowey 1987, Cross et al 1991, Hinssen et al 1978). A side-polar filament with these characteristics can be generated by packing unfolded myosin monomers with the head regions extending in opposing directions on each side of the filament (Figure 4).

Most recent electron microscopic evidence for myosin filament structure in vivo favors a non-helical side-polar arrangement of crossbridges along a rodlike myosin filament, with no central bare zone (Cooke et al 1989, Xu et al 1996). In ultrastructural studies, bare areas have been observed at the ends of myosin filaments in vivo and little evidence for a central bare zone has been obtained (Somlyo 1980, Hinssen et al 1978, Xu et al 1996). As in skeletal muscle myosin, a continuous 14 nm axial repeat arising from crossbridges distributed along the myosin filament is characteristic of native as well as synthetic myosin filaments from smooth muscle (Small 1977, Cooke et al 1989, Hinssen et al 1978, Craig and Megerman 1977, Xu et al 1996). However, the crossbridges in smooth muscle filaments are observed to project in opposite directions on opposite sides of the filament, and the tails form antiparallel interactions along the entire length of the filament (Craig and Megerman 1977, Xu et al 1996). Both of these features imply a side-polar arrangement of crossbridges and are inconsistent with a helical bipolar structure. The appearance of a helical structure of smooth muscle myosin filaments in some studies (Hinssen et al 1978) probably arose from the presence of different orientations of essentially side-polar filaments (Cooke et al 1989).

4
The thin filaments of smooth muscle cells

The thin filaments of smooth muscle cells are extremely stable during most fixation protocols and are the most abundant filament in smooth muscle. Actin is the primary protein constituent and forms the thin filament backbone. In addition, at least three other proteins, tropomyosin, caldesmon, and calponin, bind to actin and are localized to the thin filaments (Figure 5).

Actin occurs in multiple isoforms within individual smooth muscle cells. All of these isoforms are capable of forming filamentous actin that can interact with myosin to generate force. Although the functional importance of the different actin isoforms is presently unclear, there is evidence to suggest that they may serve to "customize" actin filaments to serve different functional roles within the cell by determining its interactions with different binding proteins. The first part of this section will review the molecular structure of the thin filament. The structure of actin and the relationship to the other protein constituents of the thin filament to actin

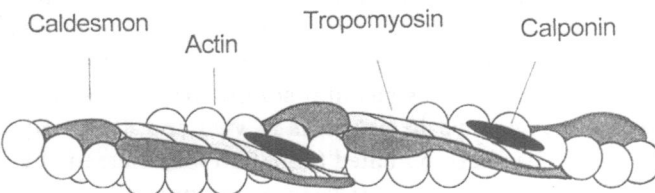

Fig. 5. Schematic representation of components of the thin filament of smooth muscle. Although calponin and caldesmon are both shown as components of the same filament, there is evidence that they may selectively associate with different actin isoforms in different populations of thin filaments

will be discussed. In subsequent sections possible differences in the functional role of different isoforms of actin within smooth muscle cells will be addressed.

4.1
Molecular organization of the thin filaments

4.1.1
Molecular structure of actin

Filamentous actin (F-actin) is a polymeric protein composed of asymmetric bi-lobed 42 kDa actin monomers. Three-dimensional maps of thin filaments obtained by cryo-electron microscopy and X-ray crystallographic data demonstrate that each actin monomer consists of two major domains, "inner" and "outer", which are unequal in size (Milligan et al 1990, Holmes et al 1990, Kabsch et al 1990). Each domain is further divided into two subdomains, resulting in subdomains 1, 2, 3, and 4 (Kabsch et al 1990). A single divalent cation binding site and an adenosine nucleotide binding site are located in the cleft between the inner and outer domains of each actin monomer (Estes et al 1992). Stoichiometric amounts of ADP are strongly bound to F-actin (Bárány et al 1992). In filamentous actin, the actin monomers form a double-stranded helical array with a right-handed long-pitch in which the two strands cross every 36 nm and a left-handed genetic helix of 5.9- nm pitch (Milligan et al 1990, Holmes et al 1990). There are about 13 actin monomers between each 36-nm crossover point. This essential structural organization of actin filaments was confirmed in thin filament samples from smooth muscle (Moody et al 1990). The binding of myosin to F-actin is currently thought to begin with an initial weak contact on subdomain-1 of the outer domain of actin followed by a strong stereo-specific interaction closer to the junction of its inner and outer domains (Rayment et al 1993a). It is this strong-binding stereospecific interaction that is thought to form the basis of the power-stroke which generates force (See Section 3.2).

The polymer state of actin is dynamic. In vitro, actin polymerizes at the filament ends until a critical concentration is reached (See Pollard and Cooper 1986). Even after a steady state is achieved in which net polymerization ceases, subunit exchange continues at both ends. In many differentiated eukaryotic cells, including the sarcomeres of striated muscle, actin filament lengths are precisely determined and maintained by capping proteins such as tropomodulin and CapZ. In addition, in skeletal muscle, the giant molecule nebulin may act as a molecular template for the regulation of actin filament length (See Fowler 1996). In many cell types, agonists can regulate the polymer state of actin through the activation and inhibition of proteins that nucleate, cap, or sever actin filaments (See Zigmond 1996). There is presently no direct evidence that the polymerization state of actin is modulated in smooth muscle cells; however smooth muscle contains large amounts of gelsolin and profilin, both of which are potent regulators of actin filament polymerization (Hinssen et al 1984, Buss and Jockusch 1989). The degree to which the dynamic remodeling of actin filaments may be regulated in smooth muscle cells and the physiologic role of actin filament regulatory proteins in these cells remains to be demonstrated.

4.1.2
Role of tropomyosin in the thin filament

Both smooth and skeletal muscle actin filaments are saturated with tropomyosin (Sobieszek and Bremel 1975). Both exhibit the same characteristic stoichiometry of binding of 1 molecule of tropomyosin interacting with 7 monomeric units of F-actin on each of the two strands of F-actin (Hartshorne 1987). The length of tropomyosin molecules (284 amino acids) and their periodicity in smooth and striated muscles is the same (Matsumura and Lin 1982). In both tissues, tropomyosin exists as a dimeric α-helical coil (Caspar et al 1969). Individual tropomyosin molecules bind in an end to end fashion to form a continuous strand on the thin filament that lies along the long-pitch of the double helix formed by the actin monomers (Moore et al 1970, OBrien et al 1971, Spudich et al 1972, Milligan et al 1990).

The functional role of tropomyosin has been extensively studied in striated muscles. According to the well-known "steric blocking" model, the Ca^{2+}-sensitive troponin complex modulates the position of tropomyosin along the actin filament to regulate myosin binding to actin and consequently the activation and inactivation of actomyosin ATPase activity. Direct structural evidence in support of this model was recently obtained by Lehman et al (1994) using skeletal muscle thin filaments from the arthropod Limulus, and subsequently confirmed using actin from vertebrate skeletal muscle (Vibert et al 1997). In these tissues, longitudinally continuous strands of density were observed along the long-pitch actin helices, that were presumed to be tropomyosin. In the absence of Ca^{2+}, the tropomyosin strands appeared to be in contact with the extreme inner edge of the outer domains of actin, coinciding with the site on actin that is thought to be the locus of the strong stereospecific actomyosin interaction. In contrast, in the presence of Ca^{2+}, the tropomyosin strands appeared to be closely associated with the inner domain of actin. These observations are consistent with a steric mechanism in which tropomyosin interferes with the transition from a weak to strong binding state between actin and myosin, thus inhibiting crossbridge cycling.

Although tropomyosin is also localized along the actin filaments of smooth muscle, its function is less clear. Electron microscopy of smooth muscle thin filament preparations isolated from chicken gizzard smooth muscle reveal continuous narrow asymmetric strands of density running helically along the filament axis composed primarily of tropomyosin or an association of tropomyosin and caldesmon (Moody et al 1990, Vibert et al 1993). Reconstructions made from caldesmon-deficient filaments confirm that a major portion of this strand density is contributed by tropomyosin. However, as the thin filaments of smooth muscle lack troponin, the troponin-tropomyosin mechanism that regulates the actomyosin interaction in skeletal muscle cannot be operative in smooth muscle. It has been proposed that the thin filament protein, caldesmon, may interact with tropomyosin to modulate actomyosin ATPase activity in smooth muscle in a manner somewhat analogous to that of troponin in skeletal muscle. However, this is not supported by recent ultrastructural analysis of the effect of caldesmon on the position of tropomyosin in smooth muscle thin filaments (See Section 4.1.3).

4.1.3
Role of caldesmon in the thin filament

Caldesmon is present ubiquitously in smooth muscle cells. It is generally agreed that it is localized to the thin filaments in regions that are replete with myosin (Sobue et al 1981, Furst et al 1986, Mabuchi et al 1996, North et al 1994a). Caldesmon is a long thin flexible asymmetric molecule that is considerably longer than tropomyosin (approximately 75–80 nm) (Lynch et al 1987, Mabuchi and Wang 1991, Bretscher 1984, Graceffa et al 1988, Furst et al 1986). The molecular mass of chicken gizzard caldesmon is 90 kDa as measured by analytical ultracentrifuge sedimentation (Graceffa et al 1988). From electron micrographic studies and nuclear magnetic resonance spectroscopy, each molecule appears to be made up of a central rigid rod section of about 30 nm with a 20-nm flexible section at each end (Mabuchi and Wang 1991). The caldesmon molecule contains binding sites for myosin, tropomyosin, actin and calmodulin (Levine et al 1990, Wang et al 1991).

Ultrastructural studies of isolated chicken gizzard thin filaments localized caldesmon on the thin filament beside tropomyosin, arranged continuously along the axis of the actin double helix (Moody et al 1990, Vibert et al 1993, Lehman et al 1997). In smooth muscle filaments derived from vascular or visceral tissue, the stoichiometry of caldesmon to tropomyosin and actin has been determined to be 1:2:14 (Lehman et al 1989, Marston 1990, Lehman et al 1993). Marston and Redwood (1991) proposed that each caldesmon molecule is placed in register with tropomyosin and extends for 78 nm, the length of two tropomyosin molecules. Each caldesmon molecule interacts with 14 actin monomers. This would result in a filament without radial symmetry such that different parts of the caldesmon molecule would appear on the same side of the actin filament.

In the presence of tropomyosin, caldesmon can inhibit actomyosin ATP-ase activity at the low ratios of caldesmon to actin likely to occur in vivo (1:14) (Marston and Redwood 1993). The affinity of caldesmon for actin is greatly attenuated by Ca^{2+}-calmodulin (Marston and Redwood 1993, Smith et al 1987, Ngai and Walsh 1984, Marston and Lehman 1985. (Reviewed by Sobue and Sellers 1991 and by Marston and Redwood 1991). These observations led to the proposal that role of caldesmon in smooth muscle may be analogous to that played by troponin in skeletal muscle (Chalovich et al 1987, Smith et al 1987, Marston and Smith 1985, Sobue et al 1982). Marston and colleagues have suggested that caldesmon acts as a Ca^{2+}-sensitive switch to regulate the interaction of actin and myosin by a steric mechanism in which the actin-tropomyosin complex is regulated between two states that have different binding affinities for myosin (Marston et al 1994, Marston and Redwood 1993). However, based on the analysis of three-dimensional image reconstructions of reconstituted smooth muscle thin filaments, Vibert et al (1993) concluded when that when caldesmon is present, tropomyosin does not cover the strong-binding sites for myosin on actin (on the inner edge of the outer domain) when actomyosin is in the inhibited state. Instead, tropomyosin appears in contact with the inner domain of each actin monomer when caldesmon is present. When caldesmon is dissociated from the thin filament by treatment with Ca^{2+}-calmodulin, tropomyosin appears to move to a position closer to the outer domain of actin, covering the myosin-binding region. This suggests that the steric-blocking mechanism in effect in striated muscle does not operate in smooth muscle, as this mechanism involves troponin fixing tropomyosin over the strong-binding sites on actin. However, the position of caldesmon on the smooth muscle thin filament may coincide with the proposed sites of weak myosin binding that are thought to be necessary for the initiation of actomyosin ATPase activity (Miller et al 1995, Rayment et al 1993). Alternatively, the presence of caldesmon and tropomyosin adjacent to the strong binding sites for myosin may interfere with myosin binding, or caldesmon may affect the structure of actin in a way that interferes with myosin binding (Lehman et al 1997).

Caldesmon also binds to smooth muscle myosin in vitro at a ratio of binding of 2–3 caldesmon molecules per myosin molecule, although a stoichiometry of 1:1 in the presence of ATP has been reported (Ikebe and Reardon 1988, Marston and Redwood 1992, Hemric and Chalovich 1990, Marston and Huber 1996). In vitro studies have shown that the thick and thin filaments can be cross-linked via caldesmon in an interaction that is independent of crossbridge activity (Marston and Redwood 1992). In vitro, the cross-linking interaction promotes myosin filament polymerization by stabilizing the myosin filaments after formation (Ikebe and Reardon 1988, Hemric et al 1994). In an vitro motility assay, caldesmon also appears to promote the interaction and movement of actin filaments over myosin filaments, possibly because of its "tethering" action linking the two filaments together (Haeberle et al 1992). Thus, it has also been proposed that the caldesmon-myosin interaction may function physiologically to promote the organization of actin and myosin filaments in smooth muscle cells and to stabilize dephosphorylated myosin filaments (Katayama et al 1995, Katayama and Ikebe 1995, Marston et al 1992, Yamashiro and Matsumura 1991, Hemric et al 1994, Ikebe and Reardon 1988). However, recent electron microscopic images of native thin filaments do not provide support for a cross-linking function of caldesmon in vivo. Moody et al 1990 examined electron microscopic images of native thin filaments from vertebrate smooth muscle, and found no evidence for lateral projections extending away from the shaft of the thin filaments, although they observed lateral projections on reconstituted thin filaments. They concluded that it was unlikely that caldesmon acts as a cross-linking protein in vivo. Their conclusions are supported by electron microscopic images of native thin filaments obtained by Mabuchi et al (1993), which demonstrated that both ends of the caldesmon molecule interact with the actin filament.

Thus, the physiologic role of caldesmon binding to myosin is presently uncertain.

4.1.4
Role of calponin in the thin filament

Calponin is a 35-kDa Ca^{2+}-calmodulin binding protein that was originally shown by Takahashi et al (1986) to bind to and sediment with F-actin. It is more or less exclusively expressed in smooth muscle tissues in vivo (Takahashi et al 1987, Gimona et al 1990, Takeuchi et al. 1991). The expression of calponin is down-regulated in smooth muscle cells cultured in vitro (Gimona et al 1990, Birukov et al 1991, Durand-Arczynska et al. 1993). In isolated toad stomach and chicken gizzard smooth muscle cells, calponin co-distributes with longitudinally oriented bundles that also contain actin and tropomyosin, indicating a close association of calponin with actin in smooth muscle cells in vivo (Walsh et al 1993, North et al 1994a).

Calponin has a very high binding affinity for smooth muscle F-actin which is 78 fold greater than for skeletal muscle actin (Winder et al 1991). The affinity of calponin for actin is independent of the presence of tropomyosin. The maximum in vitro binding stoichiometry of calponin to smooth muscle actin is 1 mol calponin /3 mol actin (Winder et al 1991), whereas stoichiometric ratios of calponin to actin in vivo have been estimated variously to be 1:7 (Takahashi et al 1986), 1:10 (Nishida et al 1990), and 1:16 (Marston 1991). Differences in vivo and in vitro stoichiometry may reflect an uneven distribution of calponin among different compartments of actin filaments (See Section 4.2).

The function of calponin and its structural relationship to other proteins on the actin filament is currently unclear. The calponin molecule is flexible and elongated with a length (16 nm) sufficient to span three actin subunits along an actin filament (Stafford et al 1995). It binds to the C terminus on actin and therefore does not block the strong-binding site for myosin (Mezgueldi et al 1992, Bonet-Kerrache and Mornet 1995, Hodgkinson et al 1997). Like caldesmon, calponin inhibits actin-activated myosin MgATPase activity in vitro (Winder and Walsh 1990, Gimona and Small 1996), and also inhibits actin filament movement in vitro motility assays (Shirinsky et al 1992, Haeberle et al 1992). However, in contrast to caldesmon, the inhibitory effect of calponin on actomyosin ATPase activity is not tropomyosin-dependent (Winder and Walsh 1990). Haeberle and Hemric

(1994) have proposed a model in which calponin interacts with caldesmon and tropomyosin to regulate the interaction of actin and myosin.

Hodgkinson et al (1997) analyzed three-dimensional reconstructions of images of reconstituted smooth muscle actin filaments and concluded that calponin is located peripherally along the long-pitch of the actin helix with the main calponin mass over subdomain 2 of actin. The positions of calponin binding are near the sites for weak myosin binding to actin but do not block the strong-binding site for myosin. When added to tropomyosin-containing actin filaments, calponin caused a shift in the position of tropomyosin which exposed the strong-binding sites for myosin previously covered by tropomyosin, indicating that the mechanism for the inhibition of actomyosin ATPase by calponin is not analogous to that by troponin in skeletal muscle. However, the binding sites for calponin on actin are near the binding sites for a number of actin-associated proteins, that could allow calponin to compete with or interact with these actin-binding proteins in vivo.

There is considerable controversy over whether calponin and caldesmon interact on the same filament in vivo. In vitro, calponin competes with caldesmon for closely spaced sites on the actin molecule and there is evidence that the two do not complex on the same filament (Makuch et al 1991, Mezgueldi et al 1992). In addition, the inhibitory effects of calponin and caldesmon on actin-activated ATPase activity appear to be unaffected by each others presence. These observations seem to support ultrastructural evidence that these proteins may localize to different actin filaments or different locations on the same filament (North et al 1994a, Makuch et al 1991) (see Section 4.2.2).

4.1.5
Other thin filament proteins

Additional proteins may also be tightly bound to actin filaments. Among these are filamin (also called actin binding protein (ABP)), the calponin-like protein SM22, and myosin light chain kinase (Lin et al 1997). Very little is currently known about the function of SM22. Filamin is particularly abundant in smooth muscle tissues, estimated as 30–40% of the myosin content. It is not specific for smooth muscle and is also present in many other tissues and cell types (Wang 1977). Filamin is a large homodimer which belongs to a class of rod-shaped Ca^{2+}-insensitive actin cross-linking proteins (Craig and Pollard 1982, Geiger 1983). Under physiologic condi-

tions filamin exists as an elongated 500 kDa dimeric molecule formed
from the antiparallel end-to-end attachment of its monomer chains (Wang
1977). In vitro, molecules of filamin spontaneously connect polymerizing
actin filaments to form an elastic gel (Stossel 1993). Although little is
known of its function in smooth muscle cells, there is evidence that filamin
associates selectively with the or cytoskeletal form of actin (Small et al
1986).

4.2
Functional specialization of thin filaments in smooth muscle cells

4.2.1
Actin isoforms

The functional role of actin in smooth muscle in active shortening and
force development is widely accepted. However, it is also well known that
actin filaments play multiple roles in non-muscle cells, where they are
responsible for the maintenance of cell shape and organization as well as
motility (Stossel 1993). There is evidence from a number of laboratories
that actin filamants of different isoforms may be functionally specialized
in smooth muscle cells and localized to different physical domains within
the cell (Lehman et al 1987, North et al 1994b, Small 1995).

At least six different isoforms of actin have been identified in vertebrate
tissues, each encoded by a different gene (Vanderkerckhove and Weber
1978, Reddy et al 1990). Four of these isoforms have been identified in the
smooth muscle tissues of warm-blooded vertebrates: α and γ smooth
muscle actin, and β and γ non-muscle actin (Kabsch and Vanderkerckhove 1992). These actin variants have approximately 95% amino acid
sequence homology, and differ primarily in their N-terminal sequence
(Pollard and Cooper 1986).

The expression of the different actin genes is tissue specific. For example, the expression of γ-actin is highest in visceral smooth muscles and the
expression of α-actin is highest in vascular smooth muscles (Fatigati and
Murphy 1984, Hartshorne 1987). Some smooth muscle tissues contain a
mixture of both α and γ isoforms; others express one or the other exclusively (Hartshorne 1987). The β or "cytoskeletal" isoform of actin appears
to be a significant component of all smooth muscle tissues. No functional
differences among these isoforms or the skeletal muscle isoform have been
discerned with respect to the activation of myosin ATPase, actin filament

sliding velocity over myosin in vitro, or actin polymerization (Gordon et al 1977, Mossakowska and Strezelecka-Golaszewska 1985 Umemoto and Sellers 1990; Harris and Warshaw 1993). However, differences in the relative levels of expression of "cytoskeletal" β actin versus α or γ "contractile" actin occur at different times during the smooth muscle development (Eddinger and Murphy 1991, Hirai and Hirabayashi 1983). Evidence that different actin isoforms are localized to different functional domains of the smooth muscle cell (See Section 4.2.2) has led to the suggestion that the differences in the N-terminal sequence of different actin isoforms may specialize subsets of actin to different functions by directing their binding to different actin-binding proteins.

4.2.2
Actin isoform specific localization of thin filament proteins

The distribution of actin and tropomyosin are generally agreed to be highly correlated throughout the cytoplasm, and in native thin filaments, tropomyosin appears to be bound to all actin filament isoforms (Mabuchi et al 1996, Small et al 1986). However, there is evidence from a number of laboratories that different actin-binding proteins associate selectively with different actin isoforms and localize to different functional domains within smooth muscle cells.

Using immunofluorescence microscopy, Small first distinguished two distinct subsets of actin filaments that were localized to different physical domains within the smooth muscle cell (Small et al 1986, Small 1995). "Cytoskeletal" actin colocalized with the actin-binding protein filamin and with desmin, the major constituent of intermediate filaments. "Contractile" actin was associated with myosin filaments in complementary positions to those occupied by "cytoskeletal" actin. The filamin-containing actin filaments abutted the dense bodies but the myosin-associated actin filaments did not. In subsequent studies of the guinea pig taenia coli and chicken gizzard, Small and colleagues reported that the "cytoskeletal" actin was of the "β" isoform, and that β actin was selectively localized to the dense bodies, the membrane-associated dense plaques, and to the longitudinal channels linking consecutive dense bodies, which are also occupied by filamin and desmin (North et al 1994b, Draeger et al 1990, Small et al 1986). However, β actin was excluded from contractile domain of the cell. In contrast, antibodies specific for the smooth muscle actin isoforms (α and γ) failed to react with dense bodies, but reacted with actin in the contractile or myosin containing domain (North et al 1994b). Small proposed that β cytoskeletal" actin and intermediate filaments link the cytoplasmic dense bodies longitudinally, whereas α and γ contractile" smooth muscle actin connect adjacent longitudinal arrays of dense bodies obliquely (North et al 1994b, Small 1995) (Figure 6). Data supporting the localization of different actin isoforms to different cellular domains has not gone unchallenged. Drew et al (1991) found that the distribution of actin isoforms within thin filaments isolated from adult swine stomach smooth muscle was random, and they detected no significant clustering of actin isoforms within or between filaments. They concluded that actin isoforms are not functionally specialized in this tissue.

There is fairly consistent ultrastructural evidence from a number of different laboratories indicating that caldesmon distributes preferentially to actin in the actomyosin domain (Furst et al 1986, Lehman et al 1987, Mabuchi et al 1996, North et al 1994a). Ultrastructural studies by Mabuchi et al (1996) identified the subset of actin filaments that colocalize with caldesmon to be γ actin; they found γ actin to be preferentially localized in the vicinity of myosin filaments.

The results of studies describing the cellular localization of calponin are less consistent. Walsh et al (1993) found calponin to be homogeneously distributed with actin and tropomyosin throughout toad stomach and chicken gizzard smooth muscle cells. North et al (1994a) reported some-what similar observations in chicken gizzard smooth muscle, although these investigators concluded that calponin was more concentrated in regions of cytoskeletal" β actin. North et al (1994a) also observed calponin in cytoplasmic dense bodies as well as at the adhesion plaques at the cell surface. Mabuchi et al (1996) observed calponin to be distributed primarily at the periphery of cytoskeletal structures in the same general region as desmin, and very often adjacent to β-actin. In these studies, the distribution of calponin bore no similarity to that of caldesmon and myosin. Mabuchi et al (1996) concluded that caldesmon was associated selectively with contractile actin, whereas calponin associated selectively with cytoskeletal actin. Biochemical data obtained by Lehman (1991) appears to support ultrastructural evidence that calponin is selectively bound to a subset of actin filaments. Using calponin and caldesmon-specific antibodies to immunoprecipitate thin filament fractions, Lehman et al obtained subsets of actin filaments which were selectively enriched in either calponin and filamin or in caldesmon.

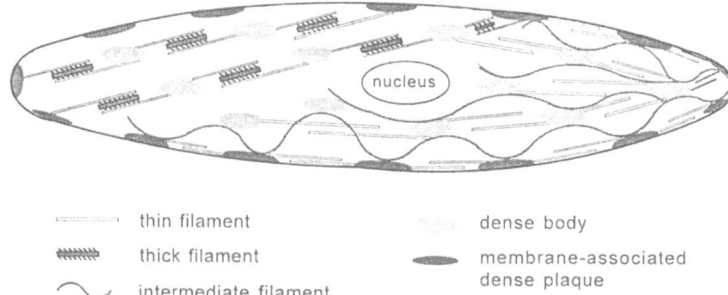

┈┈┈➤ thin filament	⬤ dense body
〰〰〰 thick filament	⬭ membrane-associated
⌒⌒ intermediate filament	dense plaque

Fig. 6. Organization of structural elements within the smooth muscle cell. For purposes of simplicity, the contractile filaments are illustrated on the left side of the cell, whereas the cytoskeletal filaments are illustrated on the right side. Thin filaments composed of "contractile" actin (α or γ isoforms) are proposed to associate with thick filaments. Thin filaments composed of "cytoskeletal" actin do not associate with myosin (as reviewed by Small (1995)). Actin filaments anchor at dense bodies in the cytosol and dense plaques at the cell membrane via linker proteins. Intermediate filaments link chains of dense bodies. Intermediate filaments are also linked to the cell surface at dense plaques

Ultrastructural evidence that calponin binds selectively to "cytoskeletal" actin has led some investigators to propose that the function of calponin in smooth muscle may be structural rather than regulatory. Based on evidence that calponin binds to desmin as well as β actin, Mabuchi et al proposed that calponin may function as a bridging protein between actin and intermediate filament networks at dense bodies (Mabuchi et al 1997) (See Section 2.4). Ultrastructural analysis of thin filaments by Hodgkinson et al (1997) indicates that the location of calponin on F-actin is similar to that of the actin cross-linking protein fimbrin, as well as to that of α-actinin and gelsolin. They suggested that a possible function of calponin may be the competitive inhibition of the binding of these or other actin-binding proteins to the actin filament. This could serve to regulate the building or remodeling of the actin cytoskeleton and thereby affect its mechanical properties (e.g. See section 5.2).

5
Ultrastructural basis for the contractile behavior of smooth muscle

The fundamental mechanism underlying force development in smooth and striated muscles is essentially the same. In both tissues, it is generally agreed that force production occurs as a result of a cyclical interaction of crossbridges that causes the sliding of adjacent actin and myosin filaments with respect to each other — the "sliding filament model" of contraction first proposed by Huxley and Niedergerke (1954) and Huxley and Hanson (1954). There are also fundamental similarities in the mechanical properties of smooth and striated muscles. A hyperbolic force-velocity relationship is characteristic of both tissues, although shortening velocities are lower in smooth than in striated muscles. Both smooth and striated muscles exhibit a length-dependence of isometric tension, in which greater isometric force is obtained as muscle length is increased until an optimal length is achieved at which force is maximal. Active tension then declines as the muscle is lengthened beyond the optimal length.

Because of the qualitative similarities of the mechanical properties in smooth and striated muscles, much research over the past several decades has focused on determining whether the functional behavior of these muscles can be explained on the basis of similar mechanisms. However, the unique structural features of smooth muscle are likely to underly some aspects of its contractile properties which differ significantly from those of striated muscles. In addition, smooth muscle tissues possess a number of distinctive functional properties that are not easily accounted for on the basis of models developed to account for the properties of striated muscles.

5.1
The sliding filament model as a basis for the length-tension properties of smooth muscle

The structure of the contractile unit is fundamental to the interpretation of the relationship between muscle length and isometric tension in skeletal muscle. Each sarcomere unit contains bipolar thick filaments with a central bare zone surrounded by pairs of actin filaments of opposite polarity. The actin filaments are anchored at Z-bands that are aligned in parallel. The length-tension behavior of skeletal muscle has been interpreted in terms of changes in the overlap between the thick and thin filaments as proposed by Gordon, Huxley and Julian (1966). In single skeletal muscle fibers, maximal active isometric tension is observed when the crossbridge arrays of the thick filaments are fully overlapped by the thin filaments. When the fiber is stretched beyond its optimum length, isometric force declines due to a reduction in the overlap of actin and myosin filaments with a consequent decrease in the number of crossbridges that can interact with actin. At fiber lengths shorter than the optimum length, the decline in active force is attributed to a decline in the interaction of actin with the myosin heads on thick filaments as the opposing thin filaments begin to overlap with the central bare zone on the thick filaments and with each other. At extremely short lengths, tension falls dramatically due to structural interference with contraction caused by myosin filaments abutting the Z lines, but these lengths are probably not reached under physiological conditions (Gordon et al 1966). The decline in force with decreasing muscle length in skeletal muscle has also been shown to be partly a function of a decrease in Ca^{2+}-activation (Schoenberg and Podolsky 1972, Edman 1980, Stephenson and Wendt 1984, Taylor and Rudel 1970).

Many aspects of this model are difficult to extrapolate to the smooth muscle cell, which does not exhibit the regular sarcomere structures of skeletal muscle. The prevailing evidence suggests that the thick filaments of smooth muscle are side-polar with no central bare zone (Section 3.4). Actin filaments are not anchored at Z-lines arranged in parallel but at elliptical dense bodies throughout the cytoplasm and at dense bands that can span the entire sarcolemma (See Sections 2.3 and 2.4 and Figure 6). Thus, a structural mechanism that can account for the length-tension properties of smooth muscle on the basis of changes in filament overlap is not evident. This is particularly true for the ascending limb of the length-tension curve, which is the physiologic range of length for most smooth muscle tissues. The structural features that have been proposed as the basis of the length-tension behavior of striated muscles in this range are specific to striated muscles. Although changes in the overlap of actin and myosin filaments in smooth muscle may contribute to its length-tension behavior, there is currently no evidence to substantiate this.

5.2
Mechanisms for the mechanical properties of smooth muscle.

The available data suggest that a number of mechanisms may contribute to the length-dependence of tension in smooth muscle tissues. These include mechanical interactions between adjacent cells, length-dependent changes in the activation of contractile filaments, and mechanosensitive alterations in the organization or length of the contractile filaments.

As the neighboring cells of smooth muscle tissues are mechanically coupled, the contractile apparatus of each individual cell exerts tension on its neighbors. Force transmission across the sarcolemma of smooth muscle cells occurs at membrane-associated dense plaques found over the entire cell surface. When isolated smooth muscle cells contract, the points of attachment of contractile filaments are drawn into the cell, resulting in out-pouching of the membrane areas between the plaques (Harris and Warshaw 1991, Draeger et al 1990).

The mechanical interactions between neighboring cells may be an important factor limiting the shortening and force development of smooth muscle at short muscle lengths (Meiss 1992, 1993, 1997). This is supported by evidence that intact smooth muscle tissues subjected to mild digestion of extracellular connective tissue with collagenase shorten more than untreated tissues (Meiss 1997, Bramley et al 1995). The observation that single smooth muscle cells, in contrast to smooth muscle tissues, develop similar levels of maximal isometric force when contractions are initiated over a wide range of cell lengths also supports the idea that a significant portion of the length-dependence of tension results from intercellular mechanical interactions (Harris and Warshaw 1991).

Some of the length-dependence of isometric force in smooth muscle appears to reflect length-dependent changes in the activation of contractile proteins (Mehta et al 1996, Hai 1991, Rembold and Murphy 1990, Wingard et al 1995). In both tracheal and arterial smooth muscle tissues, the lower levels of force associated with the stimulation of muscles under isometric conditions at muscle lengths below the optimal length (L_o) are associated with lower levels of Ca^{2+} activation and myosin light chain phosphorylation. This suggests that mechanosensitive mechanisms may modulate the signaling pathways that regulate the activation of contractile proteins; however the cellular mechanisms which mediate this property have not been determined.

Evidence from a number of laboratories suggests that there is not a fixed relationship between tissue length and active force in smooth muscle (Pratusevitch et al 1995, Meiss 1993, Gunst et al 1993 and 1995, Harris and Warshaw 1991). When isolated smooth muscle cells are subjected to stepwise shortening during contractile stimulation, the force-length curves are identical, but shifted along the length-axis in relation to the length at which the contraction is initiated (Harris and Warshaw 1991)(Figure 7). An analogous phenomenon is observed in intact trachealis smooth muscle tissue. When tracheal muscle strips are contracted isometrically, then rapidly shortened to a minimal length and slowly lengthened, the length-tension curves obtained during the slow stretch of the activated muscle are shifted in relation to the muscle length at which the contraction is initiated (Gunst 1982) (Figure 7). The relationship between muscle length and both active isometric tension and isotonic shortening velocity can be shifted similarly (Gunst 1986, Gunst et al 1993, Meiss 1993). When a muscle is contracted isometrically at one length and then shortened and allowed to redevelop force isometrically, the rate and magnitude of force redevelopment at the shorter length decreases as the length at which the contraction is initiated is increased (Gunst 1986). Shortening velocities compared at same muscle length during isotonic shortening are also lower for contractions that are initiated at longer muscle lengths (Gunst et al 1993, Meiss 1993). These observations have led to suggestions that contractile element length and smooth muscle cell length are not tightly coupled, and that smooth muscle cells may be capable of modulating the organization of the contractile apparatus in response to changes in their physical environment (Harris and Warshaw 1991, Gunst et al 1995, Gunst et al. 1993, Meiss 1993, Pratusevitch et al 1995).

There is physiologic evidence to suggest that the organization of the contractile apparatus is relatively fluid in smooth muscle cells at rest, and then becomes fixed at the onset of contractile stimulation. Isolated smooth muscle cells at rest exhibit considerable viscoelasticity (Van Dijk et al 1984). Little or no passive force can be sustained by single smooth muscle cells even when they are subjected to substantial stretch. However, this property disappears after contractile activation. The mechanical environment of the smooth muscle cell or tissue at the beginning of contractile stimulation quite reproducibly determines its response to mechanical perturbations later in the contraction, suggesting that the structural organization of the cell is "set" at the onset of contractile activation. This is also supported by observations that the shortening behavior of smooth muscles is reproducible even when significant changes in length are involved (Gunst et al 1993, Shen et al 1997, Gunst 1983).

When activated smooth muscle is allowed to shorten isotonically, the rate of shortening decreases continuously during the period of shortening until an equilibrium length is reached. This phenomenon is widely observed among different smooth muscle tissues, and occurs in single

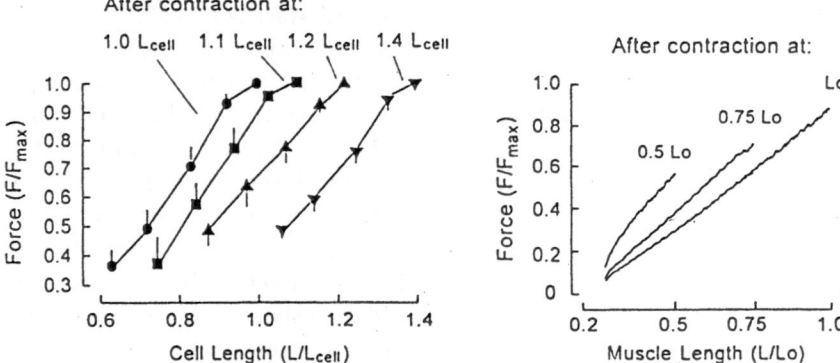

Fig.7. Plasticity of length-tension relationship in smooth muscle cells and tissues. *Left:* Active length-tension curves obtained in single isolated toad stomach cells. Cells were activated at lengths of 1.0, 1.1, 1.2, and 1.4 times the resting cell length (Lcell) after which they were subjected to step reductions in length while activated. Symbols indicate forces generated after each step reductions in length from each starting length. Although the curves are similar in shape, their position on the length axis is shifted in relation to their starting length. Redrawn from Harris and Warshaw (1991) with permission of the American Physiological Society. *Right:* Length-force curves generated by a trachealis muscle strip that was contracted isometrically at different starting lengths (as labeled) in successive contractions, and then rapidly shortened to 0.2 Lo and slowly stretched back to the starting length while still activated. The position of the force-length curves is displaced along the length axis in relation to the starting length (Gunst and Wu, unpublished data).

smooth muscle cells, indicating that it represents a property of smooth muscle that is fundamental to the organization of the contractile apparatus (Harris and Warshaw 1990, Gunst et al 1993, Meiss 1994, Meiss 1993, Arner and Hellstrand 1985). One explanation for the phenomenon is that the internal load on the contractile apparatus increases during isotonic shortening in a length-dependent manner. It is tempting to speculate that the internal load may arise from cytoskeletal connections that become fixed early in the activation of the muscle and subsequently provide an internal resistance to shortening. This explanation can account the observation that degree of resistance to shortening at any muscle length can be "reset" by the initiating the isotonic contraction at a new length (Gunst et al 1993). Thus, intracellular structures contributing the internal resistance do not appear to bear a fixed relationship to muscle length.

The cellular mechanisms underlying these unique mechanical properties of smooth muscle have not been determined. The spatial information required for reproducible shortening and tension development is preserved in glycerinated smooth muscle strips even after the dissolution of myosin filaments and the reintroduction of myosin, suggesting that the information that determines cell shape and contractile system organization is retained in the cytoskeleton (Cande et al 1983). It is possible that smooth muscle cells possess some of the mechanisms present in non-muscle cells that allow for the rapid reorganization of actin filaments. In non-muscle cells, mechanical signals sensed at integrin-containing focal adhesion sites are transmitted to the cell cytoplasm and nucleus by the cytoskeleton (Maniotus et al 1997). The membrane associated dense plaque sites in smooth muscle may play an analogous role in sensing and transmitting mechanical information from the cell surface to other cytoplasmic proteins via the cytoskeletal network, and may also initiate events that regulate cytoskeletal structure (Tang et al 1998). Although there has been little investigation of these processes in smooth muscle, the contractile activation of smooth muscle stimulates the phosphorylation of talin and paxillin, proteins localized to the membrane-associated dense plaque sites (Pavalko et al 1995, Wang et al 1996). Talin is a linker protein that attaches actin filaments to the membrane at dense plaque sites and that is also capable of nucleating actin filament formation (Burridge and Chrzanowska-Wodnicka 1996, Yamada and Geiger 1997). The phosphorylation of paxillin is associated with the formation of actin stress fibers and the assembly of focal adhesion sites in non-muscle cells (e.g., Seufferlein and Rozengurt, 1994).

Alterations in the length or number of myosin filaments have also been proposed to play a role in regulating the length-tension properties of smooth muscle. Pratusevich et al (1995) proposed that the phosphorylation of myosin light chains caused by contractile activation may stimulate the polymerization of myosin filaments, lengthening myosin filaments to adapt to contractile activation at longer muscle lengths, and perhaps also inducing the formation of new myosin filaments. Although ultrastructural evidence indicates that the number of myosin filaments may increase modestly during the contraction of some smooth muscle tissues (See Section 3.3); potential effects of muscle length on myosin filament length or number have not been evaluated.

6
Conclusions

There have been very significant advances in the past decade in our understanding of the structure of many components of the contractile apparatus of smooth muscle cells. We can point to major advances in determining the molecular basis for the motor properties of myosin and for the regulation of its activity by phosphorylation. There are extensive experimental observations that define the structure and organization of myosin filaments within smooth muscle cells. Very detailed information regarding the molecular structure of actin is also available, as well as the molecular basis for its interaction with myosin during crossbridge cycling. The basic organization of the smooth muscle thin filament is known, although there are still many open questions with regard to the localization of actin binding proteins on the thin filament and the function of these proteins. There has also been an enormous growth in our knowledge of the molecular organization of membrane sites at which the contractile system is anchored, the membrane-associated dense plaques, although much work remains to be done in determining the function of the proteins localized to these sites.

The advances in knowledge of the molecular structure of the individual components of the contractile apparatus of smooth muscle provided support for a basic model of force generation based on the relative sliding of adjacent actin and myosin filaments. The tension generated by these contractile filaments appears to be transmitted throughout the cell via a network of actin filaments anchored at dense bodies within the cytosol and at dense plaques at the cell membrane. Proteins localized to the membrane-associated dense plaques mediate the transmission of force across the membrane. However, we are still unable to satisfactorily account for some of the most fundamental functional attributes of smooth muscle cells and tissues. The mechanisms underlying the length-tension behavior of smooth muscle and of its mechanosensitivity remain to be determined. More information regarding the function of the many protein constituents of the contractile filaments and cytoskeletal apparatus as well as the way in which these proteins interact within the cell may help in this endeavor. The basis for the diversity of specialized functional attributes of different smooth muscle tissue types is also not understood, but may be clarified by molecular analysis of the functional characteristics of contractile and cytoskeletal protein isoforms present in smooth muscle cells. Finally, information on the dynamic regulation of smooth muscle cell structure, an area about which there is presently very little information, may also provide some important insights into some of the more unique functional properties of smooth muscle cells.

Acknowledgments. I am indebted to Drs. Richard Meiss, David Warshaw, Roger Craig and Dolly Mehta for expert review and discussion of the manuscript. I am also very grateful to Ming-Fang Wu for his assistance in preparing the manuscript and figures.

References

Adelstein RS, Sellers JR (1996) Myosin structure and function. In: Barany M (ed) Biochemistry of Smooth Muscle Contraction. Academic Press Inc. San Diego, California, pp 3-19

Applegate D, Pardee JD (1992) Actin-facilitated assembly of smooth muscle myosin induces formation of actomyosin fibrils. J Cell Biol 117:1223-1230

Arner A, Hellstrand P (1985) Effects of calcium and substrate on force-velocity relation and energy turnover in skinned smooth muscle of the guinea-pig. J Physiol 360:347-365

Ashton FT, Somlyo AV, Somlyo AP (1975) The contractile apparatus of vascular smooth muscle: intermediate high voltage stereo electron microscopy. J Mol Biol 98:17-29

Babij P (1993) Tissue-specific and developmentally regulated alternative splicing of a visceral isoform of smooth muscle myosin heavy chain. Nucleic Acids Res 21:1467-1471

Babij P, Periasamy M (1989) Myosin heavy chain isoform diversity in smooth muscle is produced by differential RNA processing. J Mol Biol 210:673-679

Bagby RM (1990) Ultrastructure, cytochemistry, and organization of myofilaments in vertebrate smooth muscle cells. In: Motta PM (ed) Ultrastructure of Smooth Muscle. Kluwer Academic Publishers, Boston, pp 23-61

Balint M, Sreter FA, Wolf I, Nagy B, Gergely J (1975) The substructure of heavy meromyosin. The effect of Ca^{2+} and Mg^{2+} on the tryptic fragmentation of heavy meromyosin. J Biol Chem 250:6168-6177

Bárány K, Polyak E, Bárány M (1992) Protein phosphorylation in arterial muscle contracted by high concentration of phorbol dibutyrate in the presence and absence of Ca^{2+}. Biochimica et Biophysica Acta 1134:233-241

Bennett GS, Fellini SA, Croop JM, Otto JJ, Bryan J, Holtzer H (1978) Differences among 100-A filament subunits from different cell types. Proc Natl Acad Sci U S A 75:4364-4368

Birukov KG, Stepanova OV, Nanaev AK, Shirinsky VP (1991) Expression of calponin in rabbit and human aortic smooth muscle cells. Cell Tissue Res 266:579-584

Block SM (1996) Fifty ways to love your lever: myosin motors. Cell 87:151-157

Bond M, Somlyo AV (1982) Dense bodies and actin polarity in vertebrate smooth muscle. J Cell Biol 95:403413

Bonet-Kerrache A, Mornet D (1995) Importance of the C-terminal part of actin in interactions with calponin. Biochem Biophys Res Comm 206:127132

Bramley AM, Roberts CR, Schellenberg RR (1995) Collagenase increases shortening of human bronchial smooth muscle in vitro. Am J Respir Crit Care Med 152:15131517

Bretscher A (1984) Smooth muscle caldesmon. Rapid purification and F-actin cross-linking properties. J Biol Chem 259:1287312880

Burridge K, Chrzanowska-Wodnicka M (1996) Focal adhesions, contractility, and signaling. Annu Rev Cell Dev Biol 12:463518

Buss F, Jockusch BM (1989) Tissue-specific expression of profilin. FEBS Lett 249:3134

Cande WZ, Tooth PJ, Kendrick-Jones J (1983) Regulation of contraction and thick filament assembly-disassembly in glycerinated vertebrate smooth muscle cells. J Cell Biol 97:10621071

Caspar DL, Cohen C, Longley W (1969) Tropomyosin: crystal structure, polymorphism and molecular interactions. J Mol Biol 41:87107

Cavaille F, Janmot C, Ropert S, dAlbis A (1986) Isoforms of myosin and actin in human, monkey and rat myometrium. Comparison of pregnant and non-pregnant uterus proteins. Eur J Biochem 160:507513

Chalovich JM, Cornelius P, Benson CE (1987) Caldesmon inhibits skeletal actomyosin subfragment-1 ATPase activity and the binding of myosin subfragment-1 to actin. J Biol Chem 262:57115716

Cooke PH, Fay FS (1972) Correlation between fiber length, ultrastructure, and the length-tension relationship of mammalian smooth muscle. J Cell Biol 52:105116

Cooke PH, Fay FS, Craig R (1989) Myosin filaments isolated from skinned amphibian smooth muscle cells are side-polar. J Muscle Res Cell Motil 10:206220

Cooke PH, Kargacin G, Craig R, Fogarty K, Fay FS (1987) Molecular structure and organization of filaments in single, skinned smooth muscle cells. Prog Clin Biol Res 245:125

Craig R, Megerman J (1977) Assembly of smooth muscle myosin into side-polar filaments. J Cell Biol 75:990996

Craig R, Smith R, Kendrick-Jones J (1983) Light-chain phosphorylation controls the conformation of vertebrate non-muscle and smooth muscle myosin molecules. Nature 302:436439

Craig SW, Pollard TD (1982) Actin binding proteins. Trends Biochem Sci 7:8892

Cremo CR, Sellers JR, Facemyer KC (1995) Two heads are required for phosphorylation-dependent regulation of smooth muscle myosin. J Biol Chem 270:21712175

Cross RA, Geeves MA, Kendrick-Jones J (1991) A nucleationelongation mechanism for the self-assembly of side polar sheets of smooth muscle myosin. EMBO J 10:747756

Cross RA, Vandekerckhove J (1986) Solubility-determining domain of smooth muscle myosin rod. FEBS Lett 200:355360

Devine CE, Somlyo AP (1971) Thick filaments in vascular smooth muscle. J Cell Biol 49:636649

Draeger A, Amos WB, Ikebe M, Small JV (1990) The cytoskeletal and contractile apparatus of smooth muscle: contraction bands and segmentation of the contractile elements. J Cell Biol 111:24632473

Draeger A, Stelzer EH, Herzog M, Small JV (1989) Unique geometry of actin-membrane anchorage sites in avian gizzard smooth muscle cells. J Cell Sci 94:703711

Drew JS, Moos C, Murphy RA (1991) Localization of isoactins in isolated smooth muscle thin filaments by double gold immunolabeling. Am J Physiol 260:C133240

Drew JS, Murphy RA (1997) Actin isoform expression, cellular heterogeneity, and contractile function in smooth muscle. Can J Physiol Pharmacol 75:869877

Durand-Arczynska W, Marmy N, Durand J (1993) Caldesmon, calponin and alpha-smooth muscle actin expression in subcultured smooth muscle cells from human airways. Histochem 100:465471

Eddinger TJ, Murphy RA (1991) Developmental changes in actin and myosin heavy chain isoform expression in smooth muscle. Arch Biochem Biophys 284:232237

Eddinger TJ, Wolf JA (1993) Expression of four myosin heavy chain isoforms with development in mouse uterus. Cell Motil Cytoskeleton 25:358368

Edman KA (1980) Depression of mechanical performance by active shortening during twitch and tetanus of vertebrate muscle fibres. Acta Physiologica Scandinavica 109:1526

Estes JE, Selden LA, Kinosian HJ, Gershman LC (1992) Tightly-bound divalent cation of actin. J Muscle Res Cell Motil 13:272284

Fatigati V, Murphy RA (1984) Actin and tropomyosin variants in smooth muscles. Dependence on tissue type. J Biol Chem 259:1438314388

Fay FS, Fujiwara K, Rees DD, Fogarty KE (1983) Distribution of alpha-actinin in single isolated smooth muscle cells. J Cell Biol 96:783795

Fisher AJ, Smith CA, Thoden J, Smith R, Sutoh K, Holden HM, Rayment I (1995a) Structural studies of myosin:nucleotide complexes: a revised model for the molecular basis of muscle contraction. Biophys J 68:19S-26S; discussion 27S

Fisher AJ, Smith CA, Thoden JB, Smith R, Sutoh K, Holden HM, Rayment I (1995b) X-ray structures of the myosin motor domain of Dictyostelium discoideum complexed with MgADP.BeFx and MgADP.AlF4-. Biochem 34:89608972

Fowler VM (1996) Regulation of actin filament length in erythrocytes and striated muscle. Curr Opin Cell Biol 8:8696

Frank ED, Warren L (1981) Aortic smooth muscle cells contain vimentin instead of desmin. Proc Natl Acad Sci U S A 78:30203024

Fuglsang A, Khromov A, Torok K, Somlyo AV, Somlyo AP (1993) Flash photolysis studies of relaxation and cross-bridge detachment: higher sensitivity of tonic than phasic smooth muscle to MgADP. J Muscle Res Cell Motil 14:666677

Furst DO, Cross RA, De Mey J, Small JV (1986) Caldesmon is an elongated, flexible molecule localized in the actomyosin domains of smooth muscle. EMBO J 5:251257

Gabella G (1984) Structural apparatus for force transmission in smooth muscles. Physiol Rev 64:455477

Gabella G (1990) General aspects of the fine structure of smooth muscles. In: Motta PM (ed) Ultrastructure of Smooth Muscle. Kluwer Academic Publishers, Boston, pp 1-22

Geiger B (1983) Membrane-cytoskeleton interaction. Biochimica et Biophysica Acta 737:305341

Gillis JM, Cao ML, Godfraind-De Becker A (1988) Density of myosin filaments in the rat anococcygeus muscle, at rest and in contraction. II. J Muscle Res Cell Motil 9:1829

Gimona M, Herzog M, Vandekerckhove J, Small JV (1990) Smooth muscle specific expression of calponin. FEBS Lett 274:159162

Gimona M, Small JV (1996) Calponin. In: Barany M (ed) Biochemistry of smooth muscle contraction. Academic Press Inc. San Diego,California, pp 91103

Godfraind-De Becker A, Gillis JM (1988) Analysis of the birefringence of the smooth muscle anococcygeus of the rat, at rest and in contraction. I. J Muscle Res Cell Motil 9:917

Gordon AL, Huxley AF, Julian FJ (1966) The variation of isometric tension with sarcomere length in vertebrate muscle fibres. J Physiol 184:170192

Gordon DJ, Boyer JL, Korn ED (1977) Comparative biochemistry of non-muscle actins. J Biol Chem 252:83008309

Graceffa P, Wang CL, Stafford WF (1988) Caldesmon. Molecular weight and subunit composition by analytical ultracentrifugation. J Biol Chem 263:1419614202

Guilford WH, Dupuis DE, Kennedy G, Wu J, Patlak JB, Warshaw DM (1997a) Smooth muscle and skeletal muscle myosins produce similar unitary forces and displacements in the laser trap. Biophys J 72:10061021

Guilford WH, Tyska MJ, Freyzon Y, Warshaw DM, Trybus KM (1997b) Smooth muscle myosin with an elongated neck region produces greater unitary displacements in vitro. Biophys J 72:A227

Gunst SJ (1983) Contractile force of canine airway smooth muscle during cyclical length changes. J Appl Physiol: Respirat Environ Exercise Physiol 55:759769

Gunst SJ (1986) Effect of length history on contractile behavior of canine tracheal smooth muscle. Am J Physiol 250:C14654

Gunst SJ, Meiss RA, Wu MF, Rowe M (1995) Mechanisms for the mechanical plasticity of tracheal smooth muscle. Am J Physiol 268:C126776

Gunst SJ, Russell JA (1982) Contractile force of canine tracheal smooth muscle during continuous stretch. J Appl Physiol :Respirat Environ Exercise Physiol 52:655663

Gunst SJ, Wu MF, Smith DD (1993) Contraction history modulates isotonic shortening velocity in smooth muscle. Am J Physiol 265:C46776

Haeberle JR, Hemric ME (1994) A model for the coregulation of smooth muscle actomyosin by caldesmon, calponin, tropomyosin, and the myosin regulatory light chain. Can J Physiol Pharmacol 72:14001409

Haeberle JR, Trybus KM, Hemric ME, Warshaw DM (1992) The effects of smooth muscle caldesmon on actin filament motility. J Biol Chem 267:2300123006

Hai CM (1991) Length-dependent myosin phosphorylation and contraction of arterial smooth muscle. Pflugers Archiv 418:564571

Harris DE, Warshaw DM (1990) Slowing of velocity during isotonic shortening in single isolated smooth muscle cells. Evidence for an internal load. J Gen Physiol 96:581601

Harris DE, Warshaw DM (1991) Length vs. active force relationship in single isolated smooth muscle cells. Am J Physiol 260:C110412

Harris DE, Warshaw DM (1993) Smooth and skeletal muscle myosin both exhibit low duty cycles at zero load in vitro. J Biol Chem 268:1476414768

Hartshorne DJ (1987) Biochemistry of the contractile process in smooth muscle. In: Johnson L (ed) Physiology of the gastrointestinal tract. Raven press, New York, pp 423482

Helper DJ, Lash JA, Hathaway DR (1988) Distribution of isoelectric variants of the 17,000-dalton myosin light chain in mammalian smooth muscle. J Biol Chem 263:1574815753

Hemric ME, Chalovich JM (1990) Characterization of caldesmon binding to myosin. J Biol Chem 265:1967219678

Hemric ME, Tracy PB, Haeberle JR (1994) Caldesmon enhances the binding of myosin to the cytoskeleton during platelet activation. J Biol Chem 269:41254128

Hewett TE, Martin AF, Paul RJ (1993) Correlations between myosin heavy chain isoforms and mechanical parameters in rat myometrium. J Physiol 460:351364

Hinssen H, DHaese J, Small JV, Sobieszek A (1978) Mode of filament assembly of myosins from muscle and nonmuscle cells. J Ultrastruct Res 64:282302

Hinssen H, Small JV, Sobieszek A (1984) A Ca^{2+}-dependent actin modulator from vertebrate smooth muscle. FEBS Lett 166:9095

Hirai S, Hirabayashi T (1983) Developmental change of protein constituents in chicken gizzards. Developmental Biology 97:483493

Hodge TP, Cross R, Kendrick-Jones J (1992) Role of the COOH-terminal nonhelical tailpiece in the assembly of a vertebrate nonmuscle myosin rod. J Cell Biol 118:10851095

Hodgkinson JL, El-Mezgueldi M, Craig R, Vibert P, Marston SB, Lehman W (1997) 3D Image reconstruction of reconstituted smooth muscle thin filaments containing calponin. J Mol Biol 273:150159

Holmes KC, Popp D, Gebhard W, Kabsch W (1990) Atomic model of the actin filament. Nature 347:4449

Horowitz A, Trybus KM, Bowman DS, Fay FS (1994) Antibodies probe for folded monomeric myosin in relaxed and contracted smooth muscle. J Cell Biol 126:11951200

Huxley AF, Niedergerke R (1954) Structural changes in muscle during contraction. Nature 173:971972

Huxley HE, Hanson EJ (1954) Changes in cross-striations of muscle during contraction and stretch and their structural interpretation. Nature 173:973976

Ikebe M, Hartshorne DJ (1985) Proteolysis of smooth muscle myosin by Staphylococcus aureus protease: preparation of heavy meromyosin and subfragment 1 with intact 20 000-dalton light chains. Biochem 24:23802387

Ikebe M, Hewett TE, Martin AF, Chen M, Hartshorne DJ (1991) Cleavage of a smooth muscle myosin heavy chain near its C terminus by alpha-chymotrypsin. Effect on the properties of myosin. J Biol Chem 266:70307036

Ikebe M, Reardon S (1988) Binding of caldesmon to smooth muscle myosin. J Biol Chem 263:30553058

Ingber D (1991) Integrins as mechanochemical transducers. Curr Opin Cell Biol 3:841848

Inoue T (1990) The three-dimensional ultrastructure of intracellular organization of smooth muscle cells by scanning electron microscopy. In: Motta PM (ed) Ultrastructure of Smooth Muscle. Kluwer Academic Publishers, Boston, pp 6377

Itakura S, Yamakawa H, Toyoshima YY, Ishijima A, Kojima T, Harada Y, Yanagida T, Wakabayashi T, Sutoh K (1993) Force-generating domain of myosin motor. Biochem Biophys Res Commun 196:15041510

Ito M, Dabrowska R, Guerriero V, Jr., Hartshorne DJ (1989) Identification in turkey gizzard of an acidic protein related to the C-terminal portion of smooth muscle myosin light chain kinase. J Biol Chem 264:1397113974

Kabsch W, Mannherz HG, Suck D, Pai EF, Holmes KC (1990) Atomic structure of the actin:DNase I complex. Nature 347:3744

Kabsch W, Vandekerckhove J (1992) Structure and function of actin. Annu Rev Biophys Biomol Struct 21:4976

Kargacin GJ, Cooke PH, Abramson SB, Fay FS (1989) Periodic organization of the contractile apparatus in smooth muscle revealed by the motion of dense bodies in single cells. J Cell Biol 108:14651475

Katayama E, Ikebe M (1995) Mode of caldesmon binding to smooth muscle thin filament: possible projection of the amino-terminal of caldesmon from native thin filament. Biophys J 68:24192428

Katayama E, Scott-Woo G, Ikebe M (1995) Effect of caldesmon on the assembly of smooth muscle myosin. J Biol Chem 270:39193925

Kelley CA, Kawamoto S, Conti MA, Adelstein RS (1991) Phosphorylation of vertebrate smooth muscle and nonmuscle myosin heavy chains in vitro and in intact cells. J Cell Sci Suppl 14:4954

Kelley CA, Sellers JR, Goldsmith PK, Adelstein RS (1992) Smooth muscle myosin is composed of homodimeric heavy chains. J Biol Chem 267:21272130

Kelley CA, Takahashi M, Yu JH, Adelstein RS (1993) An insert of seven amino acids confers functional differences between smooth muscle myosins from the intestines and vasculature. J Biol Chem 268:1284812854

Kelly RE, Rice RV (1969) Ultrastructural studies on the contractile mechanism of smooth muscle. J Cell Biol 42:683694

Kendrick-Jones J, Smith RC, Craig R, Citi S (1987) Polymerization of vertebrate non-muscle and smooth muscle myosins. J Mol Biol 198:241252

Kumar CC, Mohan SR, Zavodny PJ, Narula SK, Leibowitz PJ (1989) Characterization and differential expression of human vascular smooth muscle myosin light chain 2 isoform in nonmuscle cells. Biochem 28:40274035

Kuro-o M, Nagai R, Tsuchimochi H, Katoh H, Yazaki Y, Ohkubo A, Takaku F (1989) Developmentally regulated expression of vascular smooth muscle myosin heavy chain isoforms. J Biol Chem 264:1827218275

Kurzawa SE, Manstein DJ, Geeves MA (1997) Dictyostelium discoideum myosin II: characterization of functional myosin motor fragments. Biochem 36:317323

Lash JA, Helper DJ, Klug M, Nicolozakes AW, Hathaway DR (1990) Nucleotide and deduced amino acid sequence of cDNAs encoding two isoforms for the 17,000 dalton myosin light chain in bovine aortic smooth muscle. Nucleic Acids Res 18:7176

Lehman W (1991) Calponin and the composition of smooth muscle thin filaments. J Muscle Res Cell Motil 12:221224

Lehman W, Craig R, Lui J, Moody C (1989) Caldesmon and the structure of smooth muscle thin filaments: immunolocalization of caldesmon on thin filaments. J Muscle Res Cell Motil 10:101112

Lehman W, Craig R, Vibert P (1994) Ca^{2+}-induced tropomyosin movement in Limulus thin filaments revealed by three-dimensional reconstruction. Nature 368:6567

Lehman W, Denault D, Marston S (1993) The caldesmon content of vertebrate smooth muscle. Biochimica et Biophysica Acta 1203:5359

Lehman W, Sheldon A, Madonia W (1987) Diversity in smooth muscle thin filament composition. Biochimica et Biophysica Acta 914:3539

Lehman W, Vibert P, Craig R (1997) Visualization of caldesmon on smooth muscle thin filaments. J Mol Biol 274:310317

Levine BA, Moir AJ, Audemard E, Mornet D, Patchell VB, Perry SV (1990) Structural study of gizzard caldesmon and its interaction with actin. Binding involves residues of actin also recognised by myosin subfragment 1. Eur J Biochem 193:687696

Lewis JM, Schwartz MA (1995) Mapping in vivo associations of cytoplasmic proteins with integrin beta 1 cytoplasmic domain mutants. Mol Biol Cell 6:151160

Lin PJ, Luby-Phelps K, Stull JT (1997) Binding of myosin light chain kinase to cellular actin-myosin filaments. J Biol Chem 272:74127420

Lowey S, Trybus KM (1995) Role of skeletal and smooth muscle myosin light chains. Biophys J 68:120S126S; discussion 126

Lynch WP, Riseman VM, Bretscher A (1987) Smooth muscle caldesmon is an extended flexible monomeric protein in solution that can readily undergo reversible intra-and intermolecular sulfhydryl cross-linking. A mechanism for caldesmons F-actin bundling activity. J Biol Chem 262:74297437

Mabuchi K, Li B, Ip W, Tao T (1997) Association of calponin with desmin intermediate filaments. J Biol Chem 272:2266222666

Mabuchi K, Li Y, Tao T, Wang CL (1996) Immunocytochemical localization of caldesmon and calponin in chicken gizzard smooth muscle. J Muscle Res Cell Motil 17:243260

Mabuchi K, Lin JJ, Wang CL (1993) Electron microscopic images suggest both ends of caldesmon interact with actin filaments. J Muscle Res Cell Motil 14:5464

Mabuchi K, Wang CL (1991) Electron microscopic studies of chicken gizzard caldesmon and its complex with calmodulin. J Muscle Res Cell Motil 12:145151

Macklin CC (1929) The Musculature of the Bronchi and Lungs. Physiol Rev 9: 160

Makuch R, Birukov K, Shirinsky V, Dabrowska R (1991) Functional interrelationship between calponin and caldesmon. Biochem J 280:3338

Malmqvist U, Arner A (1991) Correlation between isoform composition of the 17 kDa myosin light chain and maximal shortening velocity in smooth muscle. Pflugers Archiv 418:523530

Maniotis AJ, Chen CS, Ingber DE (1997) Demonstration of mechanical connections between integrins, cytoskeletal filaments, and nucleoplasm that stabilize nuclear structure. Proc Natl Acad Sci U S A 94:849854

Marston S (1990) Stoichiometry and stability of caldesmon in native thin filaments from sheep aorta smooth muscle. Biochem J 272:305310

Marston S, Pinter K, Bennett P (1992) Caldesmon binds to smooth muscle myosin and myosin rod and crosslinks thick filaments to actin filaments. J Muscle Res Cell Motil 13:206218

Marston SB (1991) Properties of calponin isolated from sheep aorta thin filaments. FEBS Lett 292:179182

Marston SB, Fraser ID, Huber PA (1994) Smooth muscle caldesmon controls the strong binding interaction between actin-tropomyosin and myosin. J Biol Chem 269:3210432109

Marston SB, Huber PA (1996) Caldesmon. In: Barany M (ed)Biochemistry of smooth muscle contraction. Academic Press Inc. San Diego,California, pp 7790

Marston SB, Lehman W (1985) Caldesmon is a Ca^{2+}-regulatory component of native smooth-muscle thin filaments. Biochem J 231:517522

Marston SB, Redwood CS (1991) The molecular anatomy of caldesmon. Biochem J 279:116

Marston SB, Redwood CS (1992) Inhibition of actin-tropomyosin activation of myosin MgATPase activity by the smooth muscle regulatory protein caldesmon. J Biol Chem 267:1679616800

Marston SB, Redwood CS (1993) The essential role of tropomyosin in cooperative regulation of smooth muscle thin filament activity by caldesmon. J Biol Chem 268:1231712320

Marston SB, Smith CW (1985) The thin filaments of smooth muscles. J Muscle Res Cell Motil 6:669708

Marston SB, Taylor EW (1980) Comparison of the myosin and actomyosin ATPase mechanisms of the four types of vertebrate muscles. J Mol Biol 139:573600

Matsu-ura M, Ikebe M (1995) Requirement of the two -headed structure for the phosphorylation dependent regulation of smooth muscle myosin. FEBS Lett 363:246250

Matsumura F, Lin JJ (1982) Visualization of monoclonal antibody binding to tropomyosin on native smooth muscle thin filaments by electron microscopy. J Mol Biol 157:163171

Meer DP, Eddinger TJ (1996) Heterogeneity of smooth muscle myosin heavy chain expression at the single cell level. Am J Physiol 270:C181924

Mehta D, Wu MF, Gunst SJ (1996) Role of contractile protein activation in the length-dependent modulation of tracheal smooth muscle force. Am J Physiol 270:C24352

Meiss RA (1992) Limits to shortening in smooth muscle tissues. J Muscle Res Cell Motil 13:190198

Meiss RA (1993) Persistent mechanical effects of decreasing length during isometric contraction of ovarian ligament smooth muscle. J Muscle Res Cell Motil 14:205218

Meiss RA (1994) Transient length-related mechanical states in smooth muscle. Can J Physiol Pharmacol 72:13251333

Meiss RA (1997) The influence of intercellular connective tissue on smooth muscle shortening. Biophys J 72:A277(Abstract)

Mezgueldi M, Fattoum A, Derancourt J, Kassab R (1992) Mapping of the functional domains in the amino-terminal region of calponin. J Biol Chem 267:1594315951

Miller CJ, Cheung P, White P, Reisler E (1995) Actin's view of the actomyosin interface. Biophys J 68:50s56s

Milligan RA, Whittaker M, Safer D (1990) Molecular structure of F-actin and location of surface binding sites. Nature 348:217221

Mohammad MA, Sparrow MP (1989) The distribution of heavy-chain isoforms of myosin in airways smooth muscle from adult and neonate humans. Biochem J 260:421426

Moody C, Lehman W, Craig R (1990) Caldesmon and the structure of smooth muscle thin filaments: electron microscopy of isolated thin filaments. J Muscle Res Cell Motil 11:176185

Moore PB, Huxley HE, DeRosier DJ (1970) Three-dimensional reconstruction of F-actin, thin filaments and decorated thin filaments. J Mol Biol 50:279295

Morano I, Erb G, Sogl B (1993) Expression of myosin heavy and light chains changes during pregnancy in the rat uterus. Pflugers Archiv 423:434441

Mornet D, Pantel P, Audemard E, Kassab R (1979) The limited tryptic cleavage of chymotryptic S-1: an approach to the characterization of the actin site in myosin heads. Biochem Biophys Res Commun 89:925932

Mossakowska M, Strzelecka-Golaszewska H (1985) Identification of amino acid substitutions differentiating actin isoforms in their interaction with myosin. Eur J Biochem 153:373381

Murphy RA, Herlihy JT, Megerman J (1974) Force-generating capacity and contractile protein content of arterial smooth muscle. J Gen Physiol 64:691705

Murphy RA, Walker JS, Strauss JD (1997) Myosin isoforms and functional diversity in vertebrate smooth muscle. Comp Biochem Physiol 117B: 5160

Nabeshima Y, Nonomura Y, Fujii-Kuriyama Y (1987) Nonmuscle and smooth muscle myosin light chain mRNAs are generated from a single gene by the tissue-specific alternative RNA splicing. J Biol Chem 262:1060810612

Nagai R, Kuro-o M, Babij P, Periasamy M (1989) Identification of two types of smooth muscle myosin heavy chain isoforms by cDNA cloning and immunoblot analysis. J Biol Chem 264:97349737

Ngai PK, Walsh MP (1984) Inhibition of smooth muscle actin-activated myosin Mg^{2+}-ATPase activity by caldesmon. J Biol Chem 259:1365613659

Nishida W, Abe M, Takahashi K, Hiwada K (1990) Do thin filaments of smooth muscle contain calponin? A new method for the preparation. FEBS Lett 268:165168

Nonomura J (1976) Fine structure of myofilaments in chicken gizzard smooth muscle. In: Yamada E, Mazuhira K, Kurosumi K, et al (eds) Recent Progress In Electron Microscopy of Cells and Tissues. Thieme, Stuttgart, West Germany, pp 4048

North AJ, Gimona M, Cross RA, Small JV (1994a) Calponin is localised in both the contractile apparatus and the cytoskeleton of smooth muscle cells. J Cell Sci 107:437444

North AJ, Gimona M, Lando Z, Small JV (1994b) Actin isoform compartments in chicken gizzard smooth muscle cells. J Cell Sci 107:445455

OBrien EJ, Bennett PM, Hanson J (1971) Optical diffraction studies of myofibrillar structure. Philos Trans R Soc Lond B Biol Sci 261:201208

Pavalko FM, Adam LP, Wu MF, Walker TL, Gunst SJ (1995) Phosphorylation of dense-plaque proteins talin and paxillin during tracheal smooth muscle contraction. Am J Physiol 268:C56371

Pease DC, Molinari S (1960) Electron microscopy of muscular arteries; pial vessels of the cat and monkey. J Ultrastruct Res 3:447468

Pollard TD, Cooper JA (1986) Actin and actin-binding proteins. A critical evaluation of mechanisms and functions. Annu Rev Biochem 55:9871035

Pratusevich VR, Seow CY, Ford LE (1995) Plasticity in canine airway smooth muscle. J Gen Physiol 105:7394

Rayment I, Holden HM, Whittaker M, Yohn CB, Lorenz M, Holmes KC, Milligan RA (1993a) Structure of the actin-myosin complex and its implications for muscle contraction. Science 261:5865

Rayment I, Rypniewski WR, Schmidt-Base K, Smith R, Tomchick DR, Benning MM, Winkelmann DA, Wesenberg G, Holden HM (1993b) Three-dimensional structure of myosin subfragment-1: a molecular motor. Science 261:5058

Rayment I, Smith C, Yount RG (1996) The active site of myosin. Annu Rev Physiol 58:671702

Reddy S, Ozgur K, Lu M, Chang W, Mohan SR, Kumar CC, Ruley HE (1990) Structure of the human smooth muscle alpha-actin gene. Analysis of a cDNA and 5 upstream region. J Biol Chem 265:16831687

Rembold CM, Murphy RA (1990) Muscle length, shortening, myoplasmic [Ca^{2+}], and activation of arterial smooth muscle. Circ Res 66:13541361

Rovner AS, Freyzon Y, Trybus KM (1997) An insert in the motor domain determines the functional properties of expressed smooth muscle myosin isoforms. J Muscle Res Cell Motil 18:103110

Rovner AS, Murphy RA, Owens GK (1986) Expression of smooth muscle and nonmuscle myosin heavy chains in cultured vascular smooth muscle cells. J Biol Chem 261:1474014745

Sata M, Matsuura M, Ikebe M (1996) Characterization of the motor and enzymatic properties of smooth muscle long S1 and short HMM: role of the two-headed structure on the activity and regulation of the myosin motor. Biochem 35:1111311118

Sata M, Stafford WF, 3rd, Mabuchi K, Ikebe M (1997) The motor domain and the regulatory domain of myosin solely dictate enzymatic activity and phosphorylation-dependent regulation, respectively. Proc Natl Acad Sci U S A 94:9196

Schoenberg M, Podolsky RJ (1972) Length-force relation of calcium activated muscle fibers. Science 176:5254

Scholey JM, Taylor KA, Kendrick-Jones J (1980) Regulation of non-muscle myosin assembly by calmodulin-dependent light chain kinase. Nature 287:233235

Scholey JM, Taylor KA, Kendrick-Jones J (1981) The role of myosin light chains in regulating actin-myosin interaction. Biochimie 63:255271

Seufferlein T, Rozengurt E (1994) Sphingosine induces p125FAK and paxillin tyrosine phosphorylation, actin stress fiber formation, and focal contact assembly in Swiss 3T3 cells. J Biol Chem 269:2761027617

Shen X, MF Wu, RS Tepper, SJ Gunst (1997) Mechanisms for the mechanical response of airway smooth muscle to length oscillation. J Appl Physiol 83:731738

Shirinsky VP, Biryukov KG, Hettasch JM, Sellers JR (1992) Inhibition of the relative movement of actin and myosin by caldesmon and calponin. J Biol Chem 267:1588615892

Shirinsky VP, Vorotnikov AV, Birukov KG, Nanaev AK, Collinge M, Lukas TJ, Sellers JR, Watterson DM (1993) A kinase-related protein stabilizes unphosphorylated smooth muscle myosin minifilaments in the presence of ATP. J Biol Chem 268:1657816583

Shoenberg CF (1969) An electron microscope study of the influence of divalent ions on myosin filament formation in chicken gizzard extracts and homogenates. Tissue and Cell 8396

Shoenberg CF, Stewart M (1980) Filament formation in smooth muscle homogenates. J Muscle Res Cell Motil 1:117126

Siemankowski RF, Wiseman MO, White HD (1985) ADP dissociation from actomyosin subfragment 1 is sufficiently slow to limit the unloaded shortening velocity in vertebrate muscle. Proc Natl Acad Sci U S A 82:658662

Sivaramakrishnan M, Burke M (1982) The free heavy chain of vertebrate skeletal myosin subfragment 1 shows full enzymatic activity. J Biol Chem 257:11021105

Small JV (1977) Studies on isolated smooth muscle cells: The contractile apparatus. J Cell Sci 24:327349

Small JV (1985) Geometry of actin-membrane attachments in the smooth muscle cell: the localisations of vinculin and alpha-actinin. EMBO J 4:4549

Small JV (1995) Structure-function relationships in smooth muscle: the missing links. Bioessays 17:785792

Small JV, Furst DO, De Mey J (1986) Localization of filamin in smooth muscle. J Cell Biol 102:210220

Small JV, Herzog M, Barth M, Draeger A (1990) Supercontracted state of vertebrate smooth muscle cell fragments reveals myofilament lengths. J Cell Biol 111:24512461

Smith CW, Pritchard K, Marston SB (1987) The mechanism of Ca^{2+} regulation of vascular smooth muscle thin filaments by caldesmon and calmodulin. J Biol Chem 262:116122

Smith RC, Cande WZ, Craig R, Tooth PJ, Scholey JM, Kendrick-Jones J (1983) Regulation of myosin filament assembly by light-chain phosphorylation. Philos Trans R Soc Lond B Biol Sci 302:7382

Sobieszek A, Bremel RD (1975) Preparation and properties of vertebrate smooth-muscle myofibrils and actomyosin. Eur J Biochem 55:4960

Sobue K, Morimoto K, Inui M, Kanda K, Kakiuchi S (1982) Biomed Res 3:188196

Sobue K, Muramoto Y, Fujita M, Kakiuchi S (1981) Purification of a calmodulin-binding protein from chicken gizzard that interacts with F-actin. Proc Natl Acad Sci U S A 78:56525655

Sobue K, Sellers JR (1991) Caldesmon, a novel regulatory protein in smooth muscle and nonmuscle actomyosin systems. J Biol Chem 266:1211512118

Somlyo AP (1993) Myosin isoforms in smooth muscle: how may they affect function and structure?. J Muscle Res Cell Motil 14:557563

Somlyo AP, Devine CE, Somlyo AV, Rice RV (1973) Filament organization in vertebrate smooth muscle. Philos Trans Royal Soc Lond Biol Sci 265:223229

Somlyo AV (1980) Ultrastructure of vascular smooth muscle. In: Bohr DF, Somlyo AP, Sparks HV (eds) The Cardiovascular System. Vol II: Vascular Smooth Muscle. American Physiological Society, Bethesda, Maryland, pp 3367

Somlyo AV, Butler TM, Bond M, Somlyo AP (1981) Myosin filaments have non-phosphorylated light chains in relaxed smooth muscle. Nature 294:567569

Sparrow MP, Mohammad MA, Arner A, Hellstrand P, Ruegg JC (1988) Myosin composition and functional properties of smooth muscle from the uterus of pregnant and non-pregnant rats. Pflugers Archiv 412:624633

Spudich JA, Huxley HE, Finch JT (1972) Regulation of skeletal muscle contraction. II. Structural studies of the interaction of the tropomyosin-troponin complex with actin. J Mol Biol 72:619632

Stafford WF, Mabuchi K, Takahashi K, Tao T (1995) Physical characterization of calponin. A circular dichroism, analytical ultracentrifuge, and electron microscopy study. J Biol Chem 270:1057610579

Stephenson DG, Wendt IR (1984) Length dependence of changes in sarcoplasmic calcium concentration and myofibrillar calcium sensitivity in striated muscle fibres. J Muscle Res Cell Motil 5:243272

Stossel TP (1993) On the crawling of animal cells. Science 260:10861094

Suzuki H, Onishi H, Takahashi K, Watanabe S (1978) Structure and function of chicken gizzard myosin. J Biochem 84:15291542

Takahashi K, Hiwada K, Kokubu T (1986) Isolation and characterization of a 34,000-dalton calmodulin-and F-actin-binding protein from chicken gizzard smooth muscle. Biochem Biophys Res Commun 141:2026

Takahashi K, Hiwada K, Kokubu T (1987) Occurrence of anti-gizzard P34K antibody cross-reactive components in bovine smooth muscles and non-smooth muscle tissues. Life Sci 41:291296

Takeuchi K, Takahashi K, Abe M, Nishida W, Hiwada K, Nabeya T, Maruyama K (1991) Co-localization of immunoreactive forms of calponin with actin cytoskeleton in platelets, fibroblasts, and vascular smooth muscle. J Biochem 109:311316

Tang D, Mehta D, Gunst SJ (1999). Mechano-sensitive tyrosine phosphorylation of paxillin and focal adhesion kinase in tracheal smooth muscle. Am I Physiol In press, 1999.

Taubman MB, Grant JW, Nadal-Ginard B (1987) Cloning and characterization of mammalian myosin regulatory light chain (RLC) cDNA: the RLC gene is expressed in smooth, sarcomeric, and nonmuscle tissues. J Cell Biol 104:15051513

Taylor EW (1987) Comparative studies on the mechanism of regulation of smooth and striated muscle actomyosin. Prog Clin Biol Res 245:5966

Taylor SR, Rudel R (1970) Striated muscle fibers: inactivation of contraction induced by shortening. Science 167:882884

Toyoshima YY, Kron SJ, McNally EM, Niebling KR, Toyoshima C, Spudich JA (1987) Myosin subfragment-1 is sufficient to move actin filaments in vitro. Nature 328:536539

Trybus KM (1989) Filamentous smooth muscle myosin is regulated by phosphorylation. J Cell Biol 109:28872894

Trybus KM (1994a) Regulation of expressed truncated smooth muscle myosins. Role of the essential light chain and tail length. J Biol Chem 269:2081920822

Trybus KM (1994b) Role of myosin light chains. J Muscle Res Cell Motil 15:587594

Trybus KM (1996) Myosin regulation and assembly. In: Barany M (ed)Biochemistry of smooth muscle contraction. Academic Press Inc. San Diego, California, pp 3746

Trybus KM, Freyzon Y, Faust LZ, Sweeney HL (1997) Spare the rod, spoil the regulation: necessity for a myosin rod. Proc Natl Acad Sci U S A 94:4852

Trybus KM, Huiatt TW, Lowey S (1982) A bent monomeric conformation of myosin from smooth muscle. Proc Natl Acad Sci U S A 79:61516155

Trybus KM, Lowey S (1984) Conformational states of smooth muscle myosin. Effects of light chain phosphorylation and ionic strength. J Biol Chem 259:85648571

Trybus KM, Lowey S (1987) Assembly of smooth muscle myosin minifilaments: effects of phosphorylation and nucleotide binding. J Cell Biol 105:30073019

Tsao AE, Eddinger TJ (1993) Smooth muscle myosin heavy chains combine to form three native myosin isoforms. Am J Physiol 264:H165362

Tsukita S, Ishikawa H (1983) Association of actin and 10 nm filaments with the dense body in smooth muscle cells of the chicken gizzard. Cell Tissue Res 229:233242

Tsukita S, Usukura J, Ishikawa H (1982) Myosin filaments in smooth muscle cells of the guinea pig taenia coli: a freeze-substitution study. Eur J Cell Biol 28:195201

Turner CE, Burridge K (1991) Transmembrane molecular assemblies in cell-extracellular matrix interactions. Curr Opin Cell Biol 3:849853

Umemoto S, Sellers JR (1990) Characterization of in vitro motility assays using smooth muscle and cytoplasmic myosins. J Biol Chem 265:1486414869

Uyeda TQ, Abramson PD, Spudich JA (1996) The neck region of the myosin motor domain acts as a lever arm to generate movement. Proc Natl Acad Sci U S A 93:44594464

VanBuren P, Work SS, Warshaw DM (1994) Enhanced force generation by smooth muscle myosin in vitro. Proc Natl Acad Sci U S A 91:202205

Vandekerckhove J, Weber K (1978) At least six different actins are expressed in a higher mammal: an analysis based on the amino acid sequence of the amino-terminal tryptic peptide. J Mol Biol 126:783802

VanDijk AM, Wieringa PA, van der Meer M, Laird JD (1984) Mechanics of resting isolated single vascular smooth muscle cells from bovine coronary artery. Am J Physiol 246:C27787

Vibert P, Craig R, Lehman W (1993) Three-dimensional reconstruction of caldesmon-containing smooth muscle thin filaments. J Cell Biol 123:313321

Vibert P, Craig R, Lehman W (1997) Steric-model for activation of muscle thin filaments. J Mol Biol 266:814

Wagner PD, Giniger E (1981) Hydrolysis of ATP and reversible binding to F-actin by myosin heavy chains free of all light chains. Nature 292:560562

Wagner PD, Vu ND (1986) Regulation of the actin-activated ATPase of aorta smooth muscle myosin. J Biol Chem 261:77787783

Wagner PD, Vu ND (1987) Actin-activation of unphosphorylated gizzard myosin. J Biol Chem 262:1555615562

Walsh MP, Carmichael JD, Kargacin GJ (1993) Characterization and confocal imaging of calponin in gastrointestinal smooth muscle. Am J Physiol 265:C13718

Wang CL, Chalovich JM, Graceffa P, Lu RC, Mabuchi K, Stafford WF (1991) A long helix from the central region of smooth muscle caldesmon. J Biol Chem 266:1395813963

Wang K (1977) Filamin, a new high-molecular-weight protein found in smooth muscle and nonmuscle cells. Purification and properties of chicken gizzard filamin. Biochem 16:18571865

Wang N, Butler JP, Ingber DE (1993) Mechanotransduction across the cell surface and through the cytoskeleton. Science 260:11241127

Wang Z, Pavalko FM, Gunst SJ (1996) Tyrosine phosphorylation of the dense plaque protein paxillin is regulated during smooth muscle contraction. Am J Physiol 271:C1594602

Warrick HM, Spudich JA (1987) Myosin structure and function in cell motility. Annu Rev Cell Biol 3:379421

Warshaw DM, Desrosiers JM, Work SS, Trybus KM (1990) Smooth muscle myosin cross-bridge interactions modulate actin filament sliding velocity in vitro. J Cell Biol 111:453463

White S, Martin AF, Periasamy M (1993) Identification of a novel smooth muscle myosin heavy chain cDNA: isoform diversity in the S1 head region. Am J Physiol 264:C12528

Winder SJ, Sutherland C, Walsh MP (1991) Biochemical and functional characterization of smooth muscle calponin. Adv Exp Med Biol 304:3751

Winder SJ, Walsh MP (1990) Smooth muscle calponin. Inhibition of actomyosin MgATPase and regulation by phosphorylation. J Biol Chem 265:1014810155

Wingard CJ, Browne AK, Murphy RA (1995) Dependence of force on length at constant cross-bridge phosphorylation in the swine carotid media. J Physiol 488:729739

Xu JQ, Gillis JM, Craig R (1997) Polymerization of myosin on activation of rat anococcygeus smooth muscle. J Muscle Res Cell Motil 18:381393

Xu JQ, Harder BA, Uman P, Craig R (1996) Myosin filament structure in vertebrate smooth muscle. J Cell Biol 134:5366

Yamada KM, Geiger B (1997) Molecular interactions in cell adhesion complexes. Curr Opin Cell Biol 9:7685

Yamashiro S, Matsumura F (1991) Mitosis-specific phosphorylation of caldesmon: possible molecular mechanism of cell rounding during mitosis. Bioessays 13:563568

Zigmond SH (1996) Signal transduction and actin filament organization. Curr Opin Cell Biol 8:6673

Regulation of cross-bridge cycling by Ca^{2+} in smooth muscle

Anders Arner[1] and Gabriele Pfitzer[2]

[1]Department of Physiology and Neuroscience, Lund University, Sölvegatan 19
S-22362 Lund, Sweden
[2]Institut für Vegetative Physiologie, Universität zu Köln, Robert-Kochstr. 39
D-50931 Köln, Germany

Contents

1
Introduction

The cyclic interaction of myosin cross-bridges with actin is the fundamental mechanism of contraction in muscle. The generally accepted mechanism of skeletal muscle contraction is the sliding filament model where force is generated by individual myosin cross-bridges (Huxley and Niedergerke, 1954, Huxley and Hansen, 1954). Although the structural basis for filament sliding and the details of the mechanical cross-bridge cycle are still not clarified, a sliding filament model provides the general frame for analysis also of smooth muscle contraction. The energy for contraction is derived from the hydrolysis of ATP. The energetic cost of contraction depends on the rate of ATP hydrolysis and the mechanical output of the cross-bridges. In striated muscle, these parameters can be regulated by expression of different contractile protein isoforms, i.e. long-term regulation. This property might also exist in smooth muscle, although little explored. However, an important aspect of smooth muscle contraction is that the energetic cost can decrease during tension maintenance of smooth muscle, a phenomenon denoted "latch" (Dillon et al. 1981). This shows the presence of a short-term regulation in smooth muscle, i.e. a modulation of cross-bridge cycling within an individual contraction/relaxation cycle.

An increase in the cytosolic Ca^{2+} concentration triggers contraction (Rüegg 1992). However, Ca^{2+} activates contraction in different types of muscle by fundamentally different mechanisms. In striated muscle, Ca^{2+} acts through an allosteric mechanism by binding to the Ca^{2+}-binding subunit of troponin, troponin C, located on the actin filaments. This leads to a conformational change in tropomyosin and allows the transition from weakly bound to strongly bound myosin cross-bridges (Brenner 1986, Brenner et al. 1982). In contrast smooth muscle is activated through covalent modification of myosin. This mechanism was recognised by Sobieszek (1977), who showed that the actomyosin ATPase of smooth muscle increased significantly when myosin was phosphorylated. Phosphorylation was then shown to be catalysedby a calmodulin dependent enzyme, the myosin light chain kinase (MLCK, Dabrowska et al. 1978). For review of the initial work on the phosphorylation mechanisms in smooth muscle we refer to Hartshorne and Mrwa (1982).

Based on a number of critical and conclusive experiments it is generally accepted that phosphorylation is sufficient to initiate a contraction in smooth muscle cells (e.g. Hoar et al. 1979, Walsh et al. 1982, Itoh et al. 1989).

However, phosphorylation, or perhaps more precisely an increase in phosphorylation, is not always necessary to induce a contraction, as revealed by recent experiments showing that contractions could be initiated in muscle preparations where the light chains were fully dephosphorylated (Malmqvist et al. 1997). In addition to regulation by phosphorylation, several other regulatory systems have been identified. Two thin filament associated regulatory proteins, caldesmon and calponin, have been extensively investigated and several studies have shown effects on actomyosin ATPase and mechanical properties of muscle fibres (Gimona and Small 1996, Marston and Huber 1996, Chalovich and Pfitzer 1997). The precise interaction between regulation by phosphorylation-dependent and phosphorylation-independent mechanisms is still not clearly understood.

The integrated contractile and regulatory systems, outlined above, in the smooth muscle cell are targets not only for the activator Ca^{2+} but for a number of regulatory systems and physiological alterations in the physical and chemical environment. These systems can modulate the contractile responses by altering the coupling between phosphorylation and force and between [Ca^{2+}], phosphorylation and force. Force in smooth muscle is thus not primarily determined by the absolute calcium concentration but rather by a "signalling network" created by the different interacting signalling pathways in the smooth muscle cell, activated or inhibited in response to a wide array of extracellular messages.

2
Cross-bridge interaction

The fundamental process in the generation of force and shortening of muscle cells is the cyclic "cross-bridge" interaction between the thick (myosin containing) and thin (actin containing) contractile filaments. Several of the general concepts in this biological energy conversion have developed over decades from characterisation of skeletal muscle tissues. Although studies of smooth muscle contraction are of considerable interest for understanding of physiological and patho-physiological processes the properties of the cross-bridge interaction between myosin and actin in smooth muscle are not fully understood. Recent work on smooth muscle fibre preparations have revealed unique properties and from the initial view of smooth muscle as a poorly organised contractile system, a picture of a highly specialised motor system is developing.

With regard to the contractile process in smooth muscle the main characteristics, compared to the skeletal system, are: (i) a generally slow cross-bridge turnover, and (ii) a regulation of the contractile kinetics. The shortening velocity of smooth muscle is about 10- to 50-fold slower than that of fast skeletal muscle and the economy of tension maintenance is higher (cf. review by Paul 1989, Hellstrand and Paul 1982). From studies of intact smooth muscle tissue it has been demonstrated that the turnover of actin and myosin, as reflected by shortening velocity or ATPase rate can be modulated during the course of a contraction (Arner and Hellstrand 1980, Siegman et al. 1980, Dillon et al. 1981, Klemt et al. 1981, Hellstrand and Paul 1983). A smooth muscle can thus develop force at a comparatively high rate and then switch into a more economical mode of contraction. This behaviour, the "latch concept", as discussed elsewhere in this review, has been associated with several different regulatory processes. In this chapter the actin-myosin interaction in the organised contractile system in smooth muscle cells is described with focus on some of the processes at the cross-bridge level that might be of importance in determining the slow and economical mode of contraction and be involved in the regulation of smooth muscle contractile kinetics.

Although it is clear that force and shortening in smooth muscle tissue are generated by interaction between thick and thin filaments the exact organisation of the contractile apparatus in smooth muscle is not resolved at present. In particular, the structure of the thick myosin filaments has been very elusive. The presence of myosin structures in intact smooth muscle was shown with X-ray diffraction in the 70's (Lowy et al. 1970). The early studies of myosin structure however became controversial due to the appearance of ribbon-like structures. This controversy was resolved by a careful electron microscopic examination of the thick filament structure which revealed filaments more similar to those observed in skeletal muscle (Devine and Somlyo 1971, Rice et al. 1971, Somlyo 1980 for review) possibly with the appearance of tapered ends (Ashton et al. 1975), compatible with a mode of filament assembly similar to that in skeletal muscle. Other work on synthetic filaments and isolated native filaments have however revealed a "side-polar" or "phase-polar" myosin filament structure (Craig and Megerman 1977, Small 1977, Hinssen et al. 1978, Cooke et al. 1989) which might suggest a unique organisation of smooth muscle thick filaments.

At physiologic ionic strength and in the presence of MgATP, dephosphorylated myosin filaments are disassembled in vitro, forming monomeric myosin species having a sedimentation coefficient of about 10–12S, a folded conformation and a low ATPase activity (Onishi et al. 1978, Suzuki et al. 1978, Craig et al. 1983, Cross et al. 1988). When the LC_{20} on these myosin species are phosphorylated, they unfold and are able to assemble into filaments (Craig et al. 1983, Smith et al. 1983, Cross et al. 1986). These in vitro data may suggest that a fraction of the myosin in the relaxed smooth muscle cells will exist in a depolymerised state. Electron microscopy has however clearly shown that filaments are present in relaxed smooth muscle, where the myosin is dephosphorylated (Somlyo et al. 1981, Tsukita et al. 1982). Also, the amount of monomeric myosin is low both in relaxed and contracted muscles (Horowitz et al. 1994). Contrasting to this, some structural studies have suggested that filaments may be formed during contraction in some smooth muscles (Godfraind-de Becker and Gillis, 1988, Gillis et al. 1988, Watanabe et al. 1993, Xu et al. 1997). Further work is needed to clarify whether this is a general phenomenon in smooth muscle and, if it is present, how the assembly might be regulated. Also the actin filaments exhibit dynamics in their assembly. Depolymerisation of actin filaments with cytochalasin (Adler 1983, Obara and Yabu 1994, Wright and Hurn 1994, Tseng et al. 1997) and a toxin which ADP-ribosylates G-actin thereby shifting the equilibrium between F- and G- actin in favour of the latter (Mauss et al. 1989) inhibits force. Also, stabilisation of the F-actin with phalloidin interferes with force generation (Boels and Pfitzer 1992). These results suggest that actin filament dynamics might be involved in mechanical responses of smooth muscle in a manner resembling that of motile phenomena in non-muscle cells. In non-muscle cells the dynamics of the actin cytoskeleton associated with cell motility is regulated by small GTPases, Rho and Rac, belonging to the superfamily of ras-related low molecular mass GTPases (Downward 1992). Rho has been implicated in the regulation of contraction of smooth muscle (see chapter 4, and Somlyo and Somlyo in this volume). Future work has to clarify whether reorganisation of the actin cytoskeleton is of any importance for regulation of contraction and how this is regulated.

The organisation of thick and thin filaments in the smooth muscle cells differs markedly from that in the skeletal muscle. The myosin concentration in smooth muscle is about 20% of that in skeletal muscle (Murphy et al. 1974) and based on electron microscopy the thick myosin filaments appear to be surrounded by approximately 15 actin filaments (Devine and Somlyo 1971, Somlyo 1980). At present no "sarcomere" equivalents have been identified in smooth muscle, but in order to enable shortening over longer distances at sufficient velocity, the contractile system must be organised into series coupled contractile units. It has been shown that actin filaments project from dense bodies with a specific polarity (Bond and Somlyo 1982). The dense bodies in smooth muscle might therefore be structures anchoring the thin filaments and act in a similar manner as Z-bands in the skeletal muscle. Longer actin filaments have been demonstrated (Small et al. 1990) which might increase the range over which the filaments can interact, and most likely the contractile units would exhibit a broad distribution in length. In spite of the lower myosin concentration high force per cross-sectional area can be produced by smooth muscle (Murphy et al. 1974). This difference might have structural explanations. The smooth muscle thick filaments have been found to be slightly longer than those in skeletal muscle (Ashton et al. 1975) which may increase the parallel coupling of cross-bridges by creating longer contractile units. A second factor that could contribute to increased force per unit myosin, is a parallel coupling of the contractile units, e.g. via the intermediate filament system. In a recent study using a desmin-knock out animal (Sjuve et al. 1998) it was shown that removal of the intermediate filaments markedly reduced the active force of the smooth muscle, which is consistent with a role of the intermediate filaments in transmitting force within, or between, cells. Finally, a difference in force generation, between smooth and skeletal muscle, could reside in the cross-bridge interaction itself. In principle, the unitary force per cross-bridge interaction and/or the time the cross-bridge spends in attached states (duty cycle) could be increased. Recent work using optical trap methods (Guilford et al. 1997) has suggested that unitary forces are similar in smooth and skeletal myosins and that the enhanced force generating capacity in smooth muscle could be due to an increased duty cycle (i.e. the time the cross-bridges spend in force generating state).

In view of the complex systems that interact during contraction and relaxation in smooth muscle, a characterisation of the reactions involved in the actin-myosin interaction in the organised smooth muscle contractile system requires that several parameters are controlled. This can be achieved in skinned (permeabilised) smooth muscle preparations (cf. Meisheri et al. 1985, Pfitzer 1996). Several protocols have been used, although for experiments on the cross-bridge interaction the detergent Triton X-100 (e.g. Sparrow et al. 1981) which removes cell membranes and sarcoplasmic reticulum functions and enables exchange of larger proteins into the filament lattice is particularly useful. It is important to note that this skinning procedure has limitations with regard to the activation systems, since both kinase and phosphatase activity is reduced (Takai et al. 1989, Schmidt et al. 1995). Also some contractile, structural and regulatory proteins might diffuse out and/or reorganise during the skinning procedure (Kossmann et al. 1987). However, the kinetics of the actin-myosin interaction in the Triton skinned preparation, as judged from shortening velocity determinations and ATPase (Arner 1982, Arner and Hellstrand 1983), are similar to those in the intact system. Since activation of smooth muscle through LC$_{20}$ phosphorylation is ATP dependent, a characterisation of the ATP hydrolysis of actin and myosin is difficult the perform in intact tissue. One advantage of the skinned preparations is that the extent of Ca^{2+}-activation can be controlled and a state of maximal activation can be obtained by thiophosphorylation of the LC$_{20}$ using ATP-γ-S (Cassidy et al. 1979, Arheden et al. 1988). To enable studies of kinetics in a muscle fibrerapid changes in muscle length or force are used to determine stiffness and unloaded shortening velocity. Rapid changes in chemical reactants can be achieved by the use of so called caged compounds (cf. Goldman 1987, Somlyo et al. 1987, Kaplan and Somlyo 1989, Somlyo and Somlyo 1990 for review). In caged compounds the biologically active moiety is protected by a photolabile chemical modification that can be cleaved by light to release the active compound. The caged compounds can be diffused into muscle strips and equilibrated at its site of action. Then the biologically active substance (e.g. Ca^{2+}, ATP, ADP or phosphate) is released by photolysis with a light flash and becomes available for binding with its ligands within milliseconds.

The general pathways of the actin (A) - myosin (M) interaction in smooth muscle are considered to be similar to those in skeletal muscle (Marston and Taylor 1980). The cross-bridge cycle can thus be described by a scheme derived from biochemical characterization of actin-myosin interaction in vitro and from concepts developed for skeletal muscle (cf. Goldman 1987).

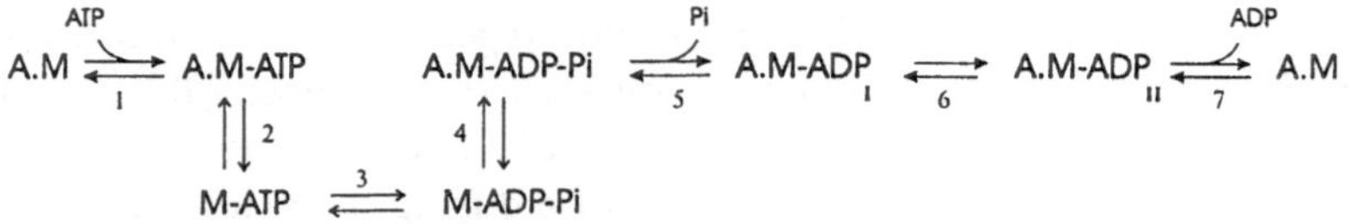

ATP   Pi   ADP
A.M $\rightleftharpoons$ A.M-ATP   A.M-ADP-Pi $\rightleftharpoons$ A.M-ADP $\rightleftharpoons$ A.M-ADP $\rightleftharpoons$ A.M
        1   5   I   6   II   7
      $\updownarrow$ 2   4 $\updownarrow$

              M-ATP $\rightleftharpoons$ M-ADP-Pi
                          3

Fig. 1. Schematic diagram for the main pathways for the interaction between actin (A), myosin (M), ATP, ADP and inorganic phosphate (Pi) during cross-bridge cycling in smooth muscle.

In the absence of ATP, actin and myosin bind strongly to form the rigor (A.M) complex (Fig. 1). A rigor state, characterised by an increased mechanical stiffness, has been proposed on the basis of experiments on ATP depleted intact smooth muscle (Lowy and Mulvany 1973, Bose and Bose, 1975, Wuytack and Casteels 1975, Butler et al. 1976) but rigor is most clearly shown in skinned fibres where an increased stiffness, a rigor force and structural X-ray diffraction and electron microscopy data show cross-bridge attachment (Arner et al. 1987a, Somlyo et al. 1988b, Arner et al. 1988). Binding of ATP rapidly dissociates the rigor state (reaction 1 and 2, Fig. 1). The second order rate constant for cross-bridge detachment by ATP has been determined to be in the range 0.4–$2.3 \ 10^4 \ M^{-1} \ s^{-1}$ which is slightly slower than in skeletal muscle (Nishiye et al. 1993). The ATP binding might also vary between smooth muscles (Khromov et al. 1996). However, this reaction is fast enough not to limit the cross-bridge turnover at physiological ATP concentrations. Following detachment, ATP is hydrolysed (reaction 3) and if the muscle is activated cross-bridges reattach (reaction 4) and release phosphate (P_i, reaction 5). In skeletal muscle the initial attachment is considered to occur in a weak binding conformation dependent on ionic strength (cf. Brenner et al. 1982, Eisenberg and Hill 1985, Brenner and Eisenberg 1987). Similar interactions have not been identified in smooth muscle fibres although lowering of the ionic strength appears to promote cross-bridge interaction (Gagelmann and Güth 1985, Arheden et al. 1988). At high phosphate concentrations active force is reduced, and the release of phosphate is considered to be associated with force generation in smooth muscle (Schneider et al. 1981, Gagelmann and Güth 1987, Österman and Arner 1995) in a similar manner as in skeletal muscle (Hibberd et al. 1985). In the skeletal system the force generation is considered to occur in two steps; an initial isomerisation followed by P_i release (Kawai and Halvorson 1991, Dantzig et al. 1992). This might occur also in smooth muscle fibres, but at present experimental data is lacking. Although the P_i

release is considered to be associated with force generation it is not rate-limiting for the rate of filament sliding, since the maximal shortening velocity (V_{max}) of muscle fibres (Österman and Arner 1995) and the filament velocity in the in vitro motility assay (Warshaw et al. 1991) are not influenced by phosphate. Following P_i release, and force generation, an isomerisation (reaction 6) is considered to precede the ADP release (reaction 7) in skeletal muscle (Sleep and Hutton 1980) and possibly also in smooth. The ADP release reaction is considered to limit the isotonic shortening velocity of muscle (Siemankowski et al. 1985). This is supported by the finding that ADP in a competitive manner inhibits the maximal shortening velocity of smooth muscle fibres (Arner et al. 1987b). ADP competes with ATP and the relaxation from rigor following release of ATP from caged-ATP is also slower in the presence of ADP (Arner et al. 1987b). The binding of ADP to the smooth muscle rigor complex was analysed by Nishiye et al. (1993) using caged-ATP experiments and was found to be very strong with an apparent dissociation constant in the micromolar range. A similarly strong ADP binding was found in smooth muscle fibres using competition with pyrophosphate (Arheden and Arner 1992) and is also found in isolated smooth muscle actomyosin (Cremo and Geeves 1998). It might be possible that ADP binding to the rigor complex in muscle fibres does not reflect processes occurring during cross-bridge cycling. It has however been shown that ADP binding is strong also to dephosphorylated cross-bridges from slow smooth muscles during relaxation induced by photolysis of a caged calcium chelator (Khromov et al. 1995). The binding of ADP to the A.M state thus appears to be stronger in smooth muscle compared to skeletal and could be a factor modulating cross-bridge turnover. Interestingly, electron microscopy (Whittaker et al. 1995) and electron paramagnetic resonance spectroscopy (Gollub et al. 1996) have recently revealed an ADP dependent structural change in the smooth muscle myosin molecule which could suggest that the ADP release reaction has a special mechanical role in the smooth muscle contractile system.

The regulation of actin myosin interaction by phosphorylation has been characterised in vitro, and the main process that is influenced by phosphorylation appears to be a step associated with the P_i release (Sellers et al. 1982, Sellers 1985, Greene and Sellers 1987). Increased levels of Ca^{2+} and phosphorylation increase active force in muscle fibres which could be consistent with an influence of phosphorylation on the P_i release reaction. By analogy with an analysis proposed for the events associated with force generation in skeletal muscle muscle (Brenner 1988), force generation in the smooth muscle can be described by a transition from weakly attached, non force generating, states to strongly attached, force generating cross-bridges. The effects of LC_{20} phosphorylation could be an increase in the apparent attachment rate, f_{app} (Rüegg and Pfitzer 1991). The non-linear relation between LC_{20} and force observed under many conditions (Siegman et al. 1989, Kenney et al. 1990, di Blasi et al. 1992, Schmidt et al. 1995) is compatible with a regulation of f_{app} through LC_{20} phosphorylation. The situation in the smooth muscle fibres appears however to be more complex. In addition to force, the maximal shortening velocity (V_{max}) is also influenced by $[Ca^{2+}]$ and phosphorylation (Arner 1983, Paul et al. 1983, Siegman et al. 1984, Barsotti et al. 1987, Malmqvist and Arner 1996); intermediate levels of Ca^{2+}/phosphorylation are associated with low force and low V_{max} compared to the situation during full activation. This could be due to phosphorylation effects on the ADP release reactions or possibly the presence of an internal load opposing shortening. The former possibility seems less likely since the ADP reaction in smooth muscle is little affected by phosphorylation both in vitro and in muscle fibres (Sellers et al. 1982, Greene and Sellers 1987, Arheden and Arner 1992). The inhibition of shortening velocity at low levels of activation could thus reflect the presence of attached cross-bridge states, possibly similar to the "latch-bridges" generated during sustained contractions in intact muscle (cf. chapter 6)

It has been shown that the rate of relaxation of skinned muscle is increased in the presence of inorganic phosphate (P_i, Schneider et al. 1981, Gagelmann and Güth 1987, Arner et al. 1987a, Somlyo et al. 1988b). Since the binding of P_i occurs early during the force generation these data suggest that unphosphorylated cross-bridges can attach into force generating states.

Interestingly when inorganic phosphate is introduced into partially activated skinned muscle fibres the force is decreased but V_{max} is increased (Österman and Arner 1995). This phenomenon thus suggests that at low levels of activation non-phosphorylated cross-bridges attach and that the resulting attached states are influenced by P_i and capable of opposing shortening. Most studies suggest that myosin has to be phosphorylated for interaction with actin (cf. Hartshorne and Mrwa 1982, Kamm and Stull 1985 for review). However, under conditions where the unphosphorylated myosin is filamentous, V_{max} of the MgATPase activity is about one-half of that of phosphorylated myosin (Wagner and Vu 1986, 1987). This observation may underly the prediction that unphosphorylated cross bridges can attach to actin.

In a recent study on isolated skinned smooth muscle cells, Malmqvist et al. (1997) extracted the regulatory protein calponin and demonstrated that in this situation, the muscles with unphosphorylated myosin could generate force, although at a slower rate than muscles with phosphorylated myosin. The data from skinned muscle thus suggest that both fully phosphorylated and dephosphorylated myosin cross-bridges can cycle and generate active force. The latter population cycle at a slower rate and could introduce a load opposing shortening at low levels of activation. Since the dephosphorylated cross-bridge states are sensitive to P_i binding, cycling of dephosphorylated cross-bridges appears to involve a comparatively large population of A.M.ADP states.

An important question is how the attached (or cycling) dephosphorylated cross-bridge states are recruited in the smooth muscle fibre. According to the "latch-bridge" hypothesis (cf. Murphy 1994, Strauss and Murphy 1996 and chapter 6) latch-bridges are formed by dephosphorylation of attached phosphorylated cross-bridges, although other mechanisms, e.g. cooperative attachment of unphosphorylated cross-bridges (Somlyo et al. 1988b, Vyas et al. 1992) possibly in association with regulation by thin filament systems (chapter 5) might also apply as discussed below.

Smooth muscles can be divided into two main groups: phasic and tonic muscle types. Phasic muscle, e.g. taenia coli, exhibits spontaneous action potentials and have faster contractile kinetics whereas tonic muscle, e.g. the aorta, do not have spontaneous activity and contract more slowly (cf. Somlyo and Somlyo 1968, Horiuti et al. 1989). Comparative studies have revealed that the shortening velocity of smooth muscles span of over a wide range (Malmqvist and Arner 1991). A fast smooth muscle, e.g. rabbit rectococcygeus, can have a shortening velocity that is about 7-fold faster than that of a slow smooth muscle, e.g. the aorta. This difference in velocity is similar to that between fast and slow skeletal muscle fibres. Thus smooth muscle is a heterogeneous group of muscles with a span of different kinetic properties in their contractile systems.

In fully activated smooth muscle the maximal shortening velocity (V_{max}) is considered to reflect the rate of cross-bridge detachment during filament sliding, whereas the rate of tension development reflects the rate at which cross-bridges enter into the force generating states. In different smooth muscles, the maximal shortening velocity appears to be correlated with the rate of tension development, although some exceptions exist

(Malmqvist and Arner 1991, Fuglsang et al. 1993, Somlyo 1993). Both parameters can be influenced by structural factors, e.g. series elasticity and sarcomere equivalent length. The initial rate of force generation after photolytic release of ATP in thiophosphorylated muscle is influenced by P_i whereas V_{max} is not (Österman and Arner 1995). Also in a recent report by He et al. (1997) it was also demonstrated, using a P_i indicator in a smooth muscle preparation, that the initial force generation after ATP release in an activated fibre preparation involved one cross-bridge turnover. These results seems to exclude that the rate of force development is simply determined by the force velocity relationship and the series elasticity. Thus the variation in contractile properties between muscles involves a regulation of both the force generating cross-bridge reactions and the reactions limiting the cycling during filament sliding. It has been shown that the ADP binding is higher in slow "tonic" smooth muscle (Fuglsang et al. 1993, Khromov et al. 1995, 1996). The ATP binding has also been reported to be weaker in slow muscles (Khromov et al. 1996). These data suggest that the muscle specific modulation of cross-bridge cycling in smooth muscle fibres occurs in the vicinity of the ADP release and cross-bridge detachment reactions. This mechanism can explain the difference in V_{max} between muscles, although it is more difficult to envision how the reactions involved in force generation may be influenced. As discussed below (chapter 5.1) strong ADP binding might also potentiate cooperative mechanisms involved in control of cross-bridge attachment.

The properties of the contractile system in a smooth muscle tissue can be altered under some (patho)physiological conditions. For example, the maximal shortening velocity increases in uterine smooth muscle oestrogen treatment or pregnancy (Hewett et al. 1993, Morano et al. 1997) and in airway smooth muscle from ragweed pollen-sensitised dogs (Jiang et al. 1995). V_{max} decreases in urinary bladder smooth muscle during hypertrophy (Sjuve et al. 1996). These alterations in the contractile system most likely have a functional role in altering e.g. the economy of force maintenance. It is possible that several changes also occur in the signalling and activation mechanisms in association with the alterations of smooth muscle phenotype: e.g. MLCK activity is increased in sensitised canine airway smooth muscle and saphenous vein (Jiang et al. 1995, Liu et al. 1996). A decrease in the activities of protein phosphatase type 1 and 2A was reported to occur in experimental vasospasm (Fukami et al. 1995). In this context it is, however, interesting to examine which contractile isoforms could be responsible for the variation in contractile properties between different smooth muscles and for the modulation of the contractile kinetics in one smooth muscle tissue.

Smooth muscle expresses four different isoforms of actin: α-, β-, and two forms of γ- actin (Vandekerckhove and Weber 1978). The isoform distribution varies during development (Eddinger and Murphy 1991) and can also change in (patho)physiological conditions, e.g. in the uterus during pregnancy (Cavaillé and Leger 1983, Cavaillé et al. 1986) or in vascular and urinary bladder smooth muscle during hypertrophy (Malmqvist and Arner 1990, Malmqvist et al. 1991a). The cellular distribution of the different actin isoforms is not clear for all muscles but it has been suggested that the β-isoform predominantly is associated with the cytoskeleton in smooth muscle (North et al. 1994b). The cellular concentration of actin varies between smooth muscle tissues and has been found to be in the approximate range 20–50 mg/g smooth muscle cell (Murphy et al. 1974, Cohen and Murphy 1979, Malmqvist and Arner 1990, Malmqvist et al. 1991a). Although the isoform distribution and content of actin appears to be modulated in smooth muscle cells the reasons and functional consequences are not clear at present. Actin is a highly conserved protein and a substantial sequence homology exists between isoforms (Pollard and Cooper 1986). Different actin isoforms do not exhibit large differences in actomyosin ATPase (Mossakowska and Strzelecka-Golaszewska 1985), the motion of actin filaments in the in vitro motility assay is little affected by the type of actin (Harris and Warshaw 1993) and data from muscle fibres suggest that the actin isoforms are functionally equivalent (Drew and Murphy 1997). These results suggest that the actin isoform distribution is not a primary determinant of kinetics of actin myosin interaction in the smooth muscle cell. Possibly, other functions of the thin filaments, e.g. depolymerisation/polymerisation processes, interaction with other parts of the cytoskeleton or with thin filament regulatory proteins, like calponin or caldesmon, might be influenced by the actin isoform distribution.

The smooth muscle myosin molecule is composed of six subunits: two heavy chains and two pairs of light chains. Adult smooth muscle expresses two main heavy chain isoforms, SM1 (molecular weight 204 kDa) and SM2 (200 kDa), which differs in their tail (light meromyosin, LMM) regions (Eddinger and Murphy 1988). These variants are formed from a single gene by alternative splicing (Nagai et al. 1989, Babij and Periasamy 1989). In addition non-muscle heavy chain (NMHC-A, 196 kDa) can be found in low amounts in tissue extracts and is expressed under certain conditions, e.g. during culture of smooth muscle cells (Rovner et al. 1986a, Kawamoto and Adelstein 1987). During development and in atherosclerotic transformation an embryonic heavy chain (SMEmb, NMHC-B, 200 kDa) can be expressed in smooth muscle (Kuro-o et al. 1989, 1991, Aikawa et al. 1993, Okamoto et al. 1992). The mechanical effects of the non-muscle and embryonic heavy chains are unknown, although the concentrations appear to be too low to enable any direct influence on mechanical performance of the muscle. The ratio of SM1/ SM2 varies between muscles, changes during development and can also be altered during hypertrophy of smooth muscle (Rovner et al. 1986b, Mohammad and Sparrow 1988, Kawamoto and Adelstein 1987, Borrione et al. 1989, Malmqvist et al. 1991a, 1991b, Malmqvist and Arner 1990, Boels et al. 1991). It has been suggested that alterations in the ratio between SM1 and SM2 can influence mechanical performance of smooth muscle in some conditions (Hewett et al. 1993). However, other comparative data suggest that the variation in smooth muscle shortening velocity, reflecting cross-bridge turnover, cannot be correlated with SM1 and SM2 (cf. Malmqvist et al. 1991a, Malmqvist and Arner 1991). The SM1 and SM2 differ in the LMM portion and these two isoforms move actin in the in vitro motility assay with similar velocities (Kelley et al. 1992) suggesting that the SM1/SM2 ratio does not have a primary function in determining the cross-bridge kinetics. More recently it has been demonstrated that an insert of 7 amino acids in the heavy meromyosin (HMM) region of the heavy chain can be present, thus generating SM1 and SM2 isoforms with and without the insert (Babij et al. 1991, Babij 1993, Kelley et al. 1993, White et al. 1993). The change is close to the ATP binding region and smooth muscle myosin with insert has a higher actin activated ATPase activity and moves actin filaments at a higher velocity in the in vitro motility assay (Kelley et al. 1993, Rovner et al. 1997). Although data from organised muscle tissues is sparse at present, variations in the amount of inserted myosin heavy chain is a strong candidate for modulating smooth muscle mechanical performance. Two types of light chains, the regulatory (20 kDa, LC_{20}) and the essential (17 kDa, LC_{17}) are present in smooth muscle (cf. review by Bárány and Bárány 1996a). Phosphorylation of the LC_{20} is the physiological activator of contraction, as described later (chapter 3), and removal of the LC_{20} results in a loss of the ability of myosin to move actin in the in vitro motility assay (Trybus et al. 1994). Isoforms of the LC_{20} have been demonstrated (cf. review by Bárány and Bárány 1996a) although their functional role in the smooth muscle contractile apparatus is not clear. The essential light chain appears to have a structural role in the neck region of myosin and removal of this component reduces velocity in the in vitro motility assay (Trybus 1994). Two isoforms of the essential light chain have been demonstrated in smooth muscle and are separated by isoelectric focusing into a more acidic LC$_{17a}$ and a more basic LC$_{17b}$ isoform (Cavaillé et al. 1986, Hasegawa et al. 1988, 1992, Helper et al. 1988, Lash et al. 1990). The composition of these isoforms has been suggested to influence the actomyosin ATPase (Helper et al. 1988, Hasegawa and Morita 1992). The relative composition of the essential light chain isoforms varies between tissues with a maximum of about 60% LC$_{17b}$ in "slow" muscles like the aorta and 0% in faster smooth muscles, like rabbit rectococcygeus (Malmqvist and Arner 1991). When different muscles are compared force development and the maximal shortening velocity are strongly correlated with the LC$_{17}$ isoform distribution, suggesting that this myosin component may have a role in determining smooth muscle contractile kinetics (Malmqvist and Arner 1991, Fuglsang et al. 1993, Morano et al. 1993, cf. Somlyo 1993 for review).

At present the data from in vitro motility assay and ATPase favour the inserted myosin as being the main factor determining the velocity of filament sliding (Rovner et al. 1997). The filament velocity in in vitro motility assays is usually considered to reflect the unloaded shortening velocity in muscle fibres. However the data cannot be directly transferred to the situation in muscle fibres where the contractile proteins are polymerised into a filament lattice, have higher effective concentrations, and contain additional regulatory proteins. It is therefore possible that contractile protein isoform might have different actions in muscle cells compared to the in vitro situation and that the different isoforms of contractile proteins might influence different functions, e.g. force generation, shortening velocity, activation mechanisms or interaction with other parts of the cytoskeleton. In several smooth muscle tissues the LC$_{17}$ isoform distribution appears to correlate with the expression of inserted myosin (cf. Malmqvist and Arner 1991, White et al. 1993, Sjuve et al. 1996) which could suggest that the expression of these two contractile protein isoforms is correlated. Muscles where the distribution of LC$_{17}$ isoforms and inserted myosin could be dissociated would be particularly interesting to clarify the role of these myosin isoforms in determining smooth muscle contractile behaviour.

3
Regulation of contraction by myosin light chain phosphorylation

It is generally believed that the reversible phosphorylation of the regulatory light chains of myosin (LC_{20}) is the key activating mechanism for
smooth muscle contraction. Phosphorylation is catalysed by the Ca^{2+}/cal-
modulin dependent myosin light chain kinase (MLCK) which is activated
when the cytosolic concentration of Ca^{2+} increases upon stimulation of
smooth muscle. A decrease in cytosolic Ca^{2+} leads to a dissociation and
inactivation of the Ca^{2+}/calmodulinll ● MLCK complex and dephosphorylation of LC_{20} by specific myosin light chain phosphatase(s) (MLCP, Fig.
2).

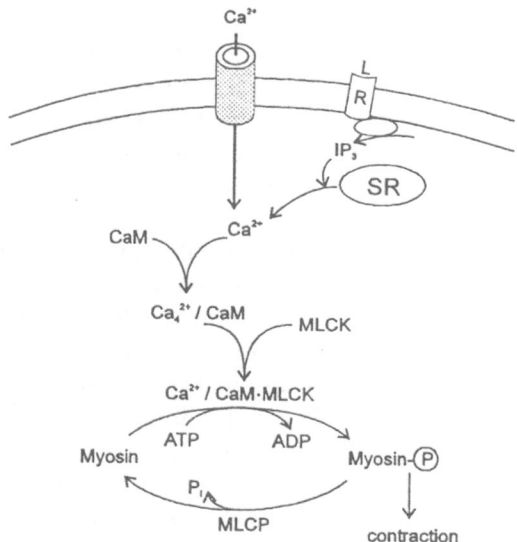

Fig. 2. Schematic diagram for the activation of smooth muscle through phosphorylation
of regulatory light chains of myosin. Upon activation of smooth muscle by membrane
depolarisation or activation of a receptor (R) through an agonist (L) the cytosolic Ca^{2+}
increases due to an influx of Ca^{2+} through plasmalemmal Ca^{2+} channels or IP_3 mediated
release from the sarcoplasmic reticulum (SR). Ca^{2+} binds to calmodulin (CaM) and
activates the myosin light chain kinase (MLCK). Myosin is dephosphorylated by
specific phosphatase(s) (MLCP).

As was initially shown by Sobieszek (Sobieszek 1977) that phosphorylation of the LC_{20} increases the actomyosin ATPase. Using more defined systems consisting of purified proteins this was then confirmed by a number of authors and it was clearly shown that the actin activated myosin MgATPase increased when LC_{20} was phosphorylated (for review Hartshorne and Mrwa 1982, Kamm and Stull 1985). Besides activation of the contractile process, phosphorylation of LC_{20} can also influence the assembly of smooth muscle myosin into filaments *in vitro* (see chapter 2).

Still, there is some controversy as to whether unphosphorylated myosin can interact with actin. Most studies suggest that myosin has to be phosphorylated for interaction with actin (Sobieszek and Small 1977, Sherry et al. 1978, Sellers et al. 1982, Chacko and Rosenfeld 1982, Sellers et al. 1985). However, under conditions where the unphosphorylated myosin is filamentous, V_{max} of the MgATPase activity is about one-half of that of phosphorylated myosin as already mentioned (Wagner and Vu 1986, 1987). This puzzle may be resolved on the basis of a recent study in which the minimal molecular requirement for myosin to be in the "off-state" was determined (Trybus et al. 1997). Mutants of myosin with different lengths of the rod showed that a length approximately equal to the myosin head was necessary to achieve a completely "off-state". It was concluded that the myosin rod mediates specific interactions with the head that are required to obtain a completely inactive state of vertebrate smooth myosins. If this interaction could be prevented, e.g. by constraints imposed by the native thick filament structure or accessory proteins, then partial activation of the actomyosin ATPase and slowly cycling of unphosphorylated cross-bridges could occur.

The predominant site of the LC_{20} being phosphorylated is serine-19 (Ikebe et al. 1986). It is thought that the effect of phosphorylation is not simply a charge effect but rather that the phosphate group in this position has a specific, steric effect (Sweeney et al. 1994). Under high activating conditions or high concentrations of MLCK, a second site becomes phosphorylated, threonine-18 (Ikebe et al. 1986). With intact myosin, phosphorylation of serine-19 is biphasic with a rapid initial phase while phosphorylation of threonine-18 is slower and follows a random process (Ikebe et al. 1986). Although phosphorylation of threonine-18 affects actomyosin ATPase activity in vitro (Ikebe and Hartshorne 1985) no effects on contractile properties of smooth muscle were detected (Haeberle et al. 1988). Since in smooth muscle, phosphorylation of threonine-18 always occurs in conjunction with phosphorylation of serine-19, the effect of threonine-18 phosphorylation alone cannot be determined in this system. Using mutants of the regulatory myosin light chains in which serine-19 and/or threonine-18 were replaced by alanine it was shown that phosphorylation of threonine-18 affected the actin-activated ATPase of myosin and translocation of actin filaments in the in vitro motility assay differently (Bresnick et al. 1995). While the actin-activated ATPase of myosin phosphorylated on threonine-18 only was approximately 15-fold lower than that of myosin phosphorylated on serine-19, the velocity with which actin filaments were moved was similar for myosins phosphorylated on threonine-18 or serine-19 or both threonine-18 and serine-19 (Bresnick et al. 1995).

LC_{20} is also phosphorylated by protein kinase C at serine-1 and 2 and threonine-9 (Bengur et al. 1987, Parente et al. 1992) which is associated with inhibition of ATPase (Ikebe et al. 1987a, de Lanerolle and Nishikawa 1988). The cellular effects of protein kinase C activation are, however, very complex since protein kinase C may affect contractile activity via several targets (cf. Somlyo and Somlyo 1994, Horowitz et al. 1996b, Singer 1996 for reviews, see chapter 4). Phosphorylation of LC_{20} of myosin II is also important for contractile phenomena in non-muscle cells such as formation of the contractile ring during cytokinesis. LC_{20} is phosphorylated by the cell cycle dependent protein kinase $p34^{cdc2}$ on the protein kinase C sites (Satterwhite et al. 1992). It was suggested that this phosphorylation which occurs during prophase and metaphase could delay cytokinesis until chromosome segregation is initiated and thus determine the timing of cytokinesis relative to earlier events in mitosis.

3.1
Myosin light chain kinase

The structure and regulation of the MLCK is extensively studied and has been reviewed in detail in recent reviews (Stull et al. 1996, Gallagher et al. 1997). Therefore, only a few aspects pertinent to the topic of this review will be covered. Unlike other protein kinases, which have a broader substrate specificity, MLCK phosphorylates only the LC_{20} of myosin II (Gallagher et al. 1997). The amount of MLCK in smooth muscle cells is approximately 3-4 µM which compares to 70–80 µM LC_{20} (Stull et al. 1996). MLCK has been cloned and sequenced from chicken (Olson et al. 1990), rabbit (Gallagher et al. 1991) and bovine smooth muscle (Kobayashi et al. 1992). Several domains can be identified in the molecule: a catalytic core, an actin binding site within the first 80 N-terminal residues, a regulatory domain with a calmodulin binding region and a putative autoregulatory site C-terminal of the catalytic core. All smooth muscle MLCKs also contain multiple structural motifs (unc motif) which are also found in twitchin, the unc 22 gene product located in muscle A bands in *C. elegans* (Benian et al. 1989), and in titin, which spans from the Z line to the M line in striated muscle (Labeit et al. 1990). Two different unc motifs have been described, motif I and II, which are members of the fibronectin and immunoglobulin superfamilies, respectively (Benian et al. 1989). Titin motif I and II bind to C protein and myosin and serve a structural role (Labeit and Kolmerer 1995). There is the intriguing possibility that MLCK in smooth muscle also serves a structural role in addition to its catalytic function (de Lanerolle et al. 1991). It was suggested that MLCK together with actin and caldesmon may form a scaffold for smooth muscle development because expression of MLCK occurred before the expression of smooth muscle myosin in the developing gizzard (Paul et al. 1995).

Interestingly one of the unc structural motifs (a motif II structure) located at the COOH-terminal end of MLCK is expressed as a separate 24 kDa acidic protein, telokin, also called kinase-related protein (Ito et al. 1989). MLCK and telokin are independently expressed proteins because the mRNAs are induced by two different promoters (Gallagher and Herring 1991). The function of telokin is not clear. Telokin may affect LC_{20} phosphorylation by modulation of the interaction of MLCK with myosin filaments (Nieznanski and Sobieszek 1997) or by enhancing MLCP activity (Wu et al. 1997). Telokin binds to myosin at the head-tail junction (Silver et al. 1997), and partially reverses the depolymerising effect of ATP on unphosphorylated myosin (Shirinsky et al. 1993). In this way it may stabilise unphosphorylated smooth muscle myosin filaments in vitro and this might be one factor explaining why unphosphorylated myosin maintains a filament structure in the presence of ATP in smooth muscle in vivo. A similar stabilising activity has also been reported for caldesmon (Katayama and Ikebe 1995, Katayama et al. 1995).

The catalytic core of MLCK, which is in the centre of the molecule, is highly homologous to other protein kinases (Stull et al. 1996) which may explain that there is no entirely specific MLCK inhibitor. The catalytic core is flanked at the C-terminal end by a regulatory segment containing a putative autoregulatory (inhibitory) and a calmodulin binding domain (Ikebe et al. 1989). Limited proteolysis at the region connecting the catalytic and regulatory domains yields a 61 kDa fragment which is constitutively active (Ikebe et al. 1987b). Within the calmodulin binding domain a pseudosubstrate structure with similarities to the consensus phosphorylation sequence of LC_{20} was detected (Kemp et al. 1987) which is probably identical to the autoregulatory domain (Olson et al. 1990, see Stull et al. 1996 for a detailed discussion). Synthetic peptides containing this sequence act as substrate antagonists (Kemp et al. 1987). Based on these observations it was proposed that in the absence of Ca^{2+}/calmodulin the autoregulatory region folds back on the catalytic site whereby the pseudo-substrate region interacts with the catalytic site thereby autoinhibiting the enzyme (Means et al. 1991, Bagchi et al. 1992). Binding of Ca^{2+}/calmodulin releases the autoinhibition and allows access of the protein substrate to the active site of the enzyme, resulting in phosphorylation of LC_{20}. Mutagenesis experiments confirmed that the pseudosubstrate has access to the catalytic site and that activation of the enzyme is accompanied from this position due to Ca^{2+}/calmodulin binding. This kind of intrasteric regulation has been termed pseudosubstrate inhibition (Kemp et al. 1994). For a detailed discussion the reader is referred to the review of Stull et al. (1996) and Gallagher et al. (1997).

It was originally recognised by Dabrowska et al. (1978) that calmodulin was part of the active holoenzyme MLCK complex. The association of calmodulin with MLCK is rapid and appears to be diffusion limited. It could be described by a two-step process, a bimolecular step and an isomerisation (Török and Trentham 1994). The time required for activating MLCK by Ca^{2+}/calmodulin may contribute to the latency of about 400–500 ms at 37°C which precedes increases in LC_{20} phosphorylation (Miller-Hance et al. 1988). The interaction of Ca^{2+}/calmodulin with MLCK may be modulated by phosphorylation of MLCK. Purified MLCK is a substrate for protein kinase A (Conti and Adelstein 1981), the multifunctional Ca^{2+}/calmodulin dependent protein kinase II (Hashimoto and Soderling 1990, Ikebe and Reardon 1990), protein kinase C (PKC) (Nishikawa et al. 1983) and mitogen activated protein kinase (MAP kinase, Klemke et al. 1997). Phosphorylation of MLCK by these protein kinases may alter the Ca^{2+}-sensitivity of the enzyme and hence of contraction, as will be discussed below.

3.2
Myosin light chain phosphatase

The dephosphorylation of the regulatory light chains of myosin is catalysed by myosin light chain phosphatase (MLCP). Compared to MLCK much less is known about MLCP. Based on their properties the protein phosphatases (PP) are categorised into two groups (Cohen 1989). Protein phosphatase-1 (PP1) dephosphorylates the ß-subunit of phosphorylase kinase specifically and is inhibited by nanomolar concentrations of two small heat- and acid-stable proteins, termed inhibitor-1 (I-1) and inhibitor-2 (I-2). The other group, protein phosphatase-2 (PP2) dephosphorylates the α-subunit of phosphorylase kinase preferentially and is insensitive to I-1 and I-2. Type 2 phosphatases could be classified into three distinct enzymes: 2A, 2B, and 2C. PP 2A is active in the absence of Ca^{2+} and Mg^{2+} while PP2B and PP2C require Ca^{2+}/calmodulin or Mg^{2+} for activity, respectively (Cohen 1989). Further means to differentiate between the phosphatases are based on the effects of polycationic macromolecules (reviewed in Erdödi et al. 1996). Another diagnostic aid is the marine toxin okadaic acid, a potent inhibitor of phosphatases with a different sensitivity against the different types of phosphatases (Takai et al. 1987). PP2A is approximately 4000 fold more sensitive than PP1 (Takai et al. 1995). The K$_i$ value for PP2A is approximately 30 pM (Takai et al. 1995). PP2B is inhibited only in the micromolar range and PP2C is not affected (Bialojan and Takai 1988). Isolated LC$_{20}$ are an excellent substrate for various forms of PP1, PP2A and PP2C (Pato and Adelstein 1983). However, when myosin as the native substrate is studied, the number of phosphatases that dephosphorylate LC$_{20}$ associated with the myosin heavy chains is considerably smaller (see Erdödi et al. 1996 for review). Use of phosphorylated myosin as a substrate is difficult because of the low solubility but myosin can be substituted for soluble heavy meromyosin (HMM), since HMM carries all the phosphatase binding properties associated with intact myosin (Erdödi et al. 1996). Using okadaic acid as probe it was suggested that the phosphatase that dephosphorylates myosin in the tissue is a type PP1 phosphatase and possibly also PP2A phosphatase (Takai et al. 1989, Ishihara et al. 1989). Type PP1 and PP2A phosphatases may also be involved in the dephosphorylation of MLCK (Nomura et al. 1992), caldesmon and calponin (Winder et al. 1992). It should be noted that in tissues often higher concentrations of okadaic acid are required than with isolated phosphatases. This difference in dose dependence could be due to preferential localisation of the inhibitor with lipids or it could reflect the intracellular concentration of the targeted phosphatase (Gong et al. 1992).

Recently, several groups have isolated a type PP1 phosphatase from chicken or turkey gizzard and pig bladder (Alessi et al. 1992, Okubo et al. 1994, Shirazi et al. 1994) with similar properties which all dephosphorylate LC$_{20}$ in intact myosin. The concentration of this enzyme was estimated to be 0.7 μM (Alessi et al. 1992) and 1 μM (Shirazi et al. 1994) and sufficient to completely dephosphorylate myosin within seconds in vivo. The type PP1 phosphatases isolated from chicken gizzard and pig bladder were trimeric and consisted of 130 kDa, 37 kDa and 20 kDa subunits (Alessi et al. 1992, Shirazi et al. 1994). Purification of a myosin-bound phosphatase from chicken gizzard actomyosin yielded a protein which consisted of two subunits, 38 kDa and 58 kDa, the latter may have been a proteolytic product of the 130-kDa subunit (Okubo et al. 1994). The 37 kDa subunit of these enzymes represents the catalytic subunit the activity of which is higher in the presence of the 130 kDa subunit than in its absence (Alessi et al. 1992). The 130 kDa subunit interacts with both the 37 kDa and 20 kDa subunit and with myosin (Alessi et al. 1992, Shirazi et al. 1994, Shimiziu et al. 1994). Therefore the 130 kDa subunit could target the phosphatase to myosin and determine its specificity. The interaction sites between myosin, the large subunit and the catalytic subunit (Ichikawa et al. 1996b, Hirano et al. 1997) and the regions of the large subunit required for regulation (Gailly et al. 1996, Johnson et al. 1996) have been mapped. Further evidence in support of the hypothesis that the 130-kDa of the trimeric phosphatase is the targeting subunit of smooth muscle phosphatase is as follows: the holoenzyme more effectively promotes relaxation of skinned smooth muscle (Shirazi et al. 1994), it is dissociated by arachidonic acid which slows relaxation and dephosphorylation of LC_{20} (Gong et al. 1992). The concept of targeting subunits which convey specificity and regulation of the protein phosphatases has been developed by Cohen and coworkers and has been exemplified with the glycogen bound form of PP1 which consists of two subunits: the catalytic subunit (37 kDa) and a 161 kDa glycogen binding subunit (Hubbard and Cohen 1989).

There are at least two possibilities to regulate the activity of PP1 in smooth muscle. Like the glycogen-bound phosphatase, the myosin bound phosphatase may be regulated by phosphorylation of the 130 kDa subunit. This concept suggests that phosphorylation of the targeting subunit causes the association or dissociation (depending on the site phosphorylated) of the subunits of holoenzyme which is associated with a change in the activity (Hubbard and Cohen 1989). The large subunit of smooth muscle myosin phosphatase has been characterised from two sources, chicken gizzard (Shimiziu et al. 1994) and rat aorta (Chen et al. 1994). cDNA clones obtained from a chicken gizzard library suggested the presence of two isoforms (Shimiziu et al. 1994) which have several potential phosphorylation sites for the cAMP-dependent protein kinase, protein kinase C, the cell cycle dependent protein kinase p34[cdc2], and glycogen synthase kinase 3 (Shimiziu et al. 1994). Subsequently it has been shown that phosphorylation of the 130 kDa subunit by a kinase which copurifies with the phosphatase (Ichikawa et al. 1996a), or by Rho-associated kinase (Kimura et al. 1996) resulted in inhibition of phosphatase activity (see chapter 4.2). An alternative mechanism of regulation could involve inhibitor-1 which has also been purified from smooth muscle tissues (Eto et al. 1995, Tokui et al. 1996). The purified inhibitor 1 was phosphorylated by protein kinase C (Eto et al. 1995) or cGMP-dependent protein kinase (Tokui et al. 1996) resulting in inhibition of MLCP. Incubation of permeabilised smooth muscle cells at submaximal Ca^{2+} with the phosphorylated inhibitor 1 induced a slow contraction indicating that phosphorylation of inhibitor 1 may also be a mechanism to increase Ca^{2+}-sensitivity of smooth muscle contraction (Tokui et al. 1996). As discussed in chapter 4.2, regulation of the activity of MLCP may be an important mechanism which regulates the Ca^{2+} sensitivity of contraction.

A further myofibrillar phosphatase which forms a multienzyme complex with myosin light chain kinase has been purified recently from chicken gizzard (Sobieszek et al. 1997b) and is composed of a 37 kDa catalytic subunit and a 67 kDa targeting subunit which bound to calmodulin in a Ca^{2+}-independent manner. The enzyme was closely associated with MLCK and myosin filaments and was called kinase- and myosin-associated protein phosphatase (KAMPPase). The enzyme was inhibited by okadaic acid (K_i = 250 nM), microcystin-LR (50 nM) and calyculin A (1.5 μM) but not by arachidonic acid or the heat-stable inhibitor-2 which suggested that this phosphatase is a type PP1 or PP2A phosphatase (Sobieszek et al. 1997a). Clearly further work is needed to identify the phosphatase specifically responsible for the dephosphorylation of myosin *in vivo*.

3.3
Regulation of contraction through LC$_{20}$ phosphorylation in intact and permeabilised smooth muscle.

Krebs and Beavo (1979) proposed several criteria which have to be fulfilled if phosphorylation-dephosphorylation of an enzyme is physiologically meaningful: (i) Demonstration in vitro that the enzyme can be phosphorylated stoichiometrically at a significant rate in reaction(s) catalysed by appropriate protein kinase(s) and dephosphorylated by a phosphoprotein phosphatase(s). (ii) Demonstration that functional properties of the enzyme undergo meaningful changes that correlate with the degree of phosphorylation. It is generally accepted that these two criteria for the regulation of smooth muscle by phosphorylation are met in vitro, as discussed above.

The third criterion of Krebs and Beavo states that it has to be demonstrated that the enzyme can be phosphorylated and dephosphorylated in vivo or in an intact cell system with accompanying functional changes. This criterion is more difficult to meet although it is accepted that there is a significant body of evidence that it is also fulfilled. One important point to consider is what is called a functional change. At first sight this may be active force of the muscle. However, as pointed out by Murphy (1994), force in smooth muscle does not necessarily represent a quantitative measure of the number of cycling cross-bridges. Perhaps under non-steady state conditions, stiffness more closely parallels LC_{20} phosphorylation (Kamm and Stull 1986). Other contractile parameters are unloaded shortening velocity and energy consumption. There is some evidence that these parameters under certain circumstances correlate much better than force with LC_{20} phosphorylation.

Contraction of a tonic type of smooth muscle is characterised by a rapid increase in tension followed by a sustained contraction. Removal of the agonist induces relaxation. Numerous studies have shown that the increase in force is typically preceded by a rapid increase in LC_{20} phosphorylation and intracellular Ca^{2+} (see Kamm and Stull 1985, Somlyo and Somlyo, 1994, Horowitz et al. 1996b, Bárány and Bárány et al. 1996b for reviews). This is also true for phasic smooth muscles such as chicken gizzard (Fischer and Pfitzer 1989), intestinal (Himpens et al. 1988), and uterine (Word et al. 1994) smooth muscle. There is also a wealth of data in support of a correlation between increases in LC_{20} phosphorylation and Ca^{2+}-induced force development in skinned smooth muscle from different origins (Hoar et al. 1979, Cassidy et al. 1979 and 1981, Chatterjee and Murphy 1983, Barsotti et al. 1987, Tanner et al. 1988, Kühn et al. 1990, Kitazawa et al. 1991, Malmqvist and Arner 1996).

The kinetics of contraction have been evaluated in electrically stimulated smooth muscle because under these conditions activation is not diffusion limited (Kamm and Stull 1985). Maximal values of unloaded shortening velocity and LC_{20} phosphorylation were attained after 5 s of stimulation and preceded maximal force. In a similar manner, maximal phosphorylation values were obtained after 3 to 4 s of electrical field stimulation of chicken gizzard strips while force was maximal after 5.5 s (Fischer and Pfitzer 1989). There is about a 500 ms lag time following stimulation and onset of LC_{20} phosphorylation in the electrically stimulated tracheal smooth muscle (Kamm and Stull 1986). At this time the fractional activation of calmodulin dependent cyclic nucleotide phosphodiesterase was maximal suggesting by extrapolation that MLCK is maximally activated within 500 ms (Miller-Hance et al. 1988). Stiffness increased in parallel with LC_{20} phosphorylation and preceded isometric force (Kamm and Stull 1986). The linear relation between stiffness and phosphorylation suggested independent attachment of each myosin head upon phosphorylation. Interestingly, the rate of LC_{20} phosphorylation has been reported to be significantly greater than the rate of increase in intracellular $[Ca^{2+}]_i$ (Word et al. 1994). It reached steady state values while $[Ca^{2+}]_i$ continued to increase. The diminished rate of LC_{20} phosphorylation coincided with an increased phosphorylation of MLCK. These experiments suggest that MLCK is very sensitive to small changes of $[Ca^{2+}]_i$ during the initiation of the contraction. Thereafter, the enzyme might be phosphorylated by another Ca^{2+}-dependent protein kinase (Ca^{2+}/calmodulin-dependent protein kinase II) and desensitised to further increases in Ca^{2+}_i (Word et al. 1994). The experiments also show that LC_{20} phosphorylation occurs fast enough to account for activation of cross-bridge cycling.

There are several steps between the beginning of stimulation and increase in contraction which contribute to the delay between activation and force. Using caged phenylephrine, the lag phase between the laser pulse and onset of contraction was 1.8 s at 20°C (Somlyo et al. 1988a, Kaplan and Somlyo 1989, Somlyo and Somlyo 1990). Of this delay approximately 0.3-0.5 s were due to steps following the increase in intracellular $[Ca^{2+}]$, e.g. binding of Ca^{2+}/calmodulin to MLCK, phosphorylation of LC_{20}, initiation of cross-bridge cycling. Using caged ATP in skinned smooth muscle, the activation kinetics of contraction depending on LC_{20} phosphorylation was analysed. The delay between the laser pulse and initiation of contraction was smaller when the light chains were prephosphorylated by treatment with ATPγS. These experiments showed that LC_{20} phosphorylation is usually slower than the cross-bridge turnover and limits the rate of force development (Somlyo and Somlyo 1990).

It is agreed by most investigators that the Ca^{2+}-dependent phosphorylation of LC_{20} is the key regulatory event which is required for cross-bridge cycling and initiating contraction. However, it was also recognised very early that there is no simple correlation between force and LC_{20} phosphorylation. In a paradigmatic experiment, Dillon et al. (1981) showed that in the tonically contracting carotid artery, phosphorylation declined together with the unloaded shortening velocity (V_{max}) while tension was maintained. The state where phosphorylation and V_{max} was low, while force was high was called "latch" state (cf. chapter 6) for discussion. In phasic smooth muscle, in which no steady state is reached, LC_{20} phosphorylation precedes relaxation and during relaxation a lower level of LC_{20} phosphorylation is required for a given level of force. Determination of the active state by imposing a quick release indicated that the active state was lower during relaxation than during contraction for a given level of force suggesting that a latch like state may also exist in phasic smooth muscles (Fischer and Pfitzer 1989).

The observation of the dissociation between steady-state force, LC_{20} phosphorylation and shortening velocity was confirmed by a number of authors (Aksoy et al. 1982, 1983, Silver and Stull 1982, Gerthoffer and Murphy 1983,) and it appeared that V_{max} correlated much better with LC_{20} phosphorylation than did isometric tension (Dillon et al. 1981, Aksoy et al. 1982, Gerthoffer and Murphy 1983, Murphy et al. 1983, Hai and Murphy 1989a for review). However, subsequent studies showed that not only force but also unloaded shortening velocity can be dissociated both in intact and permeabilised tissues (Paul et al. 1983, Siegman et al. 1984, Haeberle et al. 1985b, Barsotti et al. 1987, Moreland et al. 1987, Merkel et al. 1990, Gunst et al. 1994). As regards isometric force, different studies showed that the dependence of force on LC_{20} is steep and curvilinear, maximal force is obtained when about 20–30% of the light chains are phosphorylated (Di Blasi et al. 1992, Siegman et al. 1989, Kenney et al. 1990, Schmidt et al. 1995). This could occur either if there is a significant cooperative activation of unphosphorylated cross-bridges (Vyas et al. 1992) or if attached, dephosphorylated cross-bridges were generated from phosphorylated cross-bridges as proposed by Hai and Murphy (1988a) (see chapters 5 and 6 for detailed discussion). Under these conditions, small changes in LC_{20} phosphorylation may have large effects on force. Due to the limitations of resolution of determination of LC_{20} phosphorylation, it may be difficult at times to decide whether a change in tension is or is not associated with an appropriate change in phosphorylation.

These correlative studies neither prove nor disprove a causal link between contractile parameters and LC_{20} phosphorylation. Using permeabilised smooth muscle it was attempted to show that there is a causal relation between activation of contraction and LC_{20} phosphorylation. In permeabilised smooth muscle the medium surrounding the myofilaments can be controlled and there is free access for inhibitors and activators to the myofilaments. For functionally isolating the contractile machinery extensive permeabilisation with the detergent Triton-X-100 is generally applied as discussed above. The disadvantage of this procedure is that upstream regulatory mechanisms are attenuated and some protein components may leak out of the preparation (Kossmann et al. 1987, Takai et al. 1989, Schmidt et al. 1995). More gentle procedures use saponin, β-escin and α-toxin from *Staphylococcus aureus,* which leave intracellular membrane systems (treatment with saponin) as well as the coupling between surface membrane receptors and their effectors (treatment with β-escin or α-toxin) functionally intact (Pfitzer and Boels 1991, Pfitzer 1996 for review). With the latter two procedures it is possible to investigate the mechanisms that modulate the Ca^{2+}-sensitivity of contraction (see chapter 4).

Skinned fibres are relaxed when the free Ca^{2+}-concentration is kept below about 10^{-8} M. Increasing the Ca^{2+}-concentration in the range of 1 to 10 μM increases force in a graded manner (Filo et al. 1965) and as already mentioned, increases in LC_{20} phosphorylation were correlated with Ca^{2+}-induced force development in skinned smooth muscle from different origins (Hoar et al. 1979, Cassidy et al. 1979, Tanner et al. 1988, Chatterjee and Murphy 1983). However, these experiments did not establish a causal relationship, in particular since maximal force was observed at phosphorylation values ranging from 20 to 60% (Cassidy et al. 1981). It was therefore of prime importance to demonstrate that LC_{20} phosphorylation is sufficient to induce a contraction. This was achieved in two ways: (i) it was shown that smooth muscle strips in which LC_{20} were stably thiophosphorylated with the ATP analog, ATPγS (Sherry et al. 1978) contracted in a Ca^{2+} free solution (Hoar et al. 1979). (ii) Incubation of Triton-skinned smooth muscle with the constitutively active fragment of MLCK (Walsh et al. 1982) induced a contraction and increased LC_{20} in the absence of Ca^{2+} (Walsh et al. 1982). This contraction was inhibited by the MLCK inhibitor, ML-9 (Ishikawa et al. 1988). Injection of the constitutively active MLCK into intact smooth muscle cells produced shortening without an increase in cytosolic Ca^{2+} (Itoh et al. 1989). These experiments unequivocally demonstrated that LC_{20} phosphorylation is sufficient to initiate a contraction.

To further elucidate the activation mechanism, several inhibitors of MLCK have been applied. MLCK activity was inhibited by antagonising the binding of calmodulin to MLCK by calmodulin antagonists such as W-7 (Hidaka et al. 1978, Kanamori et al. 1981) or trifluoperazine (reviewed in Asano and Stull 1985). The disadvantage is that these inhibitors also inhibit other calmodulin-dependent enzymes in cells. ML-9 (Ishikawa et al.

1988) and wortmannin (Nakanishi et al. 1992) inhibit MLCK competitively with respect to ATP. The major disadvantage of these inhibitors is that they also inhibit other protein kinases. While wortmannin does not inhibit cAMP- and cGMP dependent protein kinases in concentrations up to 10 μM, it inhibits phosphatidylinositol 3′-kinase at lower concentrations than MLCK (Thelen et al. 1994). In an attempt to develop more specific inhibitors, peptide inhibitors have been designed. A peptide of 20 amino acid residues derived from the calmodulin-binding domain of MLCK binds calmodulin with high affinity and is a potent inhibitor of MLCK activity in vitro (Lukas et al. 1986). This peptide inhibited Ca^{2+}-induced shortening of single smooth muscle cells (Kargacin et al. 1990) and induced relaxation of Ca^{2+}-induced contractions and dephosphorylation of LC_{20} in skinned smooth muscle (Rüegg et al. 1989) . Interestingly, however, while the peptide induced complete relaxation it induced only partial dephosphorylation of LC_{20} raising the question as to a second calmodulin dependent mechanism regulating contraction (Rüegg et al. 1989). A second synthetic peptide derived from the phosphorylation site of LC_{20} (residues 11-19) inhibited the in vitro activity of MLCK (Pearson et al. 1986). It also inhibited the Ca^{2+}-induced shortening in single smooth muscle cells (Kargacin et al. 1990) and unloaded shortening in skinned smooth muscle (Strauss et al. 1992). Interestingly, it did not inhibit isometric contraction and LC_{20} phosphorylation in the skinned smooth muscle. Unexpectedly, in low Ca^{2+}-solutions the peptide significantly increased LC_{20} phosphorylation. It was concluded that this peptide also acts as a phosphatase inhibitor and while not altering the net phosphorylation it may affect the phosphate turnover. This would suggest that not only the absolute levels of LC_{20} but also the rates of dephosphorylation and phosphorylation could determine V_{max}. This was not confirmed in another study where okadaic acid was used to alter the phosphorylation/dephosphorylation rates (Malmqvist and Arner 1996). In conclusion, activation of contraction with the constitutively active fragment of MLCK has substantiated the hypothesis that LC_{20} phosphorylation is sufficient to induce a contraction. However, correlative studies in intact smooth muscle and inhibition of MLCK using synthetic peptides in skinned smooth muscle revealed the complex nature of regulation of smooth muscle contraction and call for additional regulatory mechanisms which may be linked to the thin filaments (see chapter 5).

3.4
Dephosphorylation of LC$_{20}$ as a prerequisite for relaxation

The paradigm that phosphorylation of LC$_{20}$ is the key event in activation of smooth muscle holds that dephosphorylation is a prerequisite for relaxation. When the cytosolic Ca^{2+} declines the MLCK holoenzyme complex dissociates thereby inactivating the enzyme. This permits dephosphorylation of LC$_{20}$ by myosin light chain phosphatase, deactivating the actomyosin ATPase and causing relaxation (Barron et al. 1979, Gerthoffer and Murphy 1983, Driska et al. 1989, Tanner et al. 1988, Hai and Murphy 1989b, Rembold 1991, Kühn et al. 1990, Khromov et al. 1995). These correlative studies are corroborated by experiments in skinned fibres in which it has been shown that addition of purified phosphastases induced relaxation (Rüegg et al. 1982, Haeberle et al. 1985a, Hoar et al. 1985, Shirazi et al. 1994, Gailly et al. 1996). In line with these experiments, a low phosphatase activity in skinned chicken gizzard smooth muscle was associated with impaired relaxation and dephosphorylation of LC$_{20}$. Relaxation and dephosphorylation could be rescued by incubation with a polycation modulable phosphatase (Bialojan et al. 1985). Relaxation and dephosphorylation was also inhibited by inhibitors of phosphatases such as okadaic acid (Takai et al. 1987, Bialojan et al. 1988, Erdödi et al. 1996 for review). In addition, the phosphatase inhibitors have proved to be a valuable tool in testing the hypothesis that not only the extent of LC$_{20}$ phosphorylation but also the absolute rates of the phosphorylating and dephosphorylating reactions are important in determining the contractile parameters, force and shortening velocity (Siegman et al. 1989, Malmqvist and Arner 1996, Schmidt et al. 1995).

An important question is whether dephosphorylation is the rate limiting step in relaxation. This has been addressed in different studies. Under conditions where relaxation was not diffusion limited such as termination of electrical field stimulation (Hai and Murphy 1989b), relaxation of rhythmically contracting smooth muscle (Driska et al. 1989), or rapid binding of Ca^{2+} with the caged Ca^{2+} chelator, diazo-2, in permeabilised preparations (Khromov et al. 1995) dephosphorylation preceded relaxation. From the rate of dephosphorylation a phosphatase rate constant of 0.08 sec^{-1} was estimated (Driska et al. 1989) which would be sufficient to dephosphorylate cross-bridges while attached in a strongly bound state. Following removal of Ca^{2+} with the caged chelator, diazo-2, relaxation occurred in two phases, a plateau phase followed by a monoexponential decay (Khromov et al. 1995). Thus, dephosphorylation of LC$_{20}$ does not appear to be rate limiting for relaxation. The kinetics of force relaxation may be determined, at least in tonic smooth muscle, by the dissociation of ADP from dephosphorylated attached cross-bridges (Khromov et al. 1995).

There is some evidence that relaxation may be regulated independently of LC_{20} dephosphorylation: In an early study, Gerthoffer and Murphy (1983) showed that in the tonically contracting carotid artery, relaxation following washout of high K^+ followed a dual-exponential decay (Gerthoffer and Murphy 1983). The time course of the initial rapid decay corresponded to myosin dephosphorylation which was basal after 2 min. It was hypothesised that the slow phase of relaxation (lasting up to 45 min) was due to slow inactivation of non-phosphorylated cross-bridges. The rate of decay of this slow phase depended on the concentration of extracellular Ca^{2+}. Furthermore, ß-adrenergic stimulation of a tonically contracting smooth muscle in which LC_{20} phosphorylation was basal induced a rapid relaxation (Miller et al. 1983, Gerthoffer et al. 1984).The phasic smooth muscle of chicken gizzard relaxes completely in the continued presence of stimulation which is preceded by dephosphorylation of LC_{20} (Fischer and Pfitzer 1989). Termination of stimulation at a time when phosphorylation was basal, accelerated the rate of relaxation. These observations suggest that there has to be a mechanism that regulates the net detachment rate of dephosphorylated cross-bridges.

3.5
The variable phosphorylation sensitivity of contraction

There are a number of observations that indicate that the dependence of force on phosphorylation may be modulated. Stimulation of ferret aorta with $PGF_{2\alpha}$ shifts the relation between force and LC_{20} phosphorylation to the left when compared to KCl-induced contractions (Suematsu et al. 1991). Thus, force at any one phosphorylation level was increased in the $PGF_{2\alpha}$-induced contractions. Several studies even suggested that contraction can occur without apparent increases in LC_{20} phosphorylation (Wagner and Rüegg 1986, Fulgitini et al. 1993, Sato et al. 1992). Whether this is due to disinhibiton of an inhibitory protein such as calponin (Malmqvist et al. 1997) which then allows the slow cycling of dephosphorylated cross-bridges is currently not known.

The phosphorylation sensitivity of contraction may also decrease, i.e. conditions may occur where force is low despite a high level of LC_{20} phosphorylation. Pretreatment with a low concentration of the phorbolester, PMA, which by itself had no effect on contraction and LC_{20} phosphorylation, shifted the force LC_{20} phosphorylation curve during the steady state of contractions induced by K^+ in aortic smooth muscle to the right compared to tissues which were stimulated with K^+ only (Seto et al. 1990). In tracheal smooth muscle, carbachol and serotonin induced a significant increase in LC_{20} phosphorylation in low Ca^{2+} solutions without proportional increases in force (Gerthoffer 1987). Uncoupling of relaxation from dephosphorylation was observed in arterial smooth muscle relaxed with nitrovasodilators (McDaniel et al. 1992), in intact smooth muscle incubated with okadaic acid (Tansey et al. 1990), in uterine smooth muscle (Bárány and Bárány 1993a), or in tracheal smooth muscle in which endothelin-induced contractions were relaxed with Ca^{2+} antagonists (Katoch et al. 1997). From these data one has to conclude that dephosphorylation is not under all conditions a prerequisite for relaxation. One or more additional mechanisms, possibly associated with the thin filaments, caldesmon or calponin, are therefore likely to regulate the attachment and detachment of phosphorylated and dephosphorylated cross-bridges in the smooth muscle.

4
Modulation of the Ca^{2+}-sensitivity of force and LC_{20} phosphorylation

The concept that the Ca^{2+}-sensitivity of contraction could be modulated was developed when it was shown that cAMP relaxed skinned smooth muscle strips at constant submaximal $[Ca^{2+}]$ and shifted the calcium force relation to the right towards higher values (cf. Rüegg 1992). It has also been shown that excitatory agonists that activate heterotrimeric G-proteins can increase LC_{20} phosphorylation and force at constant $[Ca^{2+}]$ (Somlyo and Somlyo 1994). In this context, a decrease in Ca^{2+}-sensitivity refers to a shift in the relation between $[Ca^{2+}]$ and phosphorylation or force towards higher $[Ca^{2+}]$; an increase in Ca^{2+}-sensitivity is a shift towards lower $[Ca^{2+}]$. The physiological relevance of the Ca^{2+}-sensitivity modulation became apparent when simultaneous measurements of intracellular $[Ca^{2+}]$ and force in intact smooth muscle showed that there was no unique relation between force and intracellular Ca^{2+} (Morgan and Morgan 1984a, 1984b) There are in principle two ways to modulate Ca^{2+}-sensitivity of contraction: (i) by altering the balance between the phosphorylating and dephosphorylating reactions which would affect the Ca^{2+}-sensitivity of LC_{20} phosphorylation but in a first approximation not the dependence of force on phosphorylation, and (ii) by LC_{20}-phosphorylation independent regulatory systems such as caldesmon or calponin which can influence force directly or alter the coupling between force and phosphorylation, i.e. change the phosphorylation sensitivity of force. It should be differentiated between non-steady state changes in Ca^{2+}-sensitivity and changes in Ca^{2+}-sensitivity during steady state. Changes in Ca^{2+}-sensitivity determined under non-steady state conditions are difficult to interpret since the apparent sensitivities of force and phosphorylation on $[Ca^{2+}]$ are influenced by the kinetics of the activation/deactivation pathways as well as by the kinetics of the force generation.

4.1
Mechanisms for decreasing the Ca^{2+}-sensitivity

Based on the phosphorylation theory of smooth muscle contraction a decrease in the Ca^{2+}-sensitivity of LC_{20} phosphorylation and hence contraction could occur either through a decrease in the activity of the myosin light chain kinase (MLCK) or an increase in the activity of the myosin light chain phosphatase (MLCP) at constant $[Ca^{2+}]$ (Fig. 3). Three main second messengers are considered to be involved in downregulation of Ca^{2+}-sensitivity: cAMP, cGMP, and Ca^{2+} itself. Incubation of purified calmodulin free MLCK with the catalytic subunit of the cAMP dependent protein kinase (Conti and Adelstein 1981), the Ca^{2+}/calmodulin-dependent protein kinase II (Hashimoto and Soderling 1990), and protein kinase C (Nishikawa et al. 1983) resulted in the phosphorylation of a serine at the C-terminus of the Ca^{2+}/calmodulin-binding sequence and in a marked lowering of the rate of MLCK activity. This was due to a 10- to 20-fold increase in the amount of calmodulin necessary for 50% activation of the kinase activity. If calmodulin is bound to MLCK *in vitro*, phosphorylation of this site by all three protein kinases is blocked.

In Triton skinned smooth muscle, cAMP or the catalytic subunit of cAMP dependent protein kinase inhibits tension development or induces relaxation at submaximal $[Ca^{2+}]$ thereby decreasing the Ca^{2+}-sensitivity of contraction (Kerrick and Hoar 1981, Rüegg et al. 1981, Rüegg and Paul 1982, Sparrow et al. 1984). Inhibition of force was associated with a decrease in LC_{20} phosphorylation (Rüegg and Pfitzer 1985). The inhibitory

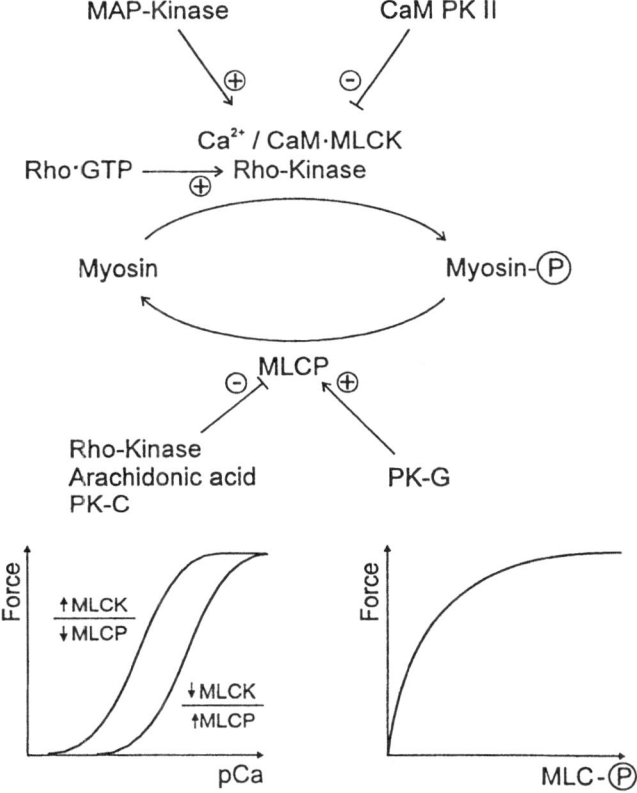

Fig. 3. Schematic diagram of pathways the modulate Ca^{2+} sensitivity by shifting the balance of the activities of myosin light chain kinase (MLCK) or myosin light chain phosphatase (MLCP) at constant $[Ca^{2+}]$. A decrease or increase in the activities of either MLCK or MLC-P changes the Ca^{2+}-sensitivity of myosin light chain phosphorylation (MLC-Ⓟ) and hence of force but does not alter the relation between force and myosin light chain phosphorylation (MLC-Ⓟ). For further details see text.

effect of cAMP or the catalytic subunit of cAMP-dependent protein kinase was reversed by high concentrations of calmodulin (Meisheri and Rüegg 1983, Pfitzer et al. 1985). These experiments in skinned smooth muscle are compatible with the hypothesis that cAMP relaxes smooth muscle by inhibition of the activity of MLCK. Initial correlative studies were also in support of this hypothesis: Relaxin, a hormone, which increases intracellular cAMP levels and inhibits uterine contractile activity, produced relaxation of precontracted uterine smooth muscle. This was associated with both dephosphorylation of LC_{20} and a decrease in the MLCK activity (Nishikori et al. 1983). Moreover, forskolin, which stimulates adenylyl cyclase, induced phosphorylation of MLCK in intact tracheal smooth muscle (de Lanerolle et al. 1984). In contrast to the experiments performed in Triton-skinned preparations, there was no decrease in the Ca^{2+}-sensitivity of mesenteric arteries permeabilised with saponin after incubation with cAMP and cAMP-dependent protein kinase (Itoh et al. 1982). This could be due to the different permeabilisation protocols (Pfitzer 1996). Others have shown that the ratio of the activities of MLCK at high and low concentrations of Ca^{2+} in the presence of calmodulin in homogenates of smooth muscle was not changed by manoeuvres that lead to an increase in intracelluar cAMP (Miller et al. 1983, van Riper et al. 1995). The activity ratio of MLCK allows the evaluation of a change in the calmodulin activation properties of MLCK in intact smooth muscle. These results have questioned the importance of inactivation of MLCK for β-adrenergic relaxation. This was substantiated by determination of the sites being phosphorylated after treatment with various pharmacological agents. It was reported that MLCK can be phosphorylated at multiple sites *in vivo* (Stull et al. 1990). However only phosphorylation of the regulatory site at the C-terminus, designed site A, led to a decreased affinity of MLCK for calmodulin (Stull et al. 1990). This site was not phosphorylated by treatment with isoproterenol in tracheal smooth muscle suggesting that cAMP-induced relaxation is not induced by desensitisation of MLCK under all conditions (Stull et al. 1990, Kotlikoff and Kamm 1996 for review). Moreover, ß-adrenergic stimulation induced relaxation at basal levels of LC_{20} phosphorylation (Miller et al. 1983). Thus, it is unlikely that inactivation of MLCK at constant Ca^{2+} contributes to ß-adrenergic relaxation. On the other hand, there is substantial experimental evidence, based on simultaneous measurements of intracellular $[Ca^{2+}]$, force, and LC_{20} phosphorylation, that desensitisation of the contractile machinery to Ca^{2+} contributes to cAMP-mediated relaxation of smooth muscle (Morgan and Morgan 1984a, 1984b, Ozaki et al. 1992, van Riper et al. 1995). If inactivation of MLCK is not responsible for the cAMP mediated decrease in force and LC_{20} phosphorylation at a constant Ca^{2+} other systems might be involved, e.g. an increase in activity of myosin light chain phosphatase (MLCP) or phosphorylation independent mechanisms. Future studies have to show the biochemical mechanism of cAMP-dependent desensitisation of the contractile apparatus.

Site A of MLCK is significantly phosphorylated under conditions where intracellular Ca^{2+} is high, i.e. during stimulation with KCl or carbachol (Stull et al. 1990, van Riper et al. 1995). The rate of Ca^{2+}-dependent phosphorylation of MLCK was slower than that of LC_{20} phosphorylation in intact smooth muscle (Tansey et al. 1994). In permeabilised smooth muscle, increasing the Ca^{2+} concentration increased the extent of both LC_{20} and MLCK phosphorylation. The latter was inhibited by inhibitors of the multifunctional Ca^{2+}/calmodulin-dependent protein kinase II activity (CaM kinase II), KN-62 or a synthetic peptide derived from the autoinhibitory sequence of CaM kinase II (Tansey et al. 1992). Under these conditions the Ca^{2+}-concentration required for half-maximal LC_{20} phosphorylation decreased from 250 nM to 170 nM. It was suggested that Ca^{2+} via binding to calmodulin plays a dual role: (i) it positively regulates LC_{20} phosphorylation via activation of MLCK (the Ca^{2+}-concentration required for half-maximal activation being 250 nM) and (ii) it negatively regulates LC_{20} phosphorylation via activation of CaM kinase II at higher Ca^{2+}-concentration (Ca^{2+}-concentration for half-maximal activation was 500 nM) leading to phosphorylation of MLCK.

Cyclic GMP has emerged as a potent, physiological second messenger involved in vasodilator activity. Nitric oxide (NO), derived from the endothelial lining of blood vessels, clinically administered nitrovasodilators, and the atrial natriuretic factor all function to widen the lumen of vessels by stimulating guanylyl cyclase in vascular smooth muscle to produce cGMP (Furchgott and Vanhoutte 1989). It was also shown that atrial natriuretic factor and nitrovasodilators relax smooth muscle by lowering intracellular Ca^{2+} and by decreasing the Ca^{2+}-sensitivity of contraction (Karaki et al. 1988, Seguchi et al. 1996). A cGMP-induced decrease in Ca^{2+}-sensitivity of contraction (Pfitzer et al. 1984, Pfitzer et al. 1986, Nishimura and van Breemen 1989, Lee et al. 1997) and of LC_{20} phosphorylation has been reported in skinned smooth muscle while the relation between force and LC_{20} phosphorylation was not altered (Pfitzer and Boels 1991, Lee et al. 1997). This suggested that cGMP either decreased the activity of MLCK or increased that of MLCP at a given concentration of Ca^{2+}. Cyclic GMP exerts its action by stimulating a cGMP-dependent protein kinase. This enzyme phosphorylated purified MLCK at a site different from site A and did not affect the activity of MLCK (Nishikawa et al. 1984, Hathaway et al. 1985) suggesting that cGMP-dependent phosphorylation of MLCK is not responsible for Ca^{2+}-desensitization. In contrast, evidence was obtained that cGMP may increase the activity of MLCP which could account for a decrease in the Ca^{2+}-sensitivity of LC_{20} phosphorylation and contraction. Under conditions where MLCK activity was blocked, the nonhydrolysable cGMP analog, 8-bromo-cGMP, increased the rate of dephosphorylation of LC_{20} and of relaxation (Lee et al. 1997). Moreover, the light chain phosphorylation at submaximal activation exhibited kinetics in the presence of 8-bromo-cGMP that were predictable from a mathematical model in which only MLCP activity is increased (Lee et al. 1997). However, cGMP-induced relaxation may be more complex. In the intact swine carotid artery, nitrovasodilators not only caused a decrease in the Ca^{2+} sensitivity of LC_{20} phosphorylation but also uncoupled force from myosin phosphorylation (McDaniel et al. 1992) suggesting that cGMP-induced desensitisation can occur by a LC_{20}-dependent and independent mechanism. The biochemical substrate of the latter mechanism has not yet been identified.

4.2
Mechanisms that increase Ca^{2+}-sensitivity

It is well established that the Ca^{2+}-sensitivity of contraction is increased in a number of different intact and permeabilised smooth muscles by different agonists (for reviews see Somlyo and Somlyo 1994, Rembold 1996, Horowitz et al. 1996b). The agonist-induced increase in Ca^{2+}-sensitivity is mimicked by the poorly hydrolyzable GTP analog, GTPγS, in smooth muscle permeabilised with β-escin or α-toxin from *S. aureus* in which the coupling between membrane bound receptors and intracellular effectors is functionally intact (Nishimura et al. 1988, Fujiwara et al. 1989, Kitazawa et al. 1989). In these experiments the $[Ca^{2+}]$ surrounding the myofilaments is kept constant by the use of high concentrations of Ca^{2+} chelators and functionally removing intracellular Ca^{2+}-stores with Ca^{2+}-ionophores which eliminates confounding effects due to Ca^{2+}-sequestration or Ca^{2+} gradients. Under many conditions, the increase in Ca^{2+}-sensitivity was not associated with a change of the dependence of force on phosphorylation (e.g. Hori et al. 1992, Rembold and Murphy 1988a) indicating that it was due to an increase in the Ca^{2+}-sensitivity of LC_{20} phosphorylation. In principle, this could be due to either an increase in the activity of MLCK or inhibition of MLCP at a given $[Ca^{2+}]$. The majority of data point to an inhibition of MLCP rather than to an activation of MLCK, as discussed below. However, it was recently shown that MLCK is phosphorylated by mitogen activated protein kinase (MAP-kinase) and that this phosphorylation stimulates MLCK (Morrison et al. 1996, Klemke et al. 1997). Expression of a mutationally active MAP kinase kinase causes activation of MAP kinase leading to phosphorylation of MLCK and LC$_{20}$ and enhanced cell migration (Klemke et al. 1997). It is not known at present whether MAP kinases phosphorylate MLCK in smooth muscle. MAP-kinases constitute a family of enzymes in the 40- to 45 kDa range (Pelech and Sanghera 1992). They are expressed and activated in smooth muscle in response to a number of stimuli and one of the substrates is caldesmon (Adam et al. 1989, 1995a, Gerthoffer et al. 1996, Katoch and Moreland 1995, Gerthoffer et al. 1997, Adam 1996 for review). *In vitro* phosphorylation of caldesmon by p44 MAP kinase slightly decreased binding of caldesmon to actin (Childs et al. 1992). Phosphorylation of serine-702 of a C-terminal fragment of caldesmon by MAP kinase reversed the inhibitory effect which the unphosphorylated fragment had on actomyosin ATPase activity (Redwood et al. 1993). This suggested that phosphorylation of caldesmon may alter its activity and this could affect Ca^{2+}-sensitivity of contraction. However, incubation of portal vein permeabilised with Triton with recombinant, activated p42 MAP kinase had no effect on Ca^{2+}-sensitivity (Nixon et al. 1995) while in a later study, activated MAP kinase potentiated Ca^{2+}-activated force (Gerthoffer et al. 1997). The potential significance of MAP kinase for regulation of contraction was also supported by the finding that a relatively specific inhibitor of MAP kinase, PD98059, partially inhibited oxytocin-induced contractions of uterine smooth muscle without affecting intracellular [Ca^{2+}] (Nohara et al. 1996). Future work will have to establish the role of MAP kinase in regulation of smooth muscle contraction which may increase the Ca^{2+}-sensitivity of contraction through two mechanisms (i) activation of MLCK, and/or (ii) phosphorylation of caldesmon.

Inhibition of MLCP appears to be a more general mechanism of increasing Ca^{2+}-sensitivity of contraction. The first experimental evidence that inhibition of MLCP can lead to an increase of Ca^{2+}-sensitivity was obtained with the phosphatase inhibitor, okadaic acid (Takai et al. 1987). Under more physiological conditions, MLCP is inhibited indirectly by a GTP dependent process (Kitazawa et al. 1991, Kubota et al. 1992) which may involve phosphorylation of the regulatory subunit of MLCP (Trinkle-Mulcahy et al. 1995).

The signalling cascade leading from the activation of a membrane bound receptor to the GTP-dependent inhibition of the activity of MLCP has not been fully elucidated. In permeabilised smooth muscle, Ca^{2+}-sensitisation can be induced by aluminium fluoride which activates heterotrimeric G protein(s) (Kawase and van Breemen 1992, Fujita et al. 1995, Gong et al. 1996). The agonist-, GTPγS-, and aluminium fluoride-induced Ca^{2+}-sensitization is inhibited by several bacterial toxins, e.g. the epidermal differentiation inhibitor (EDIN), and exoenzyme C3 from *C. botulinum* (Hirata et al. 1992, Fujita et al. 1995, Itagaki et al. 1995, Kokubu et al. 1995, Gong et al. 1996, Otto et al. 1996) which ADP-ribosylate and inactivate monomeric GTPases of the Rho subfamily of Ras related low molecular mass GTPases (Sugai et al. 1992, Aktories et al. 1992). These experiments suggest that Ca^{2+}-sensitization of force in permeabilized smooth muscle requires the activation of a heterotrimeric G protein as well as of Rho or Rho-like proteins. A note of caution should be added. It was recently reported that AlF_4^- may interact with elongation factor G (another monomeric G-protein) when it is bound to ribosomes (Mesters et al. 1993). Therefore the possibility cannot be excluded that AlF_4^- may also directly interact with Rho in smooth muscle (Gong et al. 1996). The hypothesis that proteins of the Rho family are involved in agonist-induced Ca^{2+}-sensitisation was further supported by the fact that recombinant Rho proteins, which were either permanently activated with GTPγS or constitutively active mutants, increased force at constant $[Ca^{2+}]$ in permeabilised smooth muscle (Hirata et al. 1992, Gong et al. 1996) or augment the agonist-induced Ca^{2+}-sensitisation (Otto et al. 1996). Furthermore, Ca^{2+}-sensitisation by phenylephrine, carbachol, AlF_4^-, and GTPγS was associated with translocation of RhoA from the cytosolic compartment to the plasma membrane (Gong et al. 1997). The target of Rho is Rho associated protein kinase which has been shown to phosphorylate the regulatory subunit of MLCP in cultured cells thereby decreasing its activity (Kimura et al. 1996). Moreover, Rho-kinase directly phosphorylates LC_{20} in solution (Amano et al. 1996) and induces a contraction in Triton skinned smooth muscle (Kureishi et al. 1997). Both Rho-dependent phosphorylation reactions could therefore increase LC_{20} phosphorylation at constant $[Ca^{2+}]$.

The important question whether, and to what extent, Rho proteins participate in the regulation of contractions in intact smooth muscle can be addressed by using toxin B which unlike C3 is internalised into intact cells. Toxin B monoglucosylates small GTPases of the Rho family and inactivates them (Just et al. 1995). In intact ileum longitudinal smooth muscle, carbachol induces a biphasic contraction, a fast initial increase in force, followed by a partial relaxation and a second delayed increase in force. Toxin B completely inhibited the delayed increase in force without affecting the intracellular Ca^{2+}-transient and inhibited the agonist- and GTPγS-induced increase in force at constant $[Ca^{2+}]$ after permeabilisation with β-escin (Otto et al. 1996, Lucius et al. 1998). Most interestingly, toxin B inhibited the initial increase in LC_{20} phosphorylation despite the fact that force was not significantly inhibited and despite the large increase in intracellular [Ca^{2+}]. Thus toxin B uncouples force not only from phosphorylation but also [Ca^{2+}] from phosphorylation. These data also suggest, that the activation of myosin light chain kinase by the increase in [Ca^{2+}] is not the sole factor causing the increase in LC$_{20}$ phosphorylation in the intact smooth muscle, but that additional systems that both increase LC$_{20}$ phosphorylation, e.g. through inhibition of phosphatase, and alter the phosphorylation dependence of force are activated after receptor activation.

At present it is not clear how activation of a membrane-bound receptor leads to activation of Rho. Furthermore, Rho may not under all conditions in every type of smooth muscle mediate the increase in Ca^{2+}-sensitivity (Gong et al. 1996). This indicates that there are other signalling pathways that can increase Ca^{2+}-sensitivity. For a more detailed discussion of the participation of Rho in regulating the Ca^{2+}-sensitivity see the chapter of Somlyo et al., pp. 201

Another pathway to inhibit myosin light chain phosphatase (MLCP) could involve protein kinase C and arachidonic acid. Activation of heterotrimeric G proteins that are involved in Ca^{2+}-sensitization leads to generation of diacylglycerol (e.g. Gong et al. 1995) and through activation of phospholipase A$_2$ to the formation of arachidonic acid (Gong et al. 1995, Parsons et al. 1996). The activation of protein kinase C and arachidonic acid both of which have been implicated in Ca^{2+}-sensitization may in fact act in concert to inhibit MLCP. There is a wealth of literature on the effects of phorbolesters, activators of protein kinase C, and of more or less specific inhibitors of protein kinase C, on smooth muscle contraction, intracellular [Ca^{2+}], and LC$_{20}$ phosphorylation implicating a role of protein kinase C in the activation of smooth muscle. These mechanisms have recently been reviewed in detail (see Horowitz et al. 1996b, Singer 1996, Walsh et al. 1996). However, there is still some controversy over the precise role protein kinase C plays in smooth muscle myofilament sensitisation, the discussion of which is beyond the scope to this review (Jensen et al. 1996, Gailly et al. 1997, Somlyo and Somlyo 1994 for review). In principle two types of responses to phorbol esters have been observed: (i) a slowly developing contraction at resting Ca^{2+} without an increase in LC$_{20}$ phosphorylation (e.g. Chatterjee and Tejada 1986, Jiang and Morgan 1987), and (ii) a contraction which was associated with an increase in intracellular [Ca^{2+}] and an increase in LC$_{20}$ phosphorylation (Rembold and Murphy, 1988b). The latter response could be due to activation of MLCK by Ca^{2+}. However, it was also shown that phorbolesters can increase Ca^{2+}-sensitivity of contraction (Collins et al. 1992, Itoh et al. 1993) and of LC_{20} phosphorylation in Ca^{2+} depleted smooth muscle (Singer 1990, Bárány et al. 1992) or at constant Ca^{2+} (Masuo et al. 1994). These are situations when the increase in phosphorylation could not be due to activation of MLCK. However, protein kinase C may inhibit MLCP activity. This was suggested based on the observation that half-time of LC_{20} dephosphorylation was increased in α-toxin permeabilised smooth muscle under conditions where MLCK activity was inhibited (Masuo et al. 1994). In principle there are two ways by which inhibition of MLCP could be mediated (i) indirect secondary to activation of a MLCP inhibitor, or (ii) direct through phosphorylation of a MLCP subunit or dissociation of the subunits. CPI-17 has been identified as a novel protein in vascular smooth muscle which is a substrate of protein kinase C and its phosphorylation inhibits the type 1 class of protein phosphatases (Li et al. 1998). Phosphorylated CPI-17 potentiated submaximal contractions at constant Ca^{2+} in β-escin and and Triton X-100 permeabilised smooth muscle, and decreased the $[Ca^{2+}]$ necessary for half-maximal activation by an order of magnitude. This protein could link activation of protein kinase C to inhibition of MLCP. However, there is also evidence for a direct inhibition of MLCP. It was suggested that the Ca^{2+}-sensitizing effect of excitatory agonists is mediated through activation of phospholipase A_2 and release of arachidonic acid (Gong et al. 1995, Parsons et al. 1996, Gailly et al. 1997). Two mechanisms have been described how arachidonic acid may inhibit MLCP: (i) arachidonic acid can directly inhibit MLCP in solution by dissociating the regulatory subunit from the catalytic subunit (Gong et al. 1992), or (ii) it may activate atypical protein kinase C (Gailly et al. 1997). Future studies have to identify the target of atypical protein kinase C which could be CPI-17 and/or the regulatory subunit of type 1 MLCP. In this context it is interesting to note that a partially purified myosin-bound phosphatase had an associated protein kinase which primarily phosphorylated the regulatory subunit (130 kDa) of the phosphatase resulting in inhibition of phosphatase activity (Ichikawa et al. 1996a). This kinase, which was not identified was activated by arachidonic acid and oleic acid and to a lesser extent by myristic acid suggesting that it has some characteristics of protein kinase C (Ichikawa et al. 1996a).

The mechanisms by which the phosphorylation independent Ca^{2+}-sensitization induced by protein kinase C (Jiang and Morgan 1989) occurs are presently not clear. A tentative scheme involves protein kinase C-induced activation of MAP kinase (Singer 1996, Horowitz et al. 1996b for reviews) and phosphorylation of caldesmon. This could then allow the slow cycling of unphosphorylated cross-bridges.

The pattern of activation and possible cross-talk between protein kinase C and Rho associated kinase dependent inhibition of the activity of MLCP is unknown. It appears that neither pathway can fully account for the GTP-dependent increase in Ca^{2+} sensitivity since neither inhibitors of protein kinase C including peptide inhibitors of atypical protein kinase C (Gailly et al. 1997) nor inactivation of Rho with exoenzyme C3 (Itagaki et al. 1995, Otto et al. 1996) completely inhibited the GTPγS-induced increase in Ca^{2+}-sensitivity.The different signalling pathways may be differently expressed and/or activated in different types of smooth muscle (e.g. intestinal versus arterial smooth muscle) and may also vary from species to species.

5
Thin filament associated systems

5.1
Cooperativity

Smooth muscle thin filaments contain actin and in addition tropomyosin. By analogy with the troponin-tropomyosin regulatory system in striated muscle a thin filament regulatory system might be present in smooth muscle, although at present no troponin equivalent has been clearly identified. Two thin filament associated proteins, that could have a role in a regulatory system, caldesmon and calponin, have received much attention and their functions have been the focus of several recent reviews (Sobue and Sellers 1991, Marston and Huber 1996, Dabrowska 1994, Chalovich and Pfitzer 1997, Gimona and Small 1996, Winder and Walsh, 1993, 1996). Although they have been characterised extensively in biochemical experiments their action in the smooth muscle fibre is not clear at present. In this section we will discuss some actions of the thin filament systems in the muscle cells as revealed by experiments on muscle fibre preparations.

Structural data have suggested that tropomyosin can alter its position on smooth muscle thin filaments following cross-bridge or caldesmon binding (Vibert et al. 1972, 1993, Arner et al. 1988). The structural effects of caldesmon on tropomyosin are however not identical to those induced by troponin in striated muscle thin filaments (Hodgkinson et al. 1997, Lehman et al. 1997). Biochemical data show that smooth muscle thin filaments exhibit a high degree of cooperativity in the presence of tropomyosin (Chacko and Eisenberg 1990). Cooperative cross-bridge binding has also been proposed to occur in muscle fibres (Somlyo et al. 1988b). When ATP is released in smooth muscle preparations in rigor, the relaxation involves a phase of cross-bridge reattachment prior to the final relaxation (Arner et al. 1987a, Somlyo et al. 1988b). Since the relaxation is faster in the presence of phosphate it involves cross-bridges detaching from rigor (reaction 1—2 in Fig 1 in chapter 2) and then entering the force generating reactions (steps 4 and 5) followed by a final detachment. These results are consistent with a hypothesis that rigor cross-bridges can activate a thin filament system that promotes attachment of unphosphorylated myosin. Cooperative activation has also been suggested to occur upon binding of phosphorylated myosin (Vyas et al. 1992) and it has been proposed (Somlyo et al. 1988b) that cooperative attachment could be a factor involved in recruitment of attached non-phosphorylated "latch" cross-bridges. Such a mechanism could account for the steep, curvilinear relation between force and LC_{20} phosphorylation (Di Blasi et al. 1992, Siegman et al. 1989, Schmidt et al. 1995) and would explain that 20% LC_{20} phosphorylation is sufficient for maximal contraction. As pointed out by Khromov et al. (1995) an increased ADP binding to smooth muscle would slow detachment and possibly also enhance a cooperative effect. It is possible that the thin filament associated proteins influence the cooperativity of cross-bridge binding as discussed below.

5.2
Caldesmon

Caldesmon is an actin binding molecule which is localised to the contractile apparatus of smooth muscle cells (Fürst et al. 1986, Mabuchi et al. 1996). Caldesmon is associated with thin filaments (Marston and Lehman 1985, Marston et al. 1988) and interacts with tropomyosin, myosin and Ca^{2+}-calmodulin (Sobue and Sellers 1991). In vitro, caldesmon inhibits superprecipitation of actomyosin and the actomyosin ATPase (Sobue et al. 1985, Chalovich et al. 1987, Dabrowska et al. 1985, Nagai and Walsh 1984, Smith and Marston 1985, Marston 1988) and inhibits the filament sliding in the in vitro motility assay (Shirinsky et al. 1992, Okagaki et al. 1991, Fraser and Marston 1995, Horiuchi and Chacko 1995, Wang et al. 1997). The inhibition of ATPase by caldesmon is reversed by Ca^{2+}-calmodulin (Smith et al. 1987) and caldesmon phosphorylation (Ngai and Walsh 1987).

The effects of caldesmon on actomyosin ATPase is dependent on tropomyosin (Marston 1988, Horiuchi and Chacko 1989, Horiuchi et al. 1991, Marston and Redwood 1993) and it has been proposed that caldesmon could have an inhibitory function similar to that of troponin in skeletal muscle (Fraser and Marston 1995, Marston et al. 1994, Marston and Redwood 1993). Alternatively, caldesmon might influence the binding of myosin to actin (Chalovich et al. 1987, Hemric and Chalovich 1988, Velaz et al. 1989, 1990, 1993). Using three-dimensional reconstruction of smooth muscle thin filaments (Vibert et al. 1993, Hodgkinson et al. 1997, Lehman et al. 1997) it was found that caldesmon appeared to cover weak binding sites on actin rather than moving the tropomyosin to the strong sites on actin. On the basis of these data it was suggested that a classic steric-blocking mechanism involving caldesmon and tropomyosin was not likely to operate in smooth muscle.

If caldesmon is introduced into skinned skeletal muscle fibres force and relaxed fibre stiffness is inhibited in parallel (Brenner et al. 1991) without an effect on kinetics of cross-bridge turnover reflected in a lack of an effect on the rate constant of force redevelopment or the ratio of isometric ATPase to force (Kraft et al. 1995). This is different from troponin-tropomyosin and is consistent with effects on weak cross-bridge interactions (Brenner et al. 1991, Kraft et al. 1995). In skinned smooth muscle addition of caldesmon accelerates relaxation (Szpacenko et al. 1985). A synthetic peptide against the actin and calmodulin binding sites on caldesmon causes contraction (Katsuyama et al. 1992) possibly via displacement of caldesmon from actin, although other actions of this peptide sequence might be present (Marston et al. 1994). In studies on skinned chicken gizzard fibre bundles a C-terminal 20 kDa fragment of caldesmon was found to reduce force at low myosin light chain phosphorylation levels (Pfitzer et al. 1993). Since Ca^{2+} independent myosin light chain kinase or okadaic acid, which inhibits phosphatase, was used the effects of caldesmon on force were not due to an influence on myosin light chain phosphorylation but rather to an influence on the coupling between phosphorylation and force. In skinned fibres where caldesmon had been extracted with high Mg^{2+} solutions, the relation between phosphorylation and force was almost linear with high force levels at low LC$_{20}$ phosphorylation (Malmqvist et al. 1996). In a recent study by Albrecht et al. (1997), caldesmon was found to increase the rate of relaxation from rigor suggesting that caldesmon might act via effects on cooperative mechanisms of thin filament systems. In summary, the results from skinned smooth muscle fibres show that caldesmon can influence the phosphorylation-force coupling and may therefore constitute an important component of a thin filament regulatory system. The lowered force and right-ward shift of the phosphorylation-force relationship in the presence of caldesmon suggest that either caldesmon influences the cooperative attachment of unphosphorylated cross-bridges or inhibits steps during cross-bridge binding and force generation.

Caldesmon can be phosphorylated in vitro by several kinases e.g. CaM kinase II (Scott-Woo et al. 1990, Abougou et al. 1989), protein kinase C (Adam et al. 1990, Tanaka et al. 1990) and casein kinase II (Wawrzynow et al. 1991, Vorotnikov et al. 1993, Sutherland et al. 1994) and the ERK family of mitogen activated protein (MAP) kinases (Childs et al. 1992). In vivo, several stimuli lead to caldesmon phosphorylation (Adam et al. 1989, 1990, Bárány et al. 1992, Park and Rasmussen 1986, Gerthoffer and Pohl 1994, Gerthoffer et al. 1996, 1997). However, the relation between caldesmon phosphorylation and force is not straightforward: (i) the degree of caldesmon phosphorylation and force do not correlate (Adam et al. 1989), (ii) relaxation precedes (Adam et al. 1989) or occurs without (Abe et al. 1991, Bárány et al. 1992) dephosphorylation of caldesmon, and (iii) phosphorylation is high both in a tonic and a phasic type of contraction (Adam et al. 1990). The phosphorylation pattern is consistent with the action of MAP kinase which is also present in smooth muscle (Adam et al. 1992, Adam and Hathaway 1993, Gerthoffer et al. 1996). It has been shown that the inhibitory effect of caldesmon on actomyosin ATPase is reversed when caldesmon is phosphorylated with CaM kinase II (Ngai and Walsh 1987, Sutherland and Walsh 1989). Phosphorylation of caldesmon with the possible physiological kinase, MAP kinase reverses the caldesmon induced inhibition of filament sliding in the in vitro motility assay (Gerthoffer et al. 1996) and has been shown to potentiate the Ca^{2+} sensitivity in permeabilized airway smooth muscle (Gerthoffer et al. 1997). However, negative results have been reported for skinned vascular tissue preparations (Nixon et al. 1995) which might suggest that the effects of MAP kinase phosphorylation varies between tissues.

As discussed above one important function of caldesmon appears to be in a thin filament associated regulatory system. In addition, caldesmon binds to myosin (Ikebe and Reardon 1988) and can cross-link and tether actin to myosin (Haeberle et al. 1992) and may possibly also have role in stabilising the thick filaments (Katayama et al. 1995). The regulation of these processes in the smooth muscle fibre is unknown.

5.3

Calponin

Calponin was first described by Takahashi et al. (1986) and is a protein associated with the actin filaments. Calponin binds to several contractile and regulatory proteins including actin, tropomyosin, myosin and calmodulin (Takahashi et al. 1986, 1987, 1988, Szymanski and Tao 1993, 1997). Calponin is localised both in the contractile and cytoskeletal domains in smooth muscle cells (Walsh et al. 1993, North et al. 1994a). Calponin has been suggested to translocate to the cell membrane following agonist stimulation (Parker et al. 1994) and to interact with both MAP kinase and protein kinase C-ε (Menice et al. 1997). These findings suggest that calponin may have a role as a signalling molecule, possibly an adapter protein, linking the targeting of mitogen-activated protein kinase and protein kinase C-ε to the surface membrane (Menice et al. 1997). In vitro, calponin inhibits actomyosin ATPase (Winder and Walsh 1990, Abe et al. 1990) and inhibits the filament sliding in the in vitro motility assay (Shirinsky et al., 1992, Haeberle 1994, Jaworowski et al. 1995, Borovikov et al. 1996). The inhibition by calponin is reversed by Ca^{2+}-calmodulin (Abe et al. 1990), Ca^{2+}-caltropin (Wills et al. 1994) and calponin phosphorylation (Winder and Walsh 1990, Pohl et al. 1997).

The mode of inhibition of actomyosin ATPase by calponin differs from that of caldesmon. The effect is mainly on the V_{max} with only minor effects on the K_m (Nishida et al. 1990, Horiuchi and Chacko 1991). Calponin is shown to affect strongly bound myosin (EL-Mezgueldi and Marston 1996, Borovikov et al. 1996). The inhibition is not dependent on tropomyosin (Winder and Walsh 1990) and the effects of calponin are due to changes in the actin conformation (Noda et al. 1992, Borovikov et al. 1996).

When calponin is introduced into skinned smooth muscle fibres, a reduction in active force has been observed (Itoh et al. 1994, Jaworowski et al. 1995, Uyama et al. 1996, Obara et al. 1996). Interestingly calponin influences the maximal shortening velocity of maximally activated skinned smooth muscle preparations to a larger extent than force (Jaworowski et al. 1995, Obara et al. 1996), which could suggest that calponin influences reactions associated with cross-bridge detachment or creates an internal load opposing shortening. Peptides containing the actin binding regions of calponin potentiate contraction possibly by competing with endogenous calponin (Itoh et al. 1995, Horowitz et al. 1996a). Malmqvist et al. (1997) found that extraction of calponin from skinned smooth muscle cells induces a slow contraction due to cycling of dephosphorylated myosin. It is thus possible that calponin on the thin filaments in the relaxed state exerts a tonic inhibitory role on the interaction of dephosphorylated myosin and actin and that release of this inhibitory action by regulatory mechanisms acting on calponin can lead to recruitment of slowly cycling dephosphorylated cross-bridges. At high levels of phosphorylation calponin appears to inhibit cycling of the phosphorylated cross-bridges and release of this inhibitory action would increase cross-bridge cycling rate and the maximal shortening velocity.

Although calponin can be phosphorylated in vitro and calponin phosphorylation influences its inhibitory effects, changes in calponin phosporylation in intact tissue have not been generally observed. Several groups have reported that calponin phosphorylation does not change following agonist activation (Bárány and Bárány 1993b, Gimona et al. 1992, Adam et al. 1995b). On the other hand changes in calponin phosphoryation have been described (Winder et al. 1993, Gerthoffer and Pohl 1994, Mino et al. 1995, Pohl et al. 1997).

6
The latch state

Soon after the first simultaneous measurements of force and myosin light chain (LC_{20}) phosphorylation were obtained in muscle fibres, it became obvious that there was no close correlation between force and phosphorylation but rather that the two parameters frequently dissociated from each other. In the carotid media, the initial response to stimulation was consistent with the hypothesis that activation of the contractile apparatus involves an obligatory phosphorylation of LC_{20}. However, phosphorylation started to decline during tension maintenance (Dillon et al. 1981). There was a striking correlation between LC_{20} phosphorylation and shortening velocity despite the fact that the load-bearing capacity did not change. Thus a contractile state ("latch") with high economy of force maintenance, characterised by high force, slow cross-bridge cycling and low phosphorylation levels can be induced in smooth muscle during contraction.

In the study by Dillon et al. (1981) it was proposed that force during latch was maintained by non-cycling cross-bridges (latch-bridges) formed by dephosphorylation of attached cross-bridges. The latch state was named in analogy to the catch state of molluscan smooth muscle. The striking feature of both the latch and the catch is the decreased rate of cross-bridge cycling and ATP utilisation at high levels of force maintenance. In catch, tension recovery following a quick release is absent and thus catch very much resembles a rigor like state which, however, is regulated (Rüegg 1992). Elevation of intracellular cAMP by serotonin induces a rapid relaxation (Cole and Twarog 1972, Twarog and Cole 1972, Cornelius 1982, Pfitzer and Rüegg 1982) which appears to be due to phosphorylation of a high molecular mass protein (Siegman et al. 1997, 1998). Interestingly, agonists that increase intracellular cAMP in mammalian smooth muscle relax the latch state, i.e. under conditions where phosphorylation is low (Miller et al. 1983, Gerthoffer et al. 1984). It is however still not clear, how cAMP inactivates the contractile apparatus and several other mechanisms can be involved in the regulation of the latch state in smooth muscle as discussed below.

How the latch state in smooth muscle cells is regulated, is one of the key questions for the understanding of smooth muscle contractile behaviour. It was originally suggested that there was a second mechanism, with an apparent high sensitivity for Ca^{2+}, that regulated stress maintenance by the latch-bridges in addition to regulation by LC$_{20}$ phosphorylation (Murphy 1982, Aksoy et al. 1983, Gerthoffer and Murphy 1983). In part, these suggestions initiated research into phosphorylation independent regulatory systems, and biochemical work and studies on smooth muscle fibres have identified systems that can influence the cross-bridge turnover or cytoskeletal interactions, as discussed elsewhere in this review. In a later modification of their initial latch hypothesis, Murphy and co-workers (Hai and Murphy 1988a, 1988b, 1989a, Murphy 1989) and Driska (1987) have proposed a more simple model where the only Ca^{2+}-dependent regulatory mechanism was activation of myosin light chain kinase suggesting that myosin phosphorylation was both necessary and sufficient for the development of the latch state. The central assumption of this "four-state" model is that cross-bridges can exist in four states as depicted in the Figure below; detached-phosphorylated (M$_P$), attached-phosphorylated (AM$_P$) and detached-dephosphorylated (M), attached-dephosphorylated (AM = latch-bridge).

The two attached cross-bridge states (AM and AM$_P$) were considered to generate force but had very different detachment rates; the detachment rate (k$_7$) of the dephosphorylated latch cross-bridge was proposed to be about 20% of that of the phosphorylated one (k$_4$). Hai and Murphy (1988a, 1988b) further postulated that LC$_{20}$ phosphorylation is obligatory for cross-bridge attachment, the latch state was only reached by dephosphorylation of attached phosphorylated cross-bridges (k$_8$ = 0). In the simula-

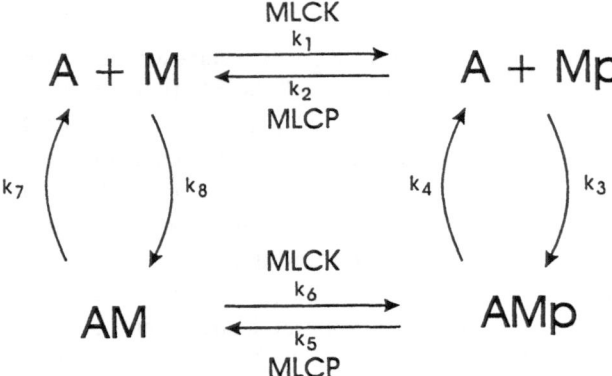

Fig. 4. Schematic diagram (modified from Rembold and Murphy 1993) for the cross-bridge cycling in smooth muscle according to the latch bridge model (Hai and Murphy 1988a, 1988b). Actin is indicated with A; M and Mp denotes the unphosphorylated and phosphorylated myosin, respectively. Note that the latch state (AM) refers to a myosin bound state which is not identical to the rigor A.M state shown in Figure 1, chapter 2.

tions (Hai and Murphy 1988a, 1988b; 1989, Rembold and Murphy 1993) it was generally also assumed that the attached phosphorylated and dephosphorylated cross-bridges produce the same amount of force and that free and attached cross-bridges are equal substrates for MLCK and MLCP.

Using the four state model Hai and Murphy (1988a, 1988b, 1989b), could simulate much of the mechanical behaviour of the swine carotid artery smooth muscle including the maintenance of force at low levels of phosphorylation (latch state) and the dependence of force on LC_{20} phosphorylation. In the intact smooth muscle force increases steeply in the suprabasal phosphorylation range 0—20% and increases only slightly with phosphorylation values over 40%. In the latch state model this non-linear quasi-hyperbolic relation is explained by recruitment of the non-phosphorylated force generating latch-bridges at the lower phosphorylation range. This behaviour is dependent on the MLCP activity; at low MLCP less latch-bridges would be formed and the dependence of force on LC_{20} phosphorylation would be less hyperbolic and approach a linear relation. The primary effect of physiological changes on MLCP activity would however be on the Ca^{2+} sensitivity of contraction; a reduction in MLCP activity would increase Ca^{2+}-sensitivity. The MLCP activity has a very high temperature dependence (Q_{10} of about 5, Mitsui et al. 1994) and it has been pointed out (Rembold and Murphy 1993, Murphy 1994) that at lower temperature latch phenomena might be less pronounced, which in part could explain that skinned muscle fibres, which usually are studied at lower temperature, do not exhibit obvious latch states.

Hai and Murphy (1988b) combined the four state model and the Huxley (1957) formalism for describing the force-velocity relationship, and could simulate also isotonic behaviour of the smooth muscle. In these simulations the slow detachment rate of latch-bridges explained a reduced shortening velocity at low levels of phosphorylation, and the linear dependence of the maximal shortening velocity on LC_{20} phosphorylation observed in intact tissue (Aksoy et al. 1982) could be reproduced. Mechanical data have suggested the presence of a cross-bridge population with slower rate of detachment during shortening in latch (Hellstrand and Nordström 1993). Using mixtures and copolymers with varying ratios of phosphorylated and unphosphorylated myosin in in vitro motility assays, Sellers et al. (1985) and Warshaw et al. (1990) could show that the in vitro velocity was dependent on the ratio of phosphorylated/unphosphorylated myosin, suggesting that unphosphorylated cross-bridges may act as a load on rapidly cycling cross-bridges. Data from skeletal muscle fibres, undergoing changes in myosin isoforms, suggest that slowly cycling cross-bridges can influence the V_{max} of muscle fibres (Reiser et al. 1988, Larsson and Moss 1993). Also in skinned fibres from smooth muscle the reduction in shortening velocity at low degrees of activation appears to be due to the presence of a population slowly cycling cross-bridges (Österman and Arner 1995). Although the mechanical and structural properties of attached dephosphorylated cross-bridges have not been clarified, it is possible that a slow rate of detachment creates an internal load that opposes shortening generated by faster phosphorylated cross-bridges. The energetic effects of such attached dephosphorylated cross-bridges during shortening might be difficult to predict, but data from muscle fibres have not found evidence for dissipation of work against an internal load during latch (Butler et al. 1986). The process of filament slowing is thus most likely more complex than the appearance of an internal resistance to shortening and it is possible that the attached dephosphorylated cross-bridges slow down general cross-bridge cycling during filament sliding.

One consequence of the simulations using the four state model (Hai and Murphy 1988a, Hai and Murphy 1992) was that the dependence of ATPase on force was non-linear and increased steeply at higher force levels, due to increased rates of both cross-bridge cycling and LC_{20} phosphorylation turnover. This result was found to be at variance with earlier data on ATP utilisation in smooth muscle (Paul 1990), although more recent measurements have demonstrated a non-linear dependence (Wingard et al. 1994) as predicted by the model. Also in skinned muscle the ATP consumption increases rapidly consistent with an increased energy cost at high activation levels (Arner and Hellstrand 1983, 1985). Another prediction of the Hai and Murphy (1988a) model simulations is that a large fraction of the ATP consumption is to be due to the phosphorylation process itself. For the swine carotid media the phosphorylation ATPase was almost equal to the ATPase of cross-bridge cycling. This muscle is a comparatively slow muscle, where the cross-bridge cycling ATPase is lower, but the calculations suggest that the cost of activation can be a significant fraction of contraction associated ATPase in smooth muscle. Using length variations in swine arterial smooth muscles and measurements of oxygen consumption is has been shown that kinetics of cross-bridge phosphorylation-dephosphorylation can rival those of cross-bridge cycling during isometric contractions (Wingard et al. 1997). A high activation associated ATPase would lower the efficiency of smooth muscle contraction (cf. Paul 1990). The almost equal energetic cost for cross-bridge cycling and activation is perhaps a necessary consequence of a regulation of cross-bridge cycling. Since smooth muscle has a comparatively high basal ATPase the contribution of these contraction associated ATPases on the total energetics of contractions is however not very large.

The four state model (cf. Murphy 1994, Strauss and Murphy 1996) thus provides a fairly simple concept, with only one regulatory system, and it can explain much of the mechanical data from smooth muscle tissue. The regulatory systems in the smooth muscle cell that influence Ca^{2+}-sensitivity of contraction via actions of MLCK and MLCP (cf. chapter 4) can be considered to operate at a higher level in the activation/deactivation cascade and their effects are usually consistent with the model predictions. However, other results in intact fibres cannot be easily explained by the model. For example the cAMP induced relaxation of latch at basal phosphorylation levels (Gerthoffer et al. 1984) or the modulation of muscle relaxation at basal phosphorylation levels (Fischer and Pfitzer 1989) cannot be explained unless one introduces an additional regulatory mechanism which regulates the net detachment rate constant. Other more severe indications that additional mechanisms, acting in parallel with LC_{20} phosphorylation, are the observations that relaxation may be induced at high levels of LC_{20} phosphorylation (Tansey et al. 1990, Bárány and Bárány 1993a, Katoch et al. 1997) and that the tissue may remain relaxed despite an increase in phosphorylation (Gerthoffer 1987). As discussed elsewhere (chapter 5) a large body of biochemical data and results from muscle fibres have shown that thin filament associated systems, e.g. caldesmon and calponin, can influence cross-bridge cycling. The action of these systems in relation to latch-bridge formation or properties is still unresolved.

The main concept of the latch-bridge model is that the attached non-phosphorylated cross-bridges are formed by dephosphorylation. An alternative to this view is attachment of unphosphorylated cross-bridges as a result of cooperative activation; in principle when a number of phosphorylated cross-bridges have attached, the unphosphorylated cross-bridges can become activated and attach into force generating states. Biochemical and structural data have shown that thin filament cooperative mechanisms can exist in smooth muscle and possibly be modulated by e.g. caldesmon (cf. chapter 5). It was originally proposed by Siegman et al. (1976a, 1976b) that attached dephosphorylated cross-bridges can exist in relaxed smooth muscle. Recent studies on skinned fibres have also shown that dephosphorylated cross-bridges can generate force and that this action was inhibited by calponin (Malmqvist et al. 1997). In skinned fibres, rigor cross-bridges have been shown to cooperatively activate attachment (Arner et al. 1987a, Somlyo et al. 1988b) and cooperative activation has been suggested to be a regulatory mechanism (Somlyo et al. 1988b). In studies by Vyas et al. (1992, 1994) it was found that a small number of thiophosphorylated cross-bridges in smooth muscle fibres increased the rate of ADP turnover and a model based on cooperative activation of cross-bridges was proposed. It thus seems possible that unphosphorylated cross-bridges can attach, and possibly cycle at a slow rate, and that this process might be regulated by a thin filament mechanism. Theoretically, cooperative activation alone would make relaxation difficult and it seems likely that such a mechanism would be sensitive to specific cross-bridge states or coupled to a regulatory system, e.g. caldesmon or calponin as suggested by results from skinned fibres (Albrecht et al. 1997, Malmqvist et al. 1997).

An important problem with any latch theory is that the "latch-bridge" in many ways remains descriptive and has not been coupled to specific biochemical actomyosin states or mode of actomyosin turnover. Skinned fibres could provide a way to explore this problem but at present there is a striking lack of a latch state in skinned smooth muscle preparations with a few exceptions. The hyperbolic phosphorylation force relationship has been demonstrated in e.g. skinned smooth muscle strips from chicken gizzard which contract maximally at a phosphorylation level between 20 and 30%. Phosphorylation can be further increased up to 60 to 100% without a major increase in tension (Hoar et al. 1979, Pfitzer et al. 1993, Schmidt et al. 1995). States with high force, low phosphorylation and low ATPase can be generated in skinned fibres (Güth and Junge 1982, Chatterjee and Murphy 1983, Gagelmann and Güth 1987, Kühn et al. 1990, Zhang and Moreland 1994), a state that ressembles the latch state in intact tissue. Although such latch like states can be produced, some results from skinned fibres are inconsistent with the latch-bridge model by Hai and Murphy (1988a, 1988b). If latch-bridges are formed by dephosphorylation of attached states the number of latch-bridges would be dependent on the absolute rates of the MLCP and MLCK activities. A prediction would be that less latch-bridges are formed when force is increased by successive inhibition of the MLCP at suboptimal $[Ca^{2+}]$ compared to the situation when force is activated by successive increases in $[Ca^{2+}]$. This could not be supported by skinned fibre data since the relation between phosphorylation and force and between force and maximal shortening velocity were similar after Ca^{2+} activation and after activation using the MLCP inhibitor okadaic acid (Siegman et al. 1989, Schmidt et al. 1995, Malmqvist and Arner 1996). Thus some of the important aspects of the Hai and Murphy model cannot be reproduced in the skinned fibre system. This can reflect that skinned fibres are studied at lower temperature where MLCP activity is lower or that MLCP and MLCK activities are lost during Triton skinning. These alterations would theoretically minimise latch-bridge formation through dephosphorylation (cf. Rembold and Murphy 1993) and perhaps favour other modes of regulation. However, the fundamental regulatory processes would still remain in the skinned fibre system and an extrapolation of skinned fibre data to the intact system could be valid for several regulatory mechanisms. The development of a skinned fibre latch state model that mimics most aspects of latch in intact tissue would be important in order to clarify the regulation of the latch state in smooth muscle and to identify the actomyosin intermediate(s) constituting the latch-bridge.

7
Conclusions and perspectives

During the last two decades significant knowledge has accumulated regarding the molecular mechanism and regulation of cross-bridge cycling in smooth muscle. There is little doubt that the phosphorylation of the regulatory light chains of myosin (LC$_{20}$) is the key regulatory event. It is also agreed upon that a sliding filament mechanism underlies contraction. However, the details of the cross-bridge cycle are not understood in the same depth as in skeletal muscle. In particular the unique feature of smooth muscle to regulate the cross-bridge cycling rate within an individual contraction/relaxation cycle is still not fully understood. This is closely related to the question of the nature of the "latch-bridge": is it entirely associated with the contractile system and if so is it a particular cross-bridge state, e.g. a strongly bound AMADP state? There is also little doubt that unphosphorylated cross-bridges may attach to actin and cycle slowly. It is not clear how the attachment or detachment of these cross-bridges are regulated. Perhaps the thin filament linked proteins, caldesmon or calponin, are involved. Just as there is no unique relation between force and LC$_{20}$ phosphorylation, the relation between cytosolic [Ca^{2+}] and force varies over a wide range of [Ca^{2+}]. Some of the signalling molecules responsible for the regulation of the Ca^{2+}-sensitivity have been identified and a picture emerges where force is regulated by interacting signalling pathways rather than by the absolute concentration of cytosolic Ca^{2+}. It is quite obvious that some of these signalling molecules such as the mitogen-activated protein kinase also regulate growth suggesting that there is a concerted regulation of cellular growth and mechanical performance of smooth muscle. The combination of the novel biophysical techniques and molecular biology including transgenic animals certainly will increase our understanding of this signalling network. This may help to design new therapeutic approaches as suggested by the recent finding that inhibition of Rho-associated kinase lowers blood pressure in several animal models of hypertension (Uehata et al. 1997).

Acknowledgements

The original studies from our laboratories cited in the text were supported by grants from the Deutsche Forschungsgemeinschaft (DFG) and from the swedish Medical Research Council (04x-8268). The collaborative work was supported by a grant from the DAAD and the Swedish Institute.

116   A. Arner and G. Pfitzer

References

Abe M, Takahashi K, Hiwada K (1990) Effect of calponin on actin-activated myosin ATPase activity. J Biochem (Tokyo) 108:835–838

Abe Y, Kasuya Y, Kudo M, Yamashita K, Goto K, Masaki T, Takuwa Y (1991) Endothelin-1-induced phosphorylation of the 20-kDa myosin light chain and caldesmon in porcine coronary artery smooth muscle. Jpn J Pharmacol 57:431–435

Abougou JC, Hagiwara M, Hachiya T, Terasawa M, Hidaka H, Hartshorne DJ (1989) Phosphorylation of caldesmon. FEBS Lett 257:408–410

Adam LP (1996) Mitogen-activated protein kinase. In: The biochemistry of smooth muscle. Bárány M (ed) Academic Press pp 167–177

Adam LP, Haeberle JR, Hathaway DR (1989) Phosphorylation of caldesmon in arterial smooth muscle. J Biol Chem 264:7698–7703

Adam LP, Milio L, Brengle B, Hathaway DR (1990) Myosin light chain and caldesmon phosphorylation in arterial muscle stimulated with endothelin-1. J Mol Cell Cardiol 22:1017–1023

Adam LP, Gapinski CJ, Hathaway DR (1992) Phosphorylation sequences in h-caldesmon from phorbol ester-stimulated canine aortas. FEBS Lett 302:223–226

Adam LP, Franklin MT, Raff GJ, Hathaway DR (1995a) Activation of mitogen-activated protein kinase in porcine carotid arteries. Circ Res 76:183–190

Adam LP, Haeberle JR, Hathaway DR (1995b) Calponin is not phosphorylated during contractions of porcine carotid arteries. Am J Physiol 268:C903–C909

Adam LP, Hathaway DR (1993) Identification of mitogen-activated protein kinase phosphorylation sequences in mammalian h-Caldesmon. FEBS Lett 322:56–60

Adler KB, Krill J, Alberghini TV, Evans JN (1983) Effect of cytochalasin D on smooth muscle contraction. Cell Motil 3:545–551

Aikawa M, Sivam PN, Kuro-o M, Kimura K, Nakahara K, Takewaki S, Ueda M, Yamaguchi H, Yazaki Y, Periasamy M, Nagai R (1993) Human smooth muscle myosin heavy chain isoforms as molecular markers for vascular development and atherosclerosis. Circ Res 73:1000–1012

Aksoy MO, Murphy RA, Kamm KE (1982) Role of Ca^{2+} and myosin light chain phosphorylation in regulation of smooth muscle. Am J Physiol 242:C109–C116

Aksoy MO, Mras S, Kamm KE, Murphy RA (1983) Ca^{2+}, cAMP, and changes in myosin phosphorylation during contraction of smooth muscle. Am J Physiol 245:C255–C270

Aktories K, Mohr C, Koch G (1992) Clostridium botulinum C3 ADP-ribosyltransferase. Current Topics in Microbiology and Immunology 175: 115–131

Albrecht K, Schneider A, Liebetrau C, Rüegg JC, Pfitzer G (1997) Exogenous caldesmon promotes relaxation of guinea-pig skinned taenia coli smooth muscles: inhibition of cooperative reattachment of latch bridges? Pflügers Arch 434:534–542

Alessi D, MacDougall LK, Sola MM, Ikebe M, Cohen P (1992) The control of protein phosphatase-1 by targetting subunits. The major myosin phosphatase in avian smooth muscle is a novel form of protein phosphatase-1. Eur J Biochem 210:1023–1035

Amano M, Ito M, Kimura K, Fukata Y, Chihara K, Nakano T, Matsuura Y, Kaibuchi K (1996) Phosphorylation and activation of myosin by Rho-associated kinase (Rho-kinase). J Biol Chem 271:20246-20249

Arheden H, Arner A (1992) Effects of magnesium pyrophosphate on mechanical properties of skinned smooth muscle from the guinea pig taenia coli. Biophys J 61:1480-1494

Arheden H, Arner A, Hellstrand P (1988) Cross-bridge behaviour in skinned smooth muscle of the guinea-pig taenia coli at altered ionic strength. J Physiol (Lond) 403:539-558

Arner A (1982) Mechanical characteristics of chemically skinned guinea-pig taenia coli. Pflügers Arch 395:277-284

Arner A (1983) Force-velocity relation in chemically skinned rat portal vein. Effects of Ca^{2+} and Mg^{2+}. Pflügers Arch 397:6-12

Arner A, Goody RS, Rapp G, Rügg JC (1987a) Relaxation of chemically skinned guinea pig taenia coli smooth muscle from rigor by photolytic release of adenosine-5'-triphosphate. J Muscle Res Cell Motil 8:377-385

Arner A, Hellstrand P, Rüegg JC (1987b) Influence of ATP, ADP and AMPPNP on the energetics of contraction in skinned smooth muscle. In: Siegman MJ, Somlyo AP, Stephens. NL (eds) Regulation and Contraction of Smooth Muscle. Alan Liss, New York pp 43-57

Arner A, Malmqvist U, Wray JS (1988) X-ray diffraction and electron microscopy of living and skinned rectococcygeus muscles of the rabbit. In: Carraro U (ed) Sarcomeric and non-sarcomeric muscles: Basic and Applied Research Prospects for the 90's. Unipress, Padova, pp 699-704

Arner A, Hellstrand P (1980) Contraction of the rat portal vein in hypertonic and isotonic medium: rates of metabolism. Acta Physiol Scand 110:69-75

Arner A, Hellstrand P (1983) Activation of contraction and ATPase activity in intact and chemically skinned smooth muscle of rat portal vein. Dependence on Ca^{++} and muscle length. Circ Res 53:695-702

Arner A, Hellstrand P (1985) Effects of calcium and substrate on force-velocity relation and energy turnover in skinned smooth muscle of the guinea-pig. J Physiol (Lond) 360:347-365

Asano M, Stull JT (1985) Effects of calmodulin antagonists on smooth muscle contraction and myosin phosphorylation. In: Hidaka H, Hartshorne DJ (eds) Calmodulin antagonists and cellular physiology. Orlando pp225-260

Ashton FT, Somlyo AV, Somlyo AP (1975) The contractile apparatus of vascular smooth muscle: intermediate high voltage stereo electron microscopy. J Mol Biol 98:17-29

Babij P (1993) Tissue-specific and developmentally regulated alternative splicing of a visceral isoform of smooth muscle myosin heavy chain. Nucleic Acids Res 21:1467-1471

Babij P, Kelly C, Periasamy M (1991) Characterization of a mammalian smooth muscle myosin heavy-chain gene: complete nucleotide and protein coding sequence and analysis of the 5' end of the gene. Proc Natl Acad Sci U S A 88:10676-10680

Babij P, Periasamy M (1989) Myosin heavy chain isoform diversity in smooth muscle is produced by differential RNA processing. J Mol Biol 210:673–679

Bagchi IC, Huang Q, Means AR (1992) Identification of amino acids essential for calmodulin binding and activation of smooth muscle myosin light chain kinase. J Biol Chem 267:3024–3029

Bárány M, Bárány K (1993a) Dissociation of relaxation and myosin light chain phosphorylation in porcine uterine muscle. Arch Biochem Biophys 305:202–204

Bárany M, Bárany K (1993b) Calponin phosphorylation does not accompany contraction of various smooth muscles. Biochim Biophys Acta 1179:229–233

Bárány K, Bárány M (1996a) Myosin light chains. In: Bárány M (ed) Biochemistry of smooth muscle contraction. Academic Press, New York, pp 21–35

Bárány K, Bárány M (1996b) Protein phosphorylation during contraction and relaxation. In: Bárány M (ed) Biochemistry of smooth muscle contraction. Academic Press, New York, pp 321–339

Bárány K, Polyak E, Bárány M (1992) Protein phosphorylation in arterial muscle contracted by high concentration of phorbol dibutyrate in the presence and absence of Ca^{2+}. Biochim Biophys Acta 1134:233–241

Barron JT, Barany M, Barany K (1979) Phosphorylation of the 20,000-dalton light chain of myosin of intact arterial smooth muscle in rest and in contraction. J Biol Chem 254:4954–4956

Barsotti RJ, Ikebe M, Hartshorne DJ (1987) Effects of Ca^{2+}, Mg^{2+}, and myosin phosphorylation on skinned smooth muscle fibers. Am J Physiol 252:C543–C554

Bengur AR, Robinson EA, Appella E, Sellers JR (1987) Sequence of the sites phosphorylated by protein kinase C in the smooth muscle myosin light chain. J Biol Chem 262:7613–7617

Benian GM, Kiff JE, Neckelmann N, Moerman DG, Waterson RH (1989) Sequence of an unusually large protein implicated in regulation of myosin activity in *C. elegans*. Nature 342:45–50

Bialojan C, Merkel L, Ruegg JC, Gifford D, Di Salvo J (1985) Prolonged relaxation of detergent-skinned smooth muscle involves decreased endogenous phosphatase activity. Proc Soc Exp Biol Med 178:648–652

Bialojan C, Takai A (1988) Inhibitory effect of a marine-sponge toxin, okadaic acid, on protein phosphatases. Specificity and kinetics. Biochem J 256:283–290

Boels PJ, Pfitzer G (1992) Relaxant effect of phalloidin on Triton-skinned microvascular and other smooth muscle preparations. J Muscle Res Cell Motil 13:71–80

Boels PJ, Troschka M, Rüegg JC, Pfitzer G (1991) Higher Ca^{2+} sensitivity of triton-skinned guinea pig mesenteric microarteries as compared with large arteries. Circ Res 69:989–996

Bond M, Somlyo AV (1982) Dense bodies and actin polarity in vertebrate smooth muscle. J Cell Biol 95:403–413

Borovikov YuS, Horiuchi KY, Avrova SV, Chacko S (1996) Modulation of actin conformation and inhibition of actin filament velocity by calponin. Biochemistry 35:13849–13857

Borrione AC, Zanellato AM, Scannapieco G, Pauletto P, Sartore S (1989) Myosin heavy-chain isoforms in adult and developing rabbit vascular smooth muscle. Eur J Biochem 183:413–417

Bose D, Bose R (1975) Mechanics of guinea pig taenia coli smooth muscle during anoxia and rigor. Am J Physiol 229:324–328

Brenner B (1986) The cross-bridge cycle in muscle. Mechanical, biochemical, and structural studies on single skinned rabbit psoas fibers to characterize cross-bridge kinetics in muscle for correlation with the actomyosin-ATPase in solution. Basic Res Cardiol 81 (Suppl 1): 1–15

Brenner B (1988) Effect of Ca^{2+} on cross-bridge turnover kinetics in skinned single rabbit psoas fibers. Proc Natl Acad Sci USA 85:3265–3269

Brenner B, Eisenberg E (1987) The mechanism of muscle contraction. Biochemical, mechanical, and structural approaches to elucidate cross-bridge action in muscle. Basic Res Cardiol 82 Suppl 2:3–16

Brenner B, Schoenberg M, Chalovich JM, Greene LE, Eisenberg E (1982) Evidence for cross-bridge attachment in relaxed muscle at low ionic strength. Proc Natl Acad Sci USA 79:7288–7291

Brenner B, Yu LC, Chalovich JM (1991) Parallel inhibition of active force and relaxed fiber stiffness in skeletal muscle by caldesmon: implications for the pathway to force generation. Proc Natl Acad Sci U S A 88:5739–5743

Bresnick AR, Wolff-Long VL, Baumann O, Pollard TD (1995) Phosphorylation on threonine-18 of the regulatory light chain dissociates the ATPase and motor properties of smooth muscle myosin II. Biochemistry 34:12576–12583

Butler TM, Siegman MJ, Davies RE (1976) Rigor and resistance to stretch in vertebrate smooth muscle. Am J Physiol 231:1509–1514

Butler TM, Siegman MJ, Mooers SU (1986) Slowing of cross-bridge cycling in smooth muscle without evidence of an internal load. Am J Physiol 251:C945–C950

Cassidy P, Hoar PE, Kerrick WG (1979) Irreversible thiophosphorylation and activation of tension in functionally skinned rabbit ileum strips by [35S]ATP gamma S. J Biol Chem 254:11148–11153

Cassidy PS, Kerrick WG, Hoar PE, Malencik DA (1981) Exogenous calmodulin increases Ca^{2+} sensitivity of isometric tension activation and myosin phosphorylation in skinned smooth muscle. Pflügers Arch 392:115–120

Cavaille F, Janmot C, Ropert S, d'Albis A (1986) Isoforms of myosin and actin in human, monkey and rat myometrium.Comparison of pregnant and non-pregnant uterus proteins. Eur J Biochem 160:507–513

Cavaillé F, Leger JJ (1983) Characterization and comparison of the contractile proteins from human gravid and non-gravid myometrium. Gynecol Obstet Invest 16:341–353

Chacko S, Eisenberg E (1990) Cooperativity of actin-activated ATPase of gizzard heavy meromyosin in the presence of gizzard tropomyosin. J Biol Chem 265:2105–2110

Chacko S, Rosenfeld A (1982) Regulation of actin-activated ATP hydrolysis by arterial myosin. Proc Natl Acad Sci U S A 79:292–296

Chalovich JM, Cornelius P, Benson CE (1987) Caldesmon inhibits skeletal actomyosin subfragment-1 ATPase activity and the binding of myosin subfragment-1 to actin. J Biol Chem 262:5711–5716

Chalovich JM, Pfitzer G (1997) Structure and function of the thin filament proteins of smooth muscle. In: Kao CY, Carsten ME (eds) Cellular aspects of smooth muscle function. Cambridge University Press Cambridge pp 253–287

Chatterjee M, Murphy RA (1983) Calcium-dependent stress maintenance without myosin phosphorylation in skinned smooth muscle. Science 221:464–466

Chatterjee M, Tejada M (1986) Phorbol ester-induced contraction in chemically skinned vascular smooth muscle. Am J Physiol 251:C356–C361

Chen YH, Chen MX, Alessi DR, Campbell DG, Shanahan C, Cohen P, Cohen PT (1994) Molecular cloning of cDNA encoding the 110 kDa and 21 kDa regulatory subunits of smooth muscle protein phosphatase 1M. FEBS Lett 356:51–55

Childs TJ, Watson MH, Sanghera JS, Campbell DL, Pelech SL, Mak AS (1992) Phosphorylation of smooth muscle caldesmon by mitogen-activated protein (MAP) kinase and expression of MAP kinase in differentiated smooth muscle cells. J Biol Chem 267:22853–22859

Cohen DM, Murphy RA (1979) Cellular thin filament protein contents and force generation in porcine arteries and veins. Circ Res 45:661–665

Cohen P (1989) The structure and regulation of protein phosphatases. Annu Rev Biochem 58:453–508

Cole RA, Twarog BM (1972) Relaxation of catch in a molluscan smooth muscle. I. Effects of drugs which act on the adenyl cyclase system. Comp Biochem Physiol A 43:321–330

Collins EM, Walsh MP, Morgan KG (1992) Contraction of single vascular smooth muscle cells by phenylephrine at constant $[Ca^{2+}]_i$. Am J Physiol 262:H754–H762

Conti MA, Adelstein RS (1981) The relationship between calmodulin binding and phosphorylation of smooth muscle kinase by the catalytic subunit of $3':5'$-cAMP dependent protein kinase. J Biol Chem 256:3178–3181

Cooke PH, Fay FS, Craig R (1989) Myosin filaments isolated from skinned amphibian smooth muscle cells are side-polar. J Muscle Res Cell Motil 10:206–220

Cornelius F (1982) Tonic contraction and the control of relaxation in a chemically skinned molluscan smooth muscle. J Gen Physiol 79:821–834

Craig R, Megerman J (1977) Assembly of smooth muscle myosin into side-polar filaments. J Cell Biol 75:990–996

Craig R, Smith R, Kendrick-Jones J (1983) Light-chain phosphorylation controls the conformation of vertebrate non-muscle and smooth muscle myosin molecules. Nature 302:436–439

Cremo CR, Geeves MA (1998) Interaction of actin and ADP with the head domain of smooth muscle myosin: implications for strain-dependent ADP release in smooth muscle. Biochemistry 37:1969–1978

Cross RA, Cross KE, Sobieszek A (1986) ATP-linked monomer-polymer equilibrium of smooth muscle myosin: the free folded monomer traps ADP.Pi. EMBO J 5:2637–2641

Cross RA, Jackson AP, Citi S, Kendrick-Jones J, Bagshaw CR (1988) Active site trapping of nucleotide by smooth and non-muscle myosins. J Mol Biol 203:173–181

Dabrowska R (1994) In: Raeburn D, Giembycz MA (eds) Airways smooth muscle: biochemical control of contraction and relaxation. Birkhäuser, Basle, pp 32–59

Dabrowska R, Goch A, Galazkiewicz B, Osinska H (1985) The influence of caldesmon on ATPase activity of the skeletal muscle actomyosin and bundling of actin filaments. Biochim Biophys Acta 842:70–75

Dabrowska R, Sherry JMF, Aromatorio DK, Hartshorne DJ (1978) Modulator protein as a component of the myosin light chain kinase from chicken gizzard. Biochemistry 17:253–258

Dantzig JA, Goldman YE, Millar NC, Lacktis J, Homsher E (1992) Reversal of the cross-bridge force-generating transition by photogeneration of phosphate in rabbit psoas muscle fibres. J Physiol (Lond) 451:247–278

de Lanerolle P, Nishikawa M (1988) Regulation of embryonic smooth muscle myosin by protein kinase C. J Biol Chem 263:9071–9074

de Lanerolle P, Nishikawa M, Yost DA, Adelstein RS (1984) Increased phosphorylation of myosin light chain kinase after an increase in cyclic AMP in intact smooth muscle. Science 223:1415–1417

de Lanerolle P, Strauss JD, Felsen R, Doerman GE, Paul RJ (1991) Effects of antibodies to myosin light chain kinase on contractility and myosin phosphorylation in chemically permeabilized smooth muscle. Circ Res 68:457–465

Devine CE, Somlyo AP (1971) Thick filaments in vascular smooth muscle. J Cell Biol 49:636–649

Di Blasi P, Van Riper D, Kaiser R, Rembold CM, Murphy RA (1992) Steady-state dependence of stress on cross-bridge phosphorylation in the swine carotid media. Am J Physiol 262:C1388–C1391

Dillon PF, Aksoy MO, Driska SP, Murphy RA (1981) Myosin phosphorylation and the cross-bridge cycle in arterial smooth muscle. Science 211:495–497

Downward J (1992) Rac and Rho in tune. Nature 359:273–274

Drew JS, Murphy RA (1997) Actin isoform expression, cellular heterogeneity, and contractile function in smooth muscle. Can J Physiol Pharmacol 75:869–877

Driska SP (1987) High myosin light chain phosphatase activity in arterial smooth muscle: can it explain the latch phenomenon? Prog Clin Biol Res 245:387–398

Driska SP, Stein PG, Porter R (1989) Myosin dephosphorylation during rapid relaxation of hog carotid artery smooth muscle. Am J Physiol 256:C315–C321

Eddinger TJ, Murphy RA (1988) Two smooth muscle myosin heavy chains differ in their light meromyosin fragment. Biochemistry 27:3807–3811

Eddinger TJ, Murphy RA (1991) Developmental changes in actin and myosin heavy chain isoform expression in smooth muscle. Arch Biochem Biophys 284:232–237

Eisenberg E, Hill TL (1985) Muscle contraction and free energy transduction in biological systems. Science 227:999–1006

EL-Mezgueldi M, Marston SB (1996) The effects of smooth muscle calponin on the strong and weak myosin binding sites of F-actin. J Biol Chem 1996 Nov 8;271(45):28161–28167

Erdödi F, Ito M, Hartshorne (1996) Myosin light chain phosphatase. In: Biochemistry of smooth muscle contraction. Bárány M (ed) Academic press pp 131–142

Eto M, Ohmori T, Suzuki M, Furuya K, Morita F (1995) A novel protein phosphatase-1 inhibitory protein potentiated by protein kinase C. Isolation from porcine aorta media and characterization. J Biochem (Tokyo) 118:1104–1107

Filo RS, Bohr DF, Rüegg JC (1965) Glycerinated skeletal and smooth muscle. Calcium and magnesium dependence. Science 147:1581–1583

Fischer W, Pfitzer G (1989) Rapid myosin phosphorylation transients in phasic contractions in chicken gizzard smooth muscle. FEBS Letters 258, 59–62

Fraser ID, Marston SB (1995) In vitro motility analysis of smooth muscle caldesmon control of actin-tropomyosin filament movement. J Biol Chem 270:19688–19693

Fuglsang A, Khromov A, Torok K, Somlyo AV, Somlyo AP (1993) Flash photolysis studies of relaxation and cross-bridge detachment: higher sensitivity of tonic than phasic smooth muscle to MgADP. J Muscle Res Cell Motil 14:666–677

Fujita A, Takeuchi T, Nakajima H, Nishio H, Hata F (1995) Involvement of heterotrimeric GTP-binding protein and rho protein, but not protein kinase C, in agonist-induced Ca^{2+}-sensitization of skinned muscle of guinea pig vas deferens. J Pharmacol Exp Ther 274:555–561

Fujiwara T, Itoh T, Kubota Y, Kuriyama H (1989) Effects of guanosine nucleotides on skinned smooth muscle tissue of the rabbit mesenteric artery. J Physiol Lond 408:535–547

Fukami M, Tani E, Takai A, Yamaura I, Minami N (1995) Activity of smooth muscle phosphatases 1 and 2A in rabbit basilar artery in vasospasm. Stroke 26:2321–2327

Fulgitini J, Singer HA, Moreland RS (1993) Phorbol ester-induced contractions of swine carotid artery are supported by slowly cycling crossbridges which are not dependent on calcium or myosin light chain phosphorylation. J Vasc Res 30:315–322

Furchgott RF, Vanhoutte PM (1989) Endothelium-derived relaxing and contracting factor. FASEB J 7:2007–2018

Furst DO, Cross RA, De Mey J, Small JV (1986) Caldesmon is an elongated, flexible molecule localized in the actomyosin domains of smooth muscle. EMBO J 5:251–257

Gagelmann M, Güth K (1985) Force generated by non-cycling crossbridges at low ionic strength in skinned smooth muscle from Taenia coli. Pflügers Arch 403:210–214

Gagelmann M, Güth K (1987) Effect of inorganic phosphate on the Ca^{2+} sensitivity in skinned Taenia coli smooth muscle fibers. Comparison of tension, ATPase activity, and phosphorylation of the regulatory myosin light chains. Biophys J 51:457–463

Gailly P, Gong MC, Somlyo AV, Somlyo AP (1997) Possible role of atypical protein kinase C activated by arachidonic acid in Ca^{2+} sensitization of rabbit smooth muscle. J Physiol Lond 500:95–109

Gailly P, Wu X, Haystead TA, Somlyo AP, Cohen PT, Cohen P, Somlyo AV (1996) Regions of the 110-kDa regulatory subunit M110 required for regulation of myosin-light-chain-phosphatase activity in smooth muscle. Eur J Biochem 239:326–332

Gallagher PJ, Hering BP, Stull JT (1997) Myosin light chain kinases. J Muscle Res Cell Mot 18:1–16

Gallagher PJ, Herring BP (1991) The carboxyl terminus of the smooth muscle myosin light chain kinase is expressed as an independent protein, telokin. J Biol Chem 266 :23945–23952

Gallagher PJ, Herring BP, Griffin SA, Stull JT (1991) Molecular characterization of a mammalian smooth muscle myosin light chain kinase. J Biol Chem 266:23936–23944

Gerthoffer WT (1987) Dissociation of myosin phosphorylation and active tension during muscarinic stimulation of tracheal smooth muscle. J Pharmacol Exp Ther 240:8–15

Gerthoffer WT, Murphy RA (1983) Ca^{2+}, myosin phosphorylation, and relaxation of arterial smooth muscle. Am J Physiol 245:C271–C277

Gerthoffer WT, Pohl J (1994) Caldesmon and calponin phosphorylation in regulation of smooth muscle contraction. Can J Physiol Pharmacol 72:1410–1414

Gerthoffer WT, Trevethick MA, Murphy RA (1984) Myosin phosphorylation and cyclic adenosine 3′,5′-monophosphate in relaxation of arterial smooth muscle by vasodilators. Circ Res 54:83–89

Gerthoffer WT, Yamboliev IA, Pohl J, Haynes R, Dang S, McHugh J (1997) Activation of MAP kinases in airway smooth muscle. Am J Physiol 272:L244–L252

Gerthoffer WT, Yamboliev IA, Shearer M, Pohl J, Haynes R, Dang S, Sato K, Sellers JR (1996) Activation of MAP kinases and phosphorylation of caldesmon in canine colonic smooth muscle. J Physiol Lond 495:597–609

Gerthoffer, WT, Murphy RA (1983) Ca^{2+}, myosin phosphorylation, and relaxation of arterial smooth muscle. Am J Physiol 245: C271–C277

Gillis JM, Cao ML, Godfraind-De Becker A (1988) Density of myosin filaments in the rat anococcygeus muscle, at rest and in contraction. II. J Muscle Res Cell Motil 9:18–29

Gimona M, Small JV (1996) Calponin. In: Bárány M (ed) The biochemistry of smooth muscle contraction. Academic Press pp 91–103

Gimona M, Sparrow MP, Strasser P, Herzog M, Small JV (1992) Calponin and SM 22 isoforms in avian and mammalian smooth muscle. Absence of phosphorylation in vivo. Eur J Biochem 205:1067–1075

Godfraind-De Becker A, Gillis JM (1988) Analysis of the birefringence of the smooth muscle anococcygeus of the rat, at rest and in contraction. I. J Muscle Res Cell Motil 1988 9:9–17

Goldman YE (1987) Kinetics of the actomyosin ATPase in muscle fibers. Annu Rev Physiol 49:637–654

Gollub J, Cremo CR, Cooke R (1996) ADP release produces a rotation of the neck region of smooth myosin but not skeletal myosin. Nat Struct Biol 3:796–802

Gong MC, Cohen P, Kitazawa T, Ikebe M, Masuo M, Somlyo AP, Somlyo AV (1992) Myosin light chain phosphatase activities and the effects of phosphatase inhibitors in tonic and phasic smooth muscle. J Biol Chem 267:14662–14668

Gong MC, Fujihara H, Somlyo AV, Somlyo AP (1997) Translocation of rhoA associated with Ca^{2+} sensitization of smooth muscle. J Biol Chem 272:10704–10709

Gong MC, Iizuka K, Nixon G, Browne JP, Hall A, Eccleston JF, Sugai M, Kobayashi S, Somlyo AV, Somlyo AP (1996) Role of guanine nucleotide-binding proteins - ras family or trimeric proteins or both - in Ca^{2+} sensitization of smooth muscle. Proc Natl Acad Sci USA 93:1340–1345

Gong MC, Kinter MT, Somlyo AV, Somlyo AP (1995) Arachidonic acid and diacylglycerol release associated with inhibition of myosin light chain dephosphorylation in rabbit smooth muscle. J Physiol Lond 486: 113–122

Greene LE, Sellers JR (1987) Effect of phosphorylation on the binding of smooth muscle heavy meromyosin ADP to actin. J Biol Chem 262:4177–4181

Guilford WH, Dupuis DE, Kennedy G, Wu J, Patlak JB, Warshaw DM (1997) Smooth muscle and skeletal muscle myosins produce similar unitary forces and displacements in the laser trap. Biophys J 72:1006–1021

Gunst SJ, al-Hassani MH, Adam LP (1994) Regulation of isotonic shortening velocity by second messengers in tracheal smooth muscle. Am J Physiol 266:C684–C691

Güth K, Junge J (1982) Low Ca^{2+} impedes cross-bridge detachment in chemically skinned Taenia coli. Nature 300:775–776

Haeberle JR (1994) Calponin decreases the rate of cross-bridge cycling and increases maximum force production by smooth muscle myosin in an in vitro motility assay. J Biol Chem 269:12424–12431

Haeberle JR, Hathaway DR, De Paoli-Roach AA (1985a) Dephosphorylation of myosin by the catalytic subunit of a type-2 phosphatase produces relaxation of chemically skinned uterine smooth muscle. J Biol Chem 260:9965–9968

Haeberle JR, Hott JW, Hathaway DR (1985b) Regulation of isometric force and isotonic shortening velocity by phosphorylation of the 20,000 dalton myosin light chain of rat uterine smooth muscle. Pflügers Arch 403:215–219

Haeberle JR, Sutton TA, Trockman BA (1988) Phosphorylation of two sites on smooth muscle myosin. J Biol Chem 263:4424–4429

Haeberle JR, Trybus KM, Hemric ME, Warshaw DM (1992) The effects of smooth muscle caldesmon on actin filament motility. J Biol Chem 267:23001–23006

Hai CM, Murphy RA (1988a) Cross-bridge phosphorylation and regulation of latch state in smooth muscle. Am J Physiol 254:C99–C106

Hai CM, Murphy RA (1988b) Regulation of shortening velocity by cross-bridge phosphorylation in smooth muscle. Am J Physiol 255:C86–C94

Hai CM, Murphy RA (1989a) Ca^{2+}, crossbridge phosphorylation, and contraction. Annu Rev Physiol 51:285–298

Hai CM, Murphy RA (1989b) Cross-bridge dephosphorylation and relaxation of vascular smooth muscle. Am J Physiol 256:C282–C287

Hai CM, Murphy RA (1992) Adenosine 5'-triphosphate consumption by smooth muscle as predicted by the coupled four-state crossbridge model. Biophys J 61: 530–541

Harris DE, Warshaw DM (1993) Smooth and skeletal muscle actin are mechanically indistinguishable in the in vitro motility assay. Circ Res 72:219–224

Hartshorne DJ, Mrwa U (1982) Regulation of smooth muscle actomyosin. Blood Vessels 19:1–18

Hasegawa Y, Morita F (1992) Role of 17-kDa essential light chain isoforms of aorta smooth muscle myosin. J Biochem (Tokyo) 111:804–809

Hasegawa Y, Ueda Y, Watanabe M, Morita F (1992) Studies on amino acid sequences of two isoforms of 17-kDa essential light chain of smooth muscle myosin from porcine aorta media. J Biochem (Tokyo) 111:798–803

Hasegawa Y, Ueno H, Horie K, Morita F (1988) Two isoforms of 17-kDa essential light chain of aorta media smooth muscle myosin. J Biochem (Tokyo) 103:15–18

Hashimoto Y, Soderling TR (1990) Phosphorylation of smooth muscle myosin light chain kinase by Ca2+/calmodulin-dependent protein kinase II: comparative study of the phosphorylation sites. Arch Biochem Biophys 278:41–45

Hathaway DR, Konicki MV, Coolican SA (1985) Phosphorylation of myosin light chain kinase from vascular smooth muscle by cAMP- and cGMP-dependent protein kinases. J Mol Cell Cardiol 17:841–850

He HZ, Ferenczi MA, Trentham DR, Webb MR, Brune M, Somlyo AP, Somlyo AV (1997) Kinetics of phosphate (Pi) release in activated isometric smooth muscle. Biophysical J 72:A177

Hellstrand P, Nordström I (1993) Cross-bridge kinetics during shortening in early and sustained contraction of intestinal smooth muscle. Am J Physiol 265:C695–C703

Hellstrand P, Paul RJ (1982) Vascular smooth muscle: relations between energy metabolism and mechanics. In: Crass ME III, Barnes CD (eds) Vascular Smooth Muscle: Metabolic, ionic and contractile mechanisms. Academic Press, New York, pp 1–35

Hellstrand P, Paul RJ (1983) Phosphagen content, breakdown during contraction, and O$_2$ consumption in rat portal vein. Am J Physiol 244:C250–C258

Helper DJ, Lash JA, Hathaway DR (1988) Distribution of isoelectric variants of the 17,000-dalton myosin light chain in mammalian smooth muscle. J Biol Chem 263:15748–15753

Hemric ME, Chalovich JM (1988) Effect of caldesmon on the ATPase activity and the binding of smooth and skeletal myosin subfragments to actin. J Biol Chem 263:1878–1885

Hewett TE, Martin AF, Paul RJ (1993) Correlations between myosin heavy chain isoforms and mechanical parameters in rat myometrium. J Physiol (Lond) 460:351–364

Hibberd MG, Dantzig JA, Trentham DR, Goldman YE (1985) Phosphate release and force generation in skeletal muscle fibers. Science 228:1317–1319

Hidaka, H, Asano M, Iwadare S, Matsumoto I, Totsuka T, Aoki N (1978) A novel vascular relaxing agent, N-(6-aminohexyl)-5-chloro-1-naphtalenesulfonamide which affects vascular smooth muscle actomyosin. J Pharmacol Exp Ther 207:8–15

Himpens B, Matthijs G, Somlyo AV, Butler TM, Somlyo AP (1988) Cytoplasmic free calcium, myosin light chain phosphorylation, and force in phasic and tonic smooth muscle. J Gen Physiol 92:713–729

Hinssen H, D'Haese J, Small JV, Sobieszek A (1978) Mode of filament assembly of myosins from muscle and nonmuscle cells. J Ultrastruct Res 64:282–302

Hirano K., Phan B.C., Hartshorne D.J. (1997) Interactions of the subunits of smooth muscle myosin phosphatase. J Biol Chem 272:3683–3688

Hirata K, Kikuchi A, Sasaki T, Kuroda S, Kaibuchi K, Matsuura Y, Seki H, Saida K, Takai Y (1992) Involvement of rho p21 in the GTP-enhanced calcium ion sensitivity of smooth muscle contraction. J Biol Chem 267:8719–8722

Hoar PE, Kerrick WG, Cassidy PS (1979) Chicken gizzard: relation between calcium-activated phosphorylation and contraction, Science 204:503–506

Hoar PE, Pato MD, Kerrick WG (1985) Myosin light chain phosphatase. Effect on the activation and relaxation of gizzard smooth muscle skinned fibers. J Biol Chem 260:8760–4876

Hodgkinson JL, Marston SB, Craig R, Vibert P, Lehman W (1997) Three-dimensional image reconstruction of reconstituted smooth muscle thin filaments: effects of caldesmon. Biophys J 72:2398–2404

Hori M, Sato K, Sakata K, Ozaki H, Takano-Ohmuro H, Tsuchiya T, Sugi H, Kato I, Karaki H (1992) Receptor agonists induce myosin phosphorylation-dependent and phosphorylation-independent contractions in vascular smooth muscle. J Pharmacol Exp Ther 261:506–512

Horiuchi KY, Chacko S (1989) Caldesmon inhibits the cooperative turning-on of the smooth muscle heavy meromyosin by tropomyosin-actin. Biochemistry 28:9111–9116

Horiuchi KY, Chacko S (1991) The mechanism for the inhibition of actin-activated ATPase of smooth muscle heavy meromyosin by calponin. Biochem Biophys Res Commun 176:1487–1493

Horiuchi KY, Chacko S (1995) Effect of unphosphorylated smooth muscle myosin on caldesmon-mediated regulation of actin filament velocity. J Muscle Res Cell Motil 16:11–19

Horiuchi KY, Samuel M, Chacko S (1991) Mechanism for the inhibition of acto-heavy meromyosin ATPase by the actin/calmodulin binding domain of caldesmon. Biochemistry 30:712–717

Horiuti K, Somlyo AV, Goldman YE, Somlyo AP (1989) Kinetics of contraction initiated by flash photolysis of caged adenosine triphosphate in tonic and phasic smooth muscles. J Gen Physiol 94:769–781

Horowitz A, Clement-Chomienne O, Walsh MP, Tao T, Katsuyama H, Morgan KG (1996a) Effects of calponin on force generation by single smooth muscle cells. Am J Physiol 270:H1858–H1863

Horowitz A, Menice CB, Laporte R, Morgan KG (1996b) Mechanisms of smooth muscle contraction. Physiol Rev 76:967–1003

Horowitz A, Trybus KM, Bowman DS, Fay FS (1994) Antibodies probe for folded monomeric myosin in relaxed and contracted smooth muscle. J Cell Biol 126:1195-1200

Hubbard MJ, Cohen P (1989) The glycogen-binding subunit of protein phosphatase-1G from rabbit skeletal muscle. Further characterisation of its structure and glycogen-binding properties. Eur J Biochem 180:457-465

Huxley AF (1957) Muscle structure and theories of contraction. Prog Biophys Biophys Chem 7:255-318

Huxley AF, Niedergerke R (1954) Interference microscopy of living muscle fibres. Nature (London) 173:971-973

Huxley HE, Hanson J (1954) Changes in the cross-striations of muscle during contraction and stretch and their structural interpretation. Nature (London) 173:973-976

Ichikawa K, Ito M, Hartshorne DJ (1996a) Phosphorylation of the large subunit of myosin phosphatase and inhibition of phosphatase activity. J Biol Chem 271:4733-4740

Ichikawa K., Hirano K., Ito M., Tanaka J., Nakano T., Hartshorne D.J. (1996b) Interactions and properties of smooth muscle myosin phosphatase. Biochemistry 35:6313-6320

Ikebe M, Hartshorne DJ (1985) Phosphorylation of smooth muscle myosin at two distinct sites by myosin light chain kinase. J Biol Chem 260:10027-10031

Ikebe M, Hartshorne DJ, Elzinga M (1986) Identification, phosphorylation, and dephosphorylation of a second site for myosin light chain kinase on the 20,000-dalton light chain of smooth muscle myosin. J Biol Chem 261:36-39

Ikebe M, Hartshorne DJ, Elzinga M (1987a) Phosphorylation of the 20,000-dalton light chain of smooth muscle myosin by the calcium-activated, phospholipid-dependent protein kinase. Phosphorylation sites and effects of phosphorylation. J Biol Chem 262:9569-9573

Ikebe M, Maruta S, Reardon S (1989) Location of the inhibitory region of smooth muscle myosin light chain kinase. J Biol Chem 264:6967-6971

Ikebe M, Reardon S (1988) Binding of caldesmon to smooth muscle myosin. J Biol Chem 263:3055-3058

Ikebe M, Reardon S (1990) Phosphorylation of smooth myosin light chain kinase by smooth muscle Ca^{2+}/calmodulin-dependent multifunctional protein kinase. J Biol Chem 265:8975-8978

Ikebe M, Stepinska M, Kemp BE, Means AR, Hartshorne DJ (1987b) Proteolysis of smooth muscle myosin light chain kinase. Formation of inactive and calmodulin-independent fragments. J Biol Chem 260:13828-13834

Ishihara H, Martin BL, Brautigan DL, Karaki H, Ozaki H, Kato Y, Fusetani N, Watabe S, Hashimoto K, Uemura D, Hartshorne DJ (1989) Calyculin A and okadaic acid: inhibitors of protein phosphatase activity. Biochem Biophys Res Commun 159:871-877

Ishikawa T, Chijiwa T, Hagiwara M, Mamiya S, Saitoh M, Hidaka H (1988) ML-9 inhibits the vascular contraction via the inhibition of myosin light chain phosphorylation. Mol Pharmacol 33:598–603

Itagaki M, Komori S, Unno T, Syuto B, Ohashi H (1995) Possible involvement of a small G-protein sensitive to exoenzyme C3 of *Clostridium botulinum* in the regulation of myofilament Ca^{2+}-sensitivity in β-escin skinned smooth muscle of guinea pig ileum. Jap J Pharmacol 67:1–7

Ito M, Dabrowska R, Guerriero V Jr, Hartshorne DJ (1989) Identification in turkey gizzard of an acidic protein related to the C-terminal portion of smooth muscle myosin light chain kinase. J Biol Chem 264:13971–13974

Itoh H, Shimomura A, Okubo S, Ichikawa K, Ito M, Konishi T, Nakano T (1993) Inhibition of myosin light chain phosphatase during Ca^{2+}-independent vasoconstriction. Am J Physiol 265:C1319–C1324

Itoh T, Hidetaka I, Kuriyama H (1982) Mechanisms of relaxation induced by activation of the β-adrenoceptors in smooth muscle cells of the guinea-pig mesenteric artery. J Physiol 326:475–493

Itoh T, Ikebe M, Kargacin GJ, Hartshorne DJ, Kemp BE, Fay FS (1989) Effects of modulators of myosin light-chain kinase activity in single smooth muscle cells. Nature 338:164–167

Itoh T, Suzuki A, Watanabe Y, Mino T, Naka M, Tanaka T (1995) A calponin peptide enhances Ca^{2+} sensitivity of smooth muscle contraction without affecting myosin light chain phosphorylation. J Biol Chem 270:20400–20403

Itoh T, Suzuki S, Suzuki A, Nakamura F, Naka M, Tanaka T (1994) Effects of exogenously applied calponin on Ca(2+)-regulated force in skinned smooth muscle of the rabbit mesenteric artery. Pflügers Arch 427:301–308

Jaworowski A, Anderson KI, Arner A, Engstrom M, Gimona M, Strasser P, Small JV (1995) Calponin reduces shortening velocity in skinned taenia coli smooth muscle fibres. FEBS Lett 365:167–171

Jensen PE, Gong MC, Somlyo AV, Somlyo AP (1996) Separate upstream and convergent downstream pathways of G-protein- and phorbol ester-mediated Ca^{2+} sensitization of myosin light chain phosphorylation in smooth muscle. Biochem J 318: 469–475

Jiang H, Rao K, Liu X, Liu G, Stephens NL (1995) Increased Ca^{2+} and myosin phosphorylation, but not calmodulin activity in sensitized airway smooth muscles. Am J Physiol 268:L739–L746

Jiang MJ, Morgan KG (1987) Intracellular calcium levels in phorbol ester-induced contractions of vascular muscle. Am J Physiol 253:H1365–H1371

Jiang MJ, Morgan KG (1989) Agonist-specific myosin phosphorylation and intracellular calcium during isometric contractions of arterial smooth muscle. Pflügers Arch 413:637–643

Johnson DF, Moorhead G, Caudwell FB, Cohen P, Chen YH, Chen MX, Cohen PT (1996) Identification of protein-phosphatase-1-binding domains on the glycogen and myofibrillar targetting subunits. Eur J Biochem 239:317–325

Just I, Selzer J, Wilm M, von Eichel-Streiber C, Mann M, Aktories K (1995) Glucosylation of Rho proteins by *Clostridium difficile* toxin B. Nature 375:500–503

Kamm KE, Stull JT (1985) The function of myosin and myosin light chain kinase phosphorylation in smooth muscle. Ann Rev Pharmacol Toxicol 25:593–620

Kamm KE, Stull JT (1986) Activation of smooth muscle contraction: relation between myosin phosphorylation and stiffness. Science 232:80–82

Kanamori M, Naka M, Asano M, Hidaka H (1981) Effects of N-(6-aminohexyl)-5-chloro-1-naphtalenesulfonamide and other calmodulin antagonists (calmodulin interacting agents) on calcium-induced contraction of rabbit aortic strips. J Pharmacol Exp Ther 217:494–499

Kaplan JH, Somlyo AP (1989) Flash photolysis of caged compounds: new tools for cellular physiology. Trends Neurosci 1989 Feb;12(2):54–59

Karaki H, Sato K, Ozaki H, Murakami K (1988) Effects of sodium nitroprusside on cytosolic calcium level in vascular smooth muscle. Eur J Pharmacol 156:259–266

Kargacin GJ, Ikebe M, Fay FS (1990) Peptide modulators of myosin light chain kinase affect smooth muscle cell contraction. Am J Physiol 259:C315–C324

Katayama E, Ikebe M (1995) Mode of caldesmon binding to smooth muscle thin filament: possible projection of the amino-terminal domain of caldesmon from native filament. Biophys J 68:2419–2428

Katayama E, Scott–Woo GC, Ikebe M (1995) Effect of caldesmon on the assembly of smooth muscle myosin. J Biol Chem 270:3919–3925

Katoch SS, Moreland RS (1995) Agonist and membrane depolarization induced activation of MAP kinase in the swine carotid artery. Am J Physiol 269:H222–H229

Katoch SS, Rüegg JC, Pfitzer G (1997) Differential effects of a K^{+} channel agonist and Ca^{2+} antagonists on myosin light chain phosphorylation in relaxation of endothelin-1 contracted tracheal smooth muscle. Pflügers Arch Eur J Physiol 433:472–477

Katsuyama H, Wang CL, Morgan KG 1(1992) Regulation of vascular smooth muscle tone by caldesmon. J Biol Chem 267:14555–14558

Kawai M, Halvorson HR (1991) Two step mechanism of phosphate release and the mechanism of force generation in chemically skinned fibers of rabbit psoas muscle. Biophys J 59:329–342

Kawamoto S, Adelstein RS (1987) Characterization of myosin heavy chains in cultured aorta smooth muscle cells. A comparative study. J Biol Chem 262:7282–7288

Kawase T, van Breemen C (1992) Aluminium fluoride induces a reversible Ca^{2+}-sensitization in alpha-toxin permeabilized vascular smooth muscle. Eur J Pharmacol 214:39–44

Kelley CA, Sellers JR, Goldsmith PK, Adelstein RS (1992) Smooth muscle myosin is composed of homodimeric heavy chains. J Biol Chem 267:2127–2130

Kelley CA, Takahashi M, Yu JH, Adelstein RS (1993) An insert of seven amino acids confers functional differences between smooth muscle myosins from the intestines and vasculature. J Biol Chem 268:12848–12854

Kemp BE, Parker MW, Hu S, Tiganis T, House C (1994) Substrate and pseudosubstrate interactions with protein kinases: determinants of specificity. Trends Biochem Sci 19:440-444

Kemp BE, Pearson RB, Guerriero V Jr, Bagchi IC, Means AR (1987) The calmodulin binding domain of chicken smooth muscle myosin light chain kinase contains a pseudosubstrate sequence. J Biol Chem 262:2542-2548

Kenney RE, Hoar PE, Kerrick WG (1990) The relationship between ATPase activity, isometric force, and myosin light-chain phosphorylation and thiophosphorylation in skinned smooth muscle fiber bundles from chicken gizzard. J Biol Chem 265:8642-8649

Kerrick WGL, Hoar PE (1981) Inhibition of smooth muscle tension by cyclic AMP-dependent protein kinase. Nature 292:253-255

Khromov A, Somlyo AV, Trentham DR, Zimmermann B, Somlyo AP (1995) The role of MgADP in force maintenance by dephosphorylated cross-bridges in smooth muscle: a flash photolysis study. Biophys J 69:2611-2622

Khromov AS, Somlyo AV, Somlyo AP (1996) Nucleotide binding by actomyosin as a determinant of relaxation kinetics of rabbit phasic and tonic smooth muscle. J Physiol (Lond) 492:669-673

Kimura K, Ito M, Amano M, Chihara K, Fukata Y, Nakafuku M, Yamamori B, Feng J, Nakano T, Okawa K, Iwamatsu A, Kaibuchi K (1996) Regulation of myosin phosphatase by Rho and Rho-associated kinase (Rho-kinase). Science 273:245-248

Kitazawa T, Gaylinn BD, Denney GH, Somlyo AP (1991) G-protein-mediated Ca^{2+} sensitization of smooth muscle contraction through myosin light chain phosphorylation. J Biol Chem 266:1708-1715

Kitazawa T, Kobayashi S, Horiuti K, Somlyo AV, Somlyo AP (1989) Receptor-coupled, permeabilized smooth muscle. Role of the phosphatidylinositol cascade, G-proteins, and modulation of the contractile response to Ca^{2+}. J Biol Chem 264:5339-5242

Klemke RL, Cai S, Giannini AL, Gallagher PJ, de Lanerolle P, Cheresh DA (1997) Regulation of cell motility by mitogen-activated protein kinase. J Cell Biol 137:481-492

Klemt P, Peiper U, Speden RN, Zilker F (1981) The kinetics of post-vibration tension recovery of the isolated rat portal vein. J Physiol (Lond) 312:281-296

Kobayashi H, Inoue A, Mikawa T, Kuwayama H, Hotta Y, Masaki T, Ebashi S (1992) Isolation of cDNA for bovine stomach 155 kDa protein exhibiting myosin light chain kinase activity. J Biochem (Tokyo) 112:786-791

Kokubu N, Satoh M, Takayanagi I (1995) Involvement of botulinum C3-sensitive GTP-binding proteins in α_1-adrenoceptor subtypes mediating Ca^{2+}-sensitization. Eur J Pharmacol 290:19-27

Kossmann T, Furst D, Small JV (1987) Structural and biochemical analysis of skinned smooth muscle preparations. J Muscle Res Cell Motil 8:135-144

Kotlikoff MI, Kamm KE (1996) Molecular mechanisms of β-adrenergic relaxation of airway smooth muscle. Annu Rev. Physiol 58:115-141

Kraft T, Chalovich JM, Yu LC, Brenner B (1995) Parallel inhibition of active force and relaxed fiber stiffness by caldesmon fragments at physiological ionic strength and temperature conditions: additional evidence that weak cross-bridge binding to actin is an essential intermediate for force generation. Biophys J 68:2404–2418

Krebs EG, Beavo JA (1979) Phosphorylation-dephosphorylation of enzymes. Annu Rev Biochem 48:923–959

Kubota Y, Nomura M, Kamm KE, Mumby MC, Stull JT (1992) GTPγS-dependent regulation of smooth muscle contractile elements. Am J Physiol 262:C405–C410

Kureishi Y, Kobayashi S, Amano M, Kimura K, Kanaide H, Nakano T, Kaibuchi K, Ito M (1997) Rho-associated kinase directly induces smooth muscle contraction through myosin light chain phosphorylation. J Biol Chem 272:12257–12260

Kuro-o M, Nagai R, Nakahara K, Katoh H, Tsai RC, Tsuchimochi H, Yazaki Y, Ohkubo A, Takaku F (1991) cDNA cloning of a myosin heavy chain isoform in embryonic smooth muscle and its expression during vascular development and in arteriosclerosis. J Biol Chem 266:3768–3773

Kuro-o M, Nagai R, Tsuchimochi H, Katoh H, Yazaki Y, Ohkubo A, Takaku F (1989) Developmentally regulated expression of vascular smooth muscle myosin heavy chain isoforms. J Biol Chem 264:18272–18275

Kühn H, Tewes A, Gagelmann M, Güth K, Arner A, Rüegg JC (1990) Temporal relationship between force, ATPase activity, and myosin phosphorylation during a contraction/relaxation cycle in a skinned smooth muscle. Pflügers Arch 416:512–518

Labeit S, Barlow DP, Gautel M, Gibson T, Holt J, Hsieh C-L, Francke U, Leonard K, Wardale J, Whiting A, Trinick J (1990) A regular pattern of two types of 100-residue motif in the sequence of titin. Nature 345:273–278

Labeit S, Kolmerer B (1995) Titins: giant proteins in charge of muscle ultrastructure and elasticity. Science 270:293–296

Larsson L, Moss RL (1993) Maximum velocity of shortening in relation to myosin isoform composition in single fibres from human skeletal muscles. J Physiol 472:595–614

Lash JA, Helper DJ, Klug M, Nicolozakes AW, Hathaway DR (1990) Nucleotide and deduced amino acid sequence of cDNAs encoding two isoforms for the 17,000 dalton myosin light chain in bovine aortic smooth muscle. Nucleic Acids Res 18:7176

Lee MR, Li L, Kitazawa T (1997) Cyclic GMP causes Ca^{2+} desensitization in vascular smooth muscle by activating the myosin light chain phosphatase. J Biol Chem 272:5963–5068

Lehman W, Vibert P, Craig R (1997) Visualization of caldesmon on smooth muscle thin filaments. J Mol Biol 274:310–317

Li L, Eto M, Lee MR, Mortia F, Yazawa M, Kitazawa T (1998) Possible involvement of the novel CPI-17 protein in protein kinase C signal transduction of rabbit arterial smooth muscle. J Physiol Lond 508:871–881

Liu G, Liu X, Rao K, Jiang H, Stephens NL (1996) Increased myosin light chain kinase content in sensitized canine saphenous vein. J Appl Physiol 80:665–669

Lowy J, Mulvany MJ (1973) Mechanical properties of guinea pig taenia coli muscles. Acta Physiol Scand 88:123–136

Lowy J, Poulsen FR, Vibert PJ (1970) Myosin filaments in vertebrate smooth muscle. Nature 225:1053–1054

Lucius C, Arner A, Steusloff A, Troschka M, Hofmann F, Aktories K, Pfitzer G (1998) Clostridium difficile toxin G inhibits carbachol-induced force and myosin light chain phosphorylation in guinea-pig smooth muscle: role of Rho proteins. J Physiol Lond 506:83–93

Lukas TJ, Burgess WH, Prendergast FG, Lau W, Watterson DM (1986) Calmodulin binding domains: characterization of a phosphorylation and calmodulin binding site from myosin light chain kinase. Biochemistry 25:1458–1464

Mabuchi K, Li Y, Tao T, Wang CL (1996) Immunocytochemical localization of caldesmon and calponin in chicken gizzard smooth muscle. J Muscle Res Cell Motil 17:243–260

Malmqvist U, Arner A (1990) Isoform distribution and tissue contents of contractile and cytoskeletal proteins in hypertrophied smooth muscle from rat portal vein. Circ Res 66:832–845

Malmqvist U, Arner A (1991) Correlation between isoform composition of the 17 kDa myosin light chain and maximal shortening velocity in smooth muscle. Pflügers Arch 418:523–530

Malmqvist U, Arner A (1996) Regulation of force and shortening velocity by calcium and myosin phosphorylation in chemically skinned smooth muscle. Pflügers Arch 433:42–48

Malmqvist U, Arner A, Makuch R, Dabrowska R (1996) The effects of caldesmon extraction on mechanical properties of skinned smooth muscle fibre preparations. Pflügers Arch 432:241–247

Malmqvist U, Arner A, Uvelius B (1991a) Contractile and cytoskeletal proteins in smooth muscle during hypertrophy and its reversal. Am J Physiol 260:C1085–C1093

Malmqvist U, Arner A, Uvelius B (1991b) Cytoskeletal and contractile proteins in detrusor smooth muscle from bladders with outlet obstruction—a comparative study in rat and man. Scand J Urol Nephrol 25:261–267

Malmqvist U, Trybus KM, Yagi S, Carmichael J, Fay FS (1997) Slow cycling of unphosphorylated myosin is inhibited by calponin, thus keeping smooth muscle relaxed. Proc Natl Acad Sci U S A 94:7655–7660

Marston S (1988) Aorta caldesmon inhibits actin activation of thiophosphorylated heavy meromyosin Mg^{2+}-ATPase activity by slowing the rate of product release. FEBS Lett 238:147–150

Marston SB, Fraser ID, Huber PA, Pritchard K, Gusev NB, Torok K (1994) Location of two contact sites between human smooth muscle caldesmon and Ca(2+)-calmodulin. J Biol Chem 269:8134–8139

Marston SB, Huber PAJ (1996) Caldesmon. In: Bárány M (ed) The biochemistry of smooth muscle contraction. Academic Press pp 77–90

Marston SB, Lehman W (1985) Caldesmon is a Ca^{2+}-regulatory component of native smooth-muscle thin filaments. Biochem J 231:517–522

Marston SB, Redwood CS (1993) The essential role of tropomyosin in cooperative regulation of smooth muscle thin filament activity by caldesmon. J Biol Chem 268:12317–12320

Marston SB, Redwood CS, Lehman W (1988) Reversal of caldesmon function by anti-caldesmon antibodies confirms its role in the calcium regulation of vascular smooth muscle thin filaments. Biochem Biophys Res Commun 155:197–202

Marston SB, Taylor EW (1980) Comparison of the myosin and actomyosin ATPase mechanisms of the four types of vertebrate muscles. J Mol Biol 139:573–600

Masuo M, Reardon S, Ikebe M, Kitazawa T (1994) A novel mechanism for the Ca^{2+} sensitizing effect of protein kinase C on vascular smooth muscle: inhibition of myosin light chain phosphatase. J Gen Physiol 104:265–286

Mauss S, Koch G, Kreye VA, Aktories K (1989) Inhibition of the contraction of the isolated longitudinal muscle of the guinea-pig ileum by botulinum C2 toxin: evidence for a role of G/F-actin transition in smooth muscle contraction. Naunyn Schmiedebergs Arch Pharmacol 340:345–351

McDaniel NL, Chen X-L, Singer HA, Murphy RA, Rembold CM (1992) Nitrovasodilators relax arterial smooth muscle by decreasing [Ca^{2+}]$_i$ and uncoupling stress from myosin phosphorylation. Am J Physiol 263:C461–C467

Means AR, Bagchi IC, Van Berkum MF, Kemp BE (1991) Regulation of smooth muscle myosin light chain kinase by calmodulin. Adv Exp Med Biol 304:11–24

Meisheri KD, Rüegg JC (1983) Dependence of cyclic-AMP-induced relaxation on Ca^{2+} and calmodulin in skinned smooth muscle of guinea pig taenia coli. Pflügers Arch Eur J Physiol 399:315–320

Meisheri KD, Rüegg JC, Paul RJ (1985) Studies on skinned fiber preparations. In: Grover AK, Daniel EE (eds) Calcium and Contractility, Humana Press, Clifton, NJ, pp 191–224

Menice CB, Hulvershorn J, Adam LP, Wang CA, Morgan KG (1997) Calponin and mitogen-activated protein kinase signaling in differentiated vascular smooth muscle. J Biol Chem 272:25157–25161

Merkel L, Gerthoffer WT, Torphy TJ (1990) Dissociation between myosin phosphorylation and shortening velocity in canine trachea. Am J Physiol 258:C524–C532

Mesters JR, Martien de Graaf J, Kraal B (1993) Divergent effects of fluoroaluminates on the peptide chain elongation factors EF-Tu and EF-G as members of the GTPase superfamily. FEBS Lett 321:149–152

Miller, JR, Silver, PJ, Stull, JT (1983) The role of myosin light chain kinase phosphorylation in beta-adrenergic relaxation of tracheal smooth muscle. Mol Pharmacol 24:235–242

Miller-Hance WC, Miller JR, Wells JN, Stull JT, Kamm KE (1988) Biochemical events associated with activation of smooth muscle contraction. J Biol Chem 263:13979–13982

Mino T, Yuasa U, Naka M, Tanaka T (1995) Phosphorylation of calponin mediated by protein kinase C in association with contraction in porcine coronary artery. Biochem Biophys Res Commun 208:397–404

Mitsui T, Kitazawa T, Ikebe M (1994) Correlation between high temperature dependence of smooth muscle myosin light chain phosphatase activity and muscle relaxation rate. J Biol Chem 1994 Feb 25;269(8):5842–5848

Mohammad MA, Sparrow MP (1988) Changes in myosin heavy chain stoichiometry in pig tracheal smooth muscle during development. FEBS Lett 228:109–112

Morano I, Erb G, Sogl B (1993) Expression of myosin heavy and light chains changes during pregnancy in the rat uterus. Pflügers Arch 423:434–441

Morano I, Koehlen S, Haase H, Erb G, Baltas LG, Rimbach S, Wallwiener D, Bastert G (1997) Alternative splicing and cycling kinetics of myosin change during hypertrophy of human smooth muscle cells. J Cell Biochem 64:171–181

Moreland S, Moreland RS, Singer HA (1987) Apparent dissociation between myosin light chain phosphorylation and maximal velocity of shortening in KCl depolarized swine carotid artery: effect of temperature and KCl concentration. Pflügers Arch 408:139–145

Morgan JP, Morgan KG (1984a) Stimulus-specific patterns of intracellular calcium levels in smooth muscle of ferret portal vein. J Physiol (Lond) 351:155–167

Morgan JP, Morgan KG (1984b) Alteration of cytoplasmic ionized calcium levels in smooth muscle by vasodilators in the ferret. J Physiol 357:539–551

Morrison DL, Sanghera JS, Stewart J, Sutherland C, Walsh MP, Pelech SL (1996) Phosphorylation and activation of smooth muscle myosin light chain kinase by MAP kinase and cyclin-dependent kinase-1. Biochem Cell Biol 74:549–557

Mossakowska M, Strzelecka-Golaszewska H (1985) Identification of amino acid substitutions differentiating actin isoforms in their interaction with myosin. Eur J Biochem 153:373–381

Murphy RA (1982) Myosin phosphorylation and crossbridge regulation in arterial smooth muscle. State-of-the-art review. Hypertension 4:3–7

Murphy RA (1989) Contraction in smooth muscle cells. Annu Rev Physiol 51:275–283

Murphy RA (1994) What is special about smooth muscle? The significance of covalent crossbridge regulation. FASEB J 8:311–318

Murphy RA, Aksoy MO, Dillon PF, Gerthoffer WT, Kamm KE (1983) The role of myosin light chain phosphorylation in regulation of the cross-bridge cycle. Fed Proc 42:51–56

Murphy RA, Herlihy JT, Megerman J (1974) Force-generating capacity and contractile protein content of arterial smooth muscle. J Gen Physiol 64:691–705

Nagai R, Kuro-o M, Babij P, Periasamy M (1989) Identification of two types of smooth muscle myosin heavy chain isoforms by cDNA cloning and immunoblot analysis. J Biol Chem 264:9734–9737

Nakanishi S, Kakita S, Takahashi I, Kawahara K, Tsukuda E, Sano T, Yamada K, Yoshida M, Kase H, Matsuda Y, et al (1992) Wortmannin, a microbial product inhibitor of myosin light chain kinase. J Biol Chem 267:2157–2163

Ngai PK, Walsh MP (1984) Inhibition of smooth muscle actin-activated myosin Mg^{2+}-ATPase activity by caldesmon. J Biol Chem 259:13656–13659

Ngai PK, Walsh MP (1987) The effects of phosphorylation of smooth-muscle caldesmon. Biochem J 244:417–425

Nieznanski K, Sobieszek A (1997) Telokin (kinase-related protein) modulates the oligomeric state of smooth-muscle myosin light-chain kinase and its interaction with myosin filaments. Biochem J 322:65–71

Nishida W, Abe M, Takahashi K, Hiwada K (1990) Do thin filaments of smooth muscle contain calponin? A new method for the preparation. FEBS Lett 268:165–168

Nishikawa M, de Lanerolle P, Lincoln TM, Adelstein RS (1984) Phosphorylation of mammalian myosin light chain kinases by the catalytic subunit of cyclic AMP-dependent protein kinase and by cyclic GMP-dependent protein kinase. J Biol Chem 259:8429–8436

Nishikawa M, Hidaka H, Adelstein RS (1983) Phosphorylation of smooth muscle heavy meromyosin by calcium-activated, phospholipid-dependent protein kinase. The effect on actin-activated MgATPase activity. J Biol Chem 258:14069–14072

Nishikori K, Weisbrodt NW, Sherwood OD, Sanborn BM (1983) Effects of relaxin on rat uterine myosin light chain kinase activity and myosin light chain phosphorylation. J Biol Chem 258:2468–2474

Nishimura J, van Breemen C (1989) Direct regulation of smooth muscle contractile elements by second messengers. Biochem Biophys Res Commun 163:929–935

Nishimura J, Kolber M, van Breemen C (1988) Norepinephrine and GTP-γ-S increase myofilament Ca^{2+} sensitivity in α-toxin permeabilized arterial smooth muscle. Biochem Biophys Res Commun 157:677–683

Nishiye E, Somlyo AV, Torok K, Somlyo AP (1993) The effects of MgADP on cross-bridge kinetics: a laser flash photolysis study of guinea-pig smooth muscle. J Physiol (Lond) 460:247–271

Nixon GF, Iizuka K, Haystead CM, Haystead TA, Somlyo AP, Somlyo AV (1995) Phosphorylation of caldesmon by mitogen-activated protein kinase with no effect on Ca^{2+} sensitivity in rabbit smooth muscle. J Physiol (Lond) 487:283–289

Noda S, Ito M, Watanabe S, Takahashi K, Maruyama K (1992) Conformational changes of actin induced by calponin. Biochem Biophys Res Commun 1992 May 29;185(1):481–487

Nohara A, Ohmichi M, Koike K, Masumoto N, Kobayashi M, Akahane M, Ikegami H, Hirota K, Miyake A, Murata Y (1996) The role of mitogen-activated protein kinase in oxytocin-induced contraction of uterine smooth muscle in pregnant rat. Biochem Biophys Res Commun 229:938–944

Nomura M, Stull JT, Kamm KE, Mumby MC (1992) Site-specific dephosphorylation of smooth muscle myosin light chain kinase by protein phosphatases 1 and 2A. Biochemistry 31:11915–11920

North AJ, Gimona M, Cross RA, Small JV (1994a) Calponin is localised in both the contractile apparatus and the cytoskeleton of smooth muscle cells. J Cell Sci 107:437–444

North AJ, Gimona M, Lando Z, Small JV (1994b) Actin isoform compartments in chicken gizzard smooth muscle cells. J Cell Sci 107:445–455

Obara K, Szymanski PT, Tao T, Paul RJ (1996) Effects of calponin on isometric force and shortening velocity in permeabilized taenia coli smooth muscle. Am J Physiol 270:C481–C487

Obara K, Yabu H (1994) Effect of cytochalasin B on intestinal smooth muscle cells. Eur J Pharmacol 255:139–147

Okagaki T, Higashi-Fujime S, Ishikawa R, Takano-Ohmuro H, Kohama K (1991) In vitro movement of actin filaments on gizzard smooth muscle myosin:requirement of phosphorylation of myosin light chain and effects of tropomyosin and caldesmon. J Biochem (Tokyo) 109:858–866

Okamoto E, Imataka K, Fujii J, Kuro-o M, Nakahara K, Nishimura H, Yazaki Y, Nagai R (1992) Heterogeneity in smooth muscle cell population accumulating in the neointimas and the media of poststenotic dilatation of the rabbit carotid artery. Biochem Biophys Res Commun 185:459–464

Okubo S, Ito M, Takashiba Y, Ichikawa K, Miyahara M, Shimizu H, Konishi T, Shima H, Nagao M, Hartshorne DJ, et al (1994) A regulatory subunit of smooth muscle myosin bound phosphatase. Biochem Biophys Res Commun 200:429–434

Olson NJ, Pearson RB, Needleman DS, Hurwitz MY, Kemp BE, Means AR (1990) Regulatory and structural motifs of chicken gizzard myosin light chain kinase. Proc Natl Acad Sci USA 87:2284–2288

Onishi H, Suzuki H, Nakamura K, Takahashi K, Watanabe S (1978) Adenosine triphosphatase activity and "thick filament" formation of chicken gizzard myosin in low salt media. J Biochem (Tokyo) 83:835–847

Österman A, Arner A (1995) Effects of inorganic phosphate on cross-bridge kinetics at different activation levels in skinned guinea-pig smooth muscle. J Physiol (Lond) 484:369–383

Otto B, Steusloff A, Just I, Aktories K, Pfitzer G (1996) Role of Rho proteins in carbachol-induced contractions in intact and permeabilized guinea-pig intestinal smooth muscle. J Physiol Lond 496:317–329

Ozaki H, Blondfield DP, Hori M, Sanders KM, Publicover NG (1992) Cyclic AMP-mediated regulation of excitation-contraction coupling in canine gastric smooth muscle. J Physiol 447:351–372

Parente JE, Walsh MP, Kerrick WG, Hoar PE (1992) Effects of the constitutively active proteolytic fragment of protein kinase C on the contractile properties of demembranated smooth muscle fibres. J Muscle Res Cell Motil 13:90–99

Park S, Rasmussen H (1986) Carbachol-induced protein phosphorylation changes in bovine tracheal smooth muscle. J Biol Chem 1986 Nov 25;261(33):15734–15739

Parker CA, Takahashi K, Tao T, Morgan KG (1994) Agonist-induced redistribution of calponin in contractile vascular smooth muscle cells. Am J Physiol 267:C1262–C1270

Parsons SJW, Sumner MJ, Garland CJ (1996) Phospholipase A_2 and protein kinase C contribute to myofilament sensitization to 5-HT in the rabbit mesenteric artery. J Physiol Lond 491:447–453

Pato MD, Adelstein RS (1983) Purification and characterization of a multisubunit phosphatase from turkey gizzard smooth muscle. The effect of calmodulin binding to myosin light chain kinase on dephosphorylation. J Biol Chem 258:7047–7054

Paul ER, Ngai PK, Walsh MP, Gröschel-Stewart U (1995) Embryonic chicken gizzard: expression of the smooth muscle regulatory proteins caldesmon and myosin light chain kinase. Cell Tissue Res 279:331–337

Paul RJ (1989) Smooth muscle energetics. Annu Rev Physiol 51:331–349

Paul RJ (1990) Smooth muscle energetics and theories of cross-bridge regulation. Am J Physiol 258:C369–C375

Paul RJ, Doerman G, Zeugner C, Rüegg JC (1983) The dependence of unloaded shortening velocity on Ca²⁺, calmodulin, and duration of contraction in "chemically skinned" smooth muscle. Circ Res 53:342–351

Pearson RB, Misconi LY, Kemp BE (1986) Smooth muscle myosin kinase requires residues on the COOH-terminal site of the phosphorylation site. J Biol Chem 261:25–27

Pelech SL, Sanghera JS (1992) Mitogen-activated protein kinases: versatile transducers for cell signaling. Trends Biochem Sci 17:233–238

Pfitzer G (1996) Permeabilized smooth muscle. In: Bárány M (ed) The biochemistry of smooth muscle contraction. Academic Press pp 191–199

Pfitzer G, Boels P (1991) Differential skinning of smooth muscle: a new approach to excitation-contraction coupling. Blood Vessels 28:262–267

Pfitzer G, Rüegg JC (1982) Molluscan catch muscle: Regulation and mechanics in living and skinned anterior byssus retractor muscle of Mytilus edulis. J Comp Physiol 147:137–142.

Pfitzer G, Hofmann F, DiSalvo J, Rüegg JC (1984) cGMP and cAMP inhibit tension development in skinned coronary arteries. Pflügers Arch Eur J Physiol 401:277–280

Pfitzer G, Merkel L, Rüegg JC, Hofmann F (1986) Cyclic GMP-dependent protein kinase relaxes skinned fibers from guinea pig taenia coli but not from chicken gizzard. Pflügers Arch Eur J Physiol 407:87–91

Pfitzer G, Rüegg JC, Zimmer M, Hofmann (1985) Relaxation of skinned coronary arteries depends on the relative concentrations of Ca²⁺, calmodulin, and active cAMP-dependent protein kinase. Pflügers Arch Eur J Physiol 405:70–76

Pfitzer G, Zeugner C, Troschka M, Chalovich JM (1993) Caldesmon and a 20-kDa actin-binding fragment of caldesmon inhibit tension development in skinned gizzard muscle fiber bundles. Proc Natl Acad Sci U S A 90:5904–5908

Pohl J, Winder SJ, Allen BG, Walsh MP, Sellers JR, Gerthoffer WT (1997) Phosphorylation of calponin in airway smooth muscle. Am J Physiol 272:L115–L123

Pollard TD, Cooper JA (1986) Actin and actin-binding proteins. A critical evaluation of mechanisms and functions. Annu Rev Biochem 55:987–1035

Redwood CS, Marston SB, Gusev NB (1993) The functional effects of mutations Thr673—Asp and Ser702—Asp at the Pro-directed kinase phosphorylation sites in the C-terminus of chicken gizzard caldesmon. FEBS Lett 327:85–89

Reiser PJ, Kasper CE, Greaser ML, Moss RL (1988) Functional significance of myosin transitions in single fibers of developing soleus muscle. Am J Physiol 254:C605-C613

Rembold CM (1991) Relaxation, [Ca^{2+}]$_i$, and the latch-bridge hypothesis in swine arterial smooth muscle. Am J Physiol 261:C41–C50

Rembold CM (1996) Electromechanical and pharmacomechanical coupling. In: Bárány M (ed) The biochemistry of smooth muscle. Academic press pp 227–239

Rembold CM, Murphy RA (1988a) Myoplasmic [Ca^{2+}] determines myosin phosphorylation in agonist-stimulated swine arterial smooth muscle. Circ Res 63:593–603

Rembold CM, Murphy RA (1988b) [Ca^{2+}]-dependent myosin phosphorylation in phorbol diester stimulated smooth muscle contraction. Am J Physiol 255:C719–C723

Rembold CM, Murphy RA (1993) Models of the mechanism for crossbridge attachment in smooth muscle. J Muscle Res Cell Motil 14:325–334

Rice RV, McManus GM, Devine OF, Somlyo AP (1971) Regular organization of thick filaments in mammalian smooth muscle. Nat New Biol 231:242–243

Rovner AS, Freyzon Y, Trybus KM (1997) An insert in the motor domain determines the functional properties of expressed smooth muscle myosin isoforms. J Muscle Res Cell Motil 18:103–110

Rovner AS, Murphy RA, Owens GK (1986a) Expression of smooth muscle and nonmuscle myosin heavy chains in cultured vascular smooth muscle cells. J Biol Chem 261:14740–14745

Rovner AS, Thompson MM, Murphy RA (1986b) Two different heavy chains are found in smooth muscle myosin. Am J Physiol 250:C861–C870

Rüegg JC (1992) Calcium in muscle contraction. Springer, Heidelberg

Rüegg JC, DiSalvo J, Paul RJ (1982) Soluble relaxing factor from vascular smooth muscle: a myosin light chain phosphatase? Biochem Biophys Res Commun 106:1126–1133

Rüegg JC, Mrwa U, Sparrow MP (1981) Cyclic-AMP mediated relaxation of chemically skinned fibers of smooth muscle. Pflügers Arch Eur J Physiol 390:198–201

Rüegg JC, Paul RJ (1982) Vascular smooth muscle: calmodulin and cyclic AMP-dependent protein kinase alter calcium sensitivity in porcine carotid skinned fibres. Circ Res 50:394–399

Rüegg JC, Pfitzer G (1985) Modulation of calcium sensitivity in guinea pig -taenia coli: skinned fiber studies. Experientia 41:997–1001

Rüegg JC, Pfitzer G (1991) Contractile protein interactions in smooth muscle. Blood Vessels 28:159–163

Rüegg JC, Zeugner C, Strauss JD, Paul RJ, Kemp B, Chem M, Li AY, Hartshorne DJ (1989) A calmodulin-binding peptide relaxes skinned muscle from guinea-pig taenia coli. Pflügers Arch 414:282–285

Sato K, Hori M, Ozaki H, Takano-Ohmura H, Tsuchiya T, Sugi H, Karaki H (1992) Myosin phosphorylation-independent contraction induced by phorbol ester in vascular smooth muscle. J Pharmacol Exp Ther 261:495–505

Satterwhite LL, Lohka MJ, Wilson KL, Scherson TY, Cisek LJ, Corden JL, Pollard TD (1992) Phosphorylation of myosin-II regulatory light chain by cyclin-p34cdc2: a mechanism for the timing of cytokinesis. J Cell Biol 118:595-605

Schmidt U, Troschka M, Pfitzer G (1995) The variable coupling between force and myosin light chain phosphorylation in Triton-skinned chicken gizzard fibre bundles: role of myosin light chain phosphatase. Pflügers Arch Eur J Physiol 429:708-715

Schneider M, Sparrow M, Rüegg JC (1981) Inorganic phosphate promotes relaxation of chemically skinned smooth muscle of guinea-pig Taenia coli. Experientia 37:980-982

Scott-Woo GC, Sutherland C, Walsh MP (1990) Kinase activity associated with caldesmon is Ca^{2+}/calmodulin-dependent kinase II. Biochem J 268:367-370

Seguchi H, Nishimura J, Toyofuku K, Kobayashi S, Kumazawa J, Kanaide H (1996) The mechanism of relaxation induced by atrial natriuretic peptide in the porcine renal artery. British J Pharmacol 118:343-351

Sellers JR (1985) Mechanism of the phosphorylation-dependent regulation of smooth muscle heavy meromyosin. J Biol Chem 260:15815-15819

Sellers JR, Eisenberg E, Adelstein RS (1982) The binding of smooth muscle heavy meromyosin to actin in the presence of ATP. Effect of phosphorylation. J Biol Chem 257:13880-13883

Sellers JR, Spudich JA, Sheetz MP (1985) Light chain phosphorylation regulates the movement of smooth muscle myosin on actin filaments. J Cell Biol 101:1897-1902

Seto M, Sasaki Y, Sasaki Y (1990) Alteration in the myosin phosphorylation pattern of smooth muscle by phorbol ester. Am J Physiol 259:C769-C774

Sherry JMF, Gorecka A, Aksoy MO, Dabrowska R, Hartshorne DJ (1978) Roles of calcium and phosphorylation in the regulation of the activity of gizzard myosin. Biochemistry 17:4411-4418

Shimizu H, Ito M, Miyahara M, Ichikawa K, Okubo S, Konishi T, Naka M, Tanaka T, Hirano K, Hartshorne DJ, et al (1994) Characterization of the myosin-binding subunit of smooth muscle myosin phosphatase. J Biol Chem 269:30407-30411

Shirazi A, Iizuka K, Fadden P, Mosse C, Somlyo AP, Somlyo AV, Haystead TA (1994) Purification and characterization of the mammalian myosin light chain phosphatase holoenzyme. The differential effects of the holoenzyme and its subunits on smooth muscle. J Biol Chem 269:31598

Shirinsky VP, Biryukov KG, Hettasch JM, Sellers JR (1992) Inhibition of the relative movement of actin and myosin by caldesmon and calponin. J Biol Chem 267:15886-15892

Shirinsky VP, Vorotnikov AV, Birukov KG, Nanaev AK, Collinge M, Lukas TJ, Sellers JR, Watterson DM (1993) A kinase-related protein stabilizes unphosphorylated smooth muscle myosin minifilaments in the presence of ATP. J Biol Chem 268:16578-16583

Siegman MJ, Butler TM, Mooers SU (1989) Phosphatase inhibition with okadaic acid does not alter the relationship between force and myosin light chain phosphorylation in permeabilized smooth muscle. Biochem Biophys Res Commun 161:838-842

Siegman MJ, Butler TM, Mooers SU, Davies RE (1976a) Crossbridge attachment, resistance to stretch, and viscoelasticity in resting mammalian smooth muscle. Science 191:383–385

Siegman MJ, Butler TM, Mooers SU, Davies RE (1976b) Calcium-dependent resistance to stretch and stress relaxation in resting smooth muscles. Am J Physiol 231:1501–1508

Siegman MJ, Butler TM, Mooers SU, Davies RE (1980) Chemical energetics of force development, force maintenance, and relaxation in mammalian smooth muscle. J Gen Physiol 76:609–629

Siegman MJ, Butler TM, Mooers SU, Michalek A (1984) Ca^{2+} can affect V_{max} without changes in myosin light chain phosphorylation in smooth muscle. Pflügers Arch 401:385–390

Siegman MJ, Funabara D, Kinoshita S, Watabe S, Hartshorne DJ, Butler TM (1998) Phosporylation of a twitchin-related protein contols catch and calcium sensitivity of force production in invertebrate smooth muscle. Proc Natl Acad Sci 95:5383–5388

Siegman MJ, Mooers SU, LI C, Narayan S, Trinkle-Mulcahy L, Watabe S, Hartshorne DJ, Butler TM (1997) Phosphorylation of a high molecular weight (600 kDa) protein regulates catch in invertebrate smooth muscle J Muscle Res Cell Mot 18:655–670

Siemankowski RF, Wiseman MO, White HD (1985) ADP dissociation from actomyosin subfragment 1 is sufficiently slow to limit the unloaded shortening velocity in vertebrate muscle. Proc Natl Acad Sci U S A 82:658–662

Silver PJ, Stull JT (1982) Regulation of myosin light chain and phosphorylase phosphorylation in tracheal smooth muscle. J Biol Chem 257:6145–6150

Silver DL, Vorotnikov AV, Watterson DM, Shirinsky VP, Sellers JR (1997) Sites of interaction between kinase-related protein and smooth muscle myosin. J Biol Chem 272:25353–25359

Singer HA (1990) Phorbol ester-induced stress and myosin light chain phosphorylation in swine carotid medial smooth muscle. J Pharmacol Exp Ther 252:1068–1074

Singer HA (1996) Protein kinase C. In: Bárány M (ed) Biochemistry of smooth muscle contraction. Academic press pp 155–165

Sjuve R, Arner A, Li Z, Mies B, Paulin D, Schmittner M, Small JV (1998) Mechanical alterations in smooth muscle from mice lacking desmin. J Muscle Res Cell Motil 19:415–429

Sjuve R, Haase H, Morano I, Uvelius B, Arner A (1996) Contraction kinetics and myosin isoform composition in smooth muscle from hypertrophied rat urinary bladder. J Cell Biochem 63:86–93

Sleep JA, Hutton RL (1980) Exchange between inorganic phosphate and adenosine 5'-triphosphate in the medium by actomyosin subfragment 1. Biochemistry 19:1276–1283

Small JV (1977) Studies on isolated smooth muscle cells: The contractile apparatus. J Cell Sci 24:327–349

Small JV, Herzog M, Barth M, Draeger A (1990) Supercontracted state of vertebrate smooth muscle cell fragments reveals myofilament lengths. J Cell Biol 111:2451–2461

Smith CW, Marston SB (1985) Disassembly and reconstitution of the Ca^{2+}-sensitive thin filaments of vascular smooth muscle. FEBS Lett 184:115–119

Smith CW, Pritchard K, Marston SB (1987) The mechanism of Ca^{2+} regulation of vascular smooth muscle thin filaments by caldesmon and calmodulin. J Biol Chem 262:116–122

Smith RC, Cande WZ, Craig R, Tooth PJ, Scholey JM, Kendrick-Jones J (1983) Regulation of myosin filament assembly by light-chain phosphorylation. Philos Trans R Soc Lond B Biol Sci 302:73–82

Sobieszek A (1977) Ca^{2+}-linked phosphorylation of a light chain of vertebrate smooth muscle myosin. Eur J Biochem 118:533–539

Sobieszek A, Babiychuk EB, Ortner B, Borkowski J (1997a) Purification and characterization of a kinase-associated myofibrillar smooth muscle myosin light chain phosphatase possessing a calmodulin-targeting subunit. J Biol Chem 272:7027–7033

Sobieszek A, Borkowski, J, Babiychuk VS (1997b) Purification and characterization of a smooth muscle myosin light chain kinase-phosphatase complex. J Biol Chem 272:7034–7041

Sobieszek A, Small JV (1977) Regulation of the actin-myosin interaction in vertebrate smooth muscle: activation via a myosin light-chain kinase and the effect of tropomyosin. J Molec Biol 112:559–576

Sobue K, Sellers JR (1991) Caldesmon, a novel regulatory protein in smooth muscle and nonmuscle actomyosin systems. J Biol Chem 266:12115–12118

Sobue K, Takahashi K, Wakabayashi I (1985) Caldesmon150 regulates the tropomyosin-enhanced actin-myosin interaction in gizzard smooth muscle. Biochem Biophys Res Commun 132:645–651

Somlyo AP (1993) Myosin isoforms in smooth muscle: how may they affect function and structure? J Muscle Res Cell Motil 14:557–563

Somlyo AP, Somlyo AV (1990) Flash photolysis studies of excitation-contraction coupling, regulation, and contraction in smooth muscle. Annu Rev Physiol 52:857–874

Somlyo AP, Somlyo AV (1994) Signal transduction and regulation in smooth muscle. Nature 372:231–236

Somlyo AP, Walker JW, Goldman YE, Trenham DR, Kobayashi S, Kitazawa T, Somlyo AV (1988a) Inositol trisphosphate, calcium and muscle contraction. Philos Trans R Soc London Ser B 320:399–414

Somlyo AV (1980) Ultrastructure of vascular smooth muscle. In: Bohr DF, Somlyo AP, Sparks HV (eds). Handbook of Physiology, Sect 2, Vol II. Am.Physiol. Soc., Bethesda, pp 33–67

Somlyo AV, Butler TM, Bond M, Somlyo AP (1981) Myosin filaments have non-phosphorylated light chains in relaxed smooth muscle. Nature 294:567–569

Somlyo AV, Goldman Y, Fujimori T, Bond M, Trentham D, Somlyo AP (1987) Crossbridge transients initiated by photolysis of caged nucleotides and crossbridge structure, in smooth muscle. Prog Clin Biol Res 245:27–41

Somlyo AV, Goldman YE, Fujimori T, Bond M, Trentham DR, Somlyo AP (1988b) Cross-bridge kinetics, cooperativity, and negatively strained cross-bridges in vertebrate smooth muscle. A laser-flash photolysis study. J Gen Physiol 91:165–192

Somlyo AV, Somlyo AP (1968) Electromechanical and pharmacomechanical coupling in vascular smooth muscle. J Pharmacol Exp Ther 159:129–145

Sparrow MP, Mrwa U, Hofmann F, Rüegg JC (1981) Calmodulin is essential for smooth muscle contraction. FEBS Lett 125:141–145

Sparrow MP, Pfitzer G, Gagelmann M, Rüegg JC (1984) Effect of calmodulin, Ca^{2+}, and cAMP protein kinase on skinned tracheal smooth muscle. Am J Physiol 246:C308–C314

Strauss JD, de Lanerolle P, Paul RJ (1992) Effects of myosin kinase inhibiting peptide on contractility and LC20 phosphorylation in skinned smooth muscle. Am J Physiol 262:C1437–C1445

Strauss JD, Murphy RA (1996) Regulation of cross-bridge cycling in smooth mucsle. In: Bárány M (ed) Biochemistry of smooth muscle contraction. Academic Press, New York, pp 341–353

Stull JT, Hsu L-C, Tansey MG, Kamm KE (1990) Myosin light chain kinase phosphorylation in tracheal smooth muscle. J Biol Chem 265:16683–16690

Stull JT, Krueger JK, Kamm KE, Gao Z-H, Zhi G, Padre R (1996) Myosin light chain kinase. In: Bárány M (ed) Biochemistry of smooth muscle contraction. Academic press pp pp 119–130

Suematsu E, Resnick M, Morgan KG (1991) Change of Ca^{2+} requirement for myosin phosphorylation by prostaglandin $F_{2\alpha}$. Am J Physiol 261:C253–C258

Sugai M, Hashimoto K, Kikuchi A, Inoue S, Okumura H, Matsumoto K, Goto Y, Ohgai H, Moriishi K, Syuto B, Yoshikawa K, Suginaka H, Takai Y (1992) Epidermal cell differentiation inhibitor ADP-ribosylates small GTP-binding proteins and induces hyperplasia of epidermis. J Biol Chem 267:2600–2604

Sutherland C, Renaux BS, McKay DJ, Walsh MP (1994) Phosphorylation of caldesmon by smooth-muscle casein kinase II. J Muscle Res Cell Motil 15:440–456

Sutherland C, Walsh MP (1989) Phosphorylation of caldesmon prevents its interaction with smooth muscle myosin. J Biol Chem 264:578–583

Suzuki H, Onishi H, Takahashi K, Watanabe S (1978) Structure and function of chicken gizzard myosin. J Biochem (Tokyo) 84:1529–1542

Sweeney HL, Yang Z, Zhi G, Stull JT, Trybus KM (1994) Charge replacement near the phosphorylatable serine of the myosin regulatory light chain mimics aspects of phosphorylation. Proc Natl Acad Sci U S A 91:1490–1494

Szpacenko A, Wagner J, Dabrowska R, Rüegg JC (1985) Caldesmon-induced inhibition of ATPase activity of actomyosin and contraction of skinned fibres of chicken gizzard smooth muscle. FEBS Lett 192:9–12

Szymanski PT, Tao T (1993) Interaction between calponin and smooth muscle myosin. FEBS Lett 334:379–382

Szymanski PT, Tao T (1997) Localization of protein regions involved in the interaction between calponin and myosin. J Biol Chem 272:11142–11146

Takahashi K, Hiwada K, Kokubu T (1986) Isolation and characterization of a 34,000-dalton calmodulin- and F-actin-binding protein from chicken gizzard smooth muscle. Biochem Biophys Res Commun 141:20–26

Takahashi K, Hiwada K, Kokubu T (1987) Occurrence of anti-gizzard P34K antibody cross-reactive components in bovine smooth muscles and non-smooth muscle tissues. Life Sci 41:291–296

Takahashi K, Hiwada K, Kokubu T (1988) Vascular smooth muscle calponin. A novel troponin T-like protein. Hypertension 11:620–626

Takai A, Bialojan C, Troschka M, Rüegg JC (1987) Smooth muscle myosin phosphatase inhibition and force enhancement by black sponge toxin. FEBS Lett 217:81–84

Takai A, Sasaki K, Nagai H, Mieskes G, Isobe M, Isono K, Yasumoto T (1995) Inhibition of specific binding of okadaic acid to protein phosphatase 2A by microcystin-LR, calyculin-A and tautomycin: method of analysis of interactions of tight-binding ligands with target protein. Biochem J 306: 657–665

Takai A, Troschka M, Mieskes G, Somlyo AV (1989) Protein phosphatase composition in the smooth muscle of guinea pig ileum studied with okadaic acid and inhibitor 2. Biochem J 262:617–623

Tanaka T, Ohta H, Kanda K, Tanaka T, Hidaka H, Sobue K (1990) Phosphorylation of high-Mr caldesmon by protein kinase C modulates the regulatory function of this protein on the interaction between actin and myosin. Eur J Biochem 188:495–500

Tanner JA, Haeberle JR, Meiss RA (1988) Regulation of glycerinated smooth muscle contraction and relaxation by myosin phosphorylation. Am J Physiol 255:C34–C42

Tansey MG, Hori M, Karaki K, Kamm KE, Stull JT (1990) Okadaic acid uncouples myosin light chain phosphorylation and tension in smooth muscle. FEBS Lett 270:219–221

Tansey MG, Word RA, Hidaka H, Singer HA, Schworer CM, Kamm KE, Stull JT (1992) Phosphorylation of myosin light chain kinase by the multifunctional calmodulin-dependent protein kinase II in smooth muscle cells. J Biol Chem 267:12511–12516

Tansey MG, Luby-Phelbs K, Kamm KE, Stull JT (1994) Ca^{2+}-dependent phosphorylation of myosin light chain kinase decreases the Ca^{2+}-sensitivity of light chain phosphorylation within smooth muscle cells. J Biol Chem 269:9912–9920

Thelen M, Wymann MP, Langen H (1994) Wortmannin binds specifically to 1-phosphatidylinositol 3-kinase while inhibiting guanine nucleotide-binding protein-coupled receptor signaling in neutrophil leukocytes. Proc Natl Acad Sci USA 91:4960–4964

Tokui T, Brozovich, F, Ando S, Ikebe M (1996) Enhancement of smooth muscle contraction with protein phosphatase inhibitor 1: Activation of inhibitor 1 by cGMP dependent protein kinase. Biochem Biophys Res Commun 220:777–783

Török K, Trentham DR (1994) Mechanism of 2-chloro-(epsilon-amino-Lys75)-[6-[4-(N,N-diethylamino)phenyl]-1,3,5-triazin-4-yl]calmodulin interactions with smooth muscle myosin light chain kinase and derived peptides. Biochemistry 33:12807–12820

Trinkle-Mulcahy L, Ichikawa K, Hartshorne DJ, Siegman MJ, Butler TM (1995) Thio-phosphorylation of the 130-kDa subunit associatesd with a decreased activity of myosin light chain phosphatase in α-toxin permeabilized smooth muscle. J Biol Chem 270:18191–18194

Trybus KM (1991) Assembly of cytoplasmic and smooth muscle myosins. Curr Opin Cell Biol 3:105–111

Trybus KM (1994) Regulation of expressed truncated smooth muscle myosins. Role of the essential light chain and tail length. J Biol Chem 269:20819–20822

Trybus KM, Freyzon Y, Faust LZ, Sweeney HL (1997) Spare the rod, spoil the regulation: necessity for a myosin rod. Proc Natl Acad Sci U S A 94:48–52

Trybus KM, Waller GS, Chatman TA (1994) Coupling of ATPase activity and motility in smooth muscle myosin is mediated by the regulatory light chain. J Cell Biol 124:963–969

Tseng S, Kim R, Kim T, Morgan KG, Hai CM (1997) F-actin disruption attenuates agonist-induced $[Ca^{2+}]$, myosin phosphorylation, and force in smooth muscle. Am J Physiol 272:C1960–C1967

Tsukita S, Tsukita S, Usukura J, Ishikawa H (1982) Myosin filaments in smooth muscle cells of the guinea pig taenia coli: a freeze-substitution study. Eur J Cell Biol 28:195–201

Twarog BM, Cole RA (1972) Relaxation of catch in a molluscan smooth muscle. II. Effects of serotonin, dopamine and related compounds. Comp Biochem Physiol A 43:331–335

Uehata M, Ishizaki T, Satoh H, Ono T, Kawahara T, Morishita T, Tamakawa H, Yamagami K, Inui J, Maekawa M, Narumiya S (1997) Calcium sensitization of smooth muscle mediated by a Rho-associated protein kinase in hypertension. Nature 389:990–994

Uyama Y, Imaizumi Y, Watanabe M, Walsh MP (1996) Inhibition by calponin of isometric force in demembranated vascular smooth muscle strips: the critical role of serine-175. Biochem J 319:551–558

van Riper DA, Weaver BA, Stull JT, Rembold CH (1995) Myosin light chain kinase phosphorylation in swine carotid artery contraction and relaxation. Am J Physiol 268:H2466–H2475

Vandekerckhove J, Weber K (1978) At least six different actins are expressed in a higher mammal: an analysis based on the amino acid sequence of the amino-terminal tryptic peptide. J Mol Biol 126:783–802

Velaz L, Chen YD, Chalovich JM (1993) Characterization of a caldesmon fragment that competes with myosin-ATP binding to actin. Biophys J 65:892–898

Velaz L, Hemric ME, Benson CE, Chalovich JM (1989) The binding of caldesmon to actin and its effect on the ATPase activity of soluble myosin subfragments in the presence and absence of tropomyosin. J Biol Chem 264:9602–9610

Velaz L, Ingraham RH, Chalovich JM (1990) Dissociation of the effect of caldesmon on the ATPase activity and on the binding of smooth heavy meromyosin to actin by partial digestion of caldesmon. J Biol Chem 265:2929–2934

Vibert P, Craig R, Lehman W (1993) Three-dimensional reconstruction of caldesmon-containing smooth muscle thin filaments. J Cell Biol 123:313–321
Vibert PJ, Haselgrove JC, Lowy J, Poulsen FR (1972) Structural changes in actin-containing filaments of muscle. J Mol Biol 71:757–767
Vorotnikov AV, Gusev NB, Hua S, Collins JH, Redwood CS, Marston SB (1993) Identification of casein kinase II as a major endogeneous caldesmon kinase in sheep aorta smooth muscle. FEBS Lett 334:18–22
Vyas TB, Mooers SU, Narayan SR, Siegman MJ, Butler TM (1994) Cross-bridge cycling at rest and during activation. Turnover of myosin-bound ADP in permeabilized smooth muscle. J Biol Chem 269:7316–7322
Vyas TB, Mooers SU, Narayan SR, Witherell JC, Siegman MJ, Butler TM (1992) Cooperative activation of myosin by light chain phosphorylation in permeabilized smooth muscle. Am J Physiol 263:C210–C219
Wagner J, Rüegg JC (1986) Skinned smooth muscle: calcium-calmodulin activation independent of myosin phosphorylation. Pflügers Arch 407:569–571
Wagner PD, Vu ND (1986) Regulation of the actin-activated ATPase of aorta smooth muscle myosin . J Biol Chem 261:7778–7783
Wagner PD, Vu ND (1987) Actin-activation of unphosphorylated gizzard myosin. J Biol Chem 262:15556–15562
Walsh MP, Bridenbaugh R, Hartshorne DJ, Kerrick GL (1982) Phosphorylation-dependent activated tension in skinned gizzard muscle fibers in the absence of Ca^{2+}. J Biol Chem 257:5987–5990
Walsh MP, Carmichael JD, Kargacin GJ (1993) Characterization and confocal imaging of calponin in gastrointestinal smooth muscle. Am J Physiol 265:C1371–C1378
Walsh MP, Horowitz A, Clément-Chomienn O, Andrea JE, Allen BG, Morgan KG (1996) Protein kinase C mediation of Ca^{2+}-independent contractions of vascular smooth muscle. Biochem Cell Biol 74:485–502
Wang Z, Jiang H, Yang ZQ, Chacko S (1997) Both N-terminal myosin-binding and C-terminal actin-binding sites on smooth muscle caldesmon are required for caldesmon-mediated inhibition of actin filament velocity. Proc Natl Acad Sci U S A 94:11899–11904
Warshaw DM, Desrosiers JM, Work SS, Trybus KM (1990) Smooth muscle myosin cross-bridge interactions modulate actin filament sliding velocity in vitro. J Cell Biol 111:453–463
Warshaw DM, Desrosiers JM, Work SS, Trybus KM (1991) Effects of MgATP, MgADP, and Pi on actin movement by smooth muscle myosin. J Biol Chem 266:24339–24343
Watanabe M, Takemori S, Yagi N (1993) X-ray diffraction study on mammalian visceral smooth muscles in resting and activated states. J Muscle Res Cell Motil 14:469–475
Wawrzynow A, Collins JH, Bogatcheva NV, Vorotnikov AV, Gusev NB (1991) Identification of the site phosphorylated by casein kinase II in smooth muscle caldesmon. FEBS Lett 289:213–216

White S, Martin AF, Periasamy M (1993) Identification of a novel smooth muscle myosin heavy chain cDNA: isoform diversity in the S1 head region. Am J Physiol 264:C1252–C1258

Whittaker M, Wilson-Kubalek EM, Smith JE, Faust L, Milligan RA, Sweeney HL (1995) A 35-A movement of smooth muscle myosin on ADP release. Nature 378:748–751

Wills FL, McCubbin WD, Gimona M, Strasser P, Kay CM (1994) Two domains of interaction with calcium binding proteins can be mapped using fragments of calponin. Protein Sci 3:2311–2321

Winder SJ, Allen BG, Fraser ED, Kang HM, Kargacin GJ, Walsh MP (1993) Calponin phosphorylation in vitro and in intact muscle. Biochem J 296:827–836

Winder SJ, Pato MD, Walsh MP (1992) Purification and characterization of calponin phosphatase from smooth muscle. Effect of dephosphorylation on calponin function. Biochem J 286:197–203

Winder SJ, Walsh MP (1990) Smooth muscle calponin. Inhibition of actomyosin MgATPase and regulation by phosphorylation. J Biol Chem 265:10148–10155

Winder SJ, Walsh MP (1993) Calponin: thin filament-linked regulation of smooth muscle contraction. Cell Signal 5:677–686

Winder SJ, Walsh MP (1996) Calponin. Curr Top Cell Regul 1996;34:33–61

Wingard CJ, Paul RJ, Murphy RA (1994) Dependence of ATP consumption on cross-bridge phosphorylation in swine carotid smooth muscle. J Physiol (Lond) 481:111–117

Wingard CJ, Paul RJ, Murphy RA (1997) Energetic cost of activation processes during contraction of swine arterial smooth muscle. J Physiol (Lond) 501:213–223

Word RA, Tang DC, Kamm KE (1994) Activation properties of myosin light chain kinase during contraction/relaxation cycles of tonic and phasic smooth muscles. J Biol Chem 269:21596–21602

Wright G, Hurn E (1994) Cytochalasin inhibition of slow tension increase in rat aortic rings. Am J Physiol 267:H1437–H1446

Wu X, Haystead TAJ, Somlyo AV, Somlyo AP (1997) Telokin relaxes smooth muscle and is phosphorylated, *in situ*. Biophys J 72:A178

Wuytack F, Casteels R (1975) Proceedings: Evidence for rigor in smooth muscle. Arch Int Physiol Biochim 83:340–341

Xu JQ, Gillis JM, Craig R (1997) Polymerization of myosin on activation of rat anococcygeus smooth muscle. J Muscle Res Cell Motil 18:381–393

Zhang Y, Moreland RS (1994) Regulation of Ca(2+)-dependent ATPase activity in detergent-skinned vascular smooth muscle. Am J Physiol 267:H1032–H1039

Calcium Permeant
Ion Channels in Smooth Muscle

Michael I. Kotlikoff[1], Gerry Herrera and Mark T. Nelson[2]

[1]Department of Animal Biology, School of Veterinary Medicine, University of Pennsylvania, Philadelphia, PA

[2]Department of Pharmacology, School of Medicine, University of Vermont Burlington, VT,

Contents

4.1 Introduction

The regulation of cytosolic calcium concentration in smooth muscle is characterized by numerous calcium permeant ion channels mediating calcium flux across the sarcolemma and sarcoplasmic reticulum, and by a substantial diversity between tissues with regard to the extent that individual channels contribute to excitation-contraction (E-C) coupling. The relative complexity of calcium signaling in smooth muscle is immediately apparent if one compares the processes underlying calcium transport during E-C coupling between skeletal and smooth muscle. During excitation of skeletal muscle a single neurotransmitter (acetylcholine) binds to a single type of receptor/ligand gated ion channel (nicotinic receptor), and mediates calcium flux from the sarcoplasmic reticulum to the cytosol via a single type of calcium channel-the ryanodine receptor. By contrast, smooth muscle E-C coupling is marked by redundancy at every level of activation. Multiple neurotransmitters and autocoids bind to cognate receptors that include ligand-gated (ionotropic) cation channels with variable calcium permeability, and G protein coupled receptors. The former receptor/channels are analogous to the nicotinic receptor in skeletal muscle in that they generate a postsynaptic potential that alters the membrane potential, thereby regulating the activity of voltage-dependent channels, including voltage-dependent calcium channels. In addition to this function, however, one now well characterized class of ionotropic receptors are calcium permeant, and thereby directly influence E-C coupling. G protein-coupled receptors also influence the behavior of calcium permeant channels in multiple ways. First, phospholipase C linked receptors result in the activation of inositol trisphosphate receptors and mediate the release of calcium from the sarcoplasmic reticulum. Second, the release of intracellular calcium by this mechanism subsequently alters the membrane potential by gating the opening of calcium-activated channels, thereby indirectly modulating the activity of voltage-dependent ion channels such as voltage-dependent calcium channels. Third, second messenger pathways are activated that result in the modulation of voltage-dependent ion channels. Finally, cation channels that are activated by second messenger pathways (metabotropic ion channels) alter membrane potential and may have substantial calcium permeability. These complex responses to extracellular signals are imposed upon a pre-existing calcium homeostasis that results from graded calcium influx through voltage-dependent calcium channels and spontaneous intracellular calcium release through ryanodine receptors. Thus the myocyte integrates numerous calcium inputs and the concentration of cytosolic free calcium ($[Ca^{2+}]_i$) at any given time is the net result of the activity of these calcium channels, as well as that of calcium pumps and exchangers that remove calcium ions from the cytosol.

In smooth muscle, the relationship between a rise in $[Ca^{2+}]_i$ and generation of force has been extensively characterized (see reviews (Bolton et al. 1988; Somlyo and Himpens 1989; van Breemen and Saida 1989; Missiaen et al. 1992b; Somlyo and Somlyo 1994; Jiang and Stephens 1994; Walsh et al. 1995; Karaki et al. 1997)). While increases in the Ca^{2+} sensitivity of myosin light chain phosphorylation have been demonstrated in smooth muscle (Kitazawa et al. 1989; Somlyo and Somlyo 1994), full activation of contractile proteins does not occur in the absence of a rise in $[Ca^{2+}]_i$, and following force development, decreases in $[Ca^{2+}]_i$ result in muscle relaxation (Somlyo and Himpens 1989). Moreover, whereas intracellular calcium release plays an important role in the initiation of contraction (Baron et al. 1984; Somlyo and Himpens 1989) of smooth muscle and calcium falls following initial peak transients (Morgan and Morgan 1982), sustained contractile responses require a sustained elevation of $[Ca^{2+}]_i$, which is dependent on extracellular calcium (Himpens et al. 1988).

Despite the complexity of calcium regulatory mechanisms in smooth muscle the characterization of specific calcium transport proteins has been greatly facilitated by recent technical advances in the measurement of ion channel currents and the non-invasive measurement of cytosolic calcium concentration $[Ca^{2+}]_i$ in cells. The former techniques include whole cell (Hamill et al. 1981) and single channel (Neher and Sakmann 1976) recording methods as well as the development of "permeabilized-patch" techniques that maintain physiologic cytosolic calcium buffering (Horn and Marty 1988; Korn and Horn 1989). Similarly, the development of a diverse array of fluorescent calcium indicators such as Fura 2, Indo, and Fluo-3 have enabled the relatively non-invasive measurement of $[Ca^{2+}]_i$ under a variety of conditions (Tsien et al. 1985; Grynkiewicz et al. 1985). The simultaneous experimental use of these techniques allows the dissection of specific transport processes that contribute to rises in cytosolic calcium. More recently the measurement of calcium release processes in muscle has been further advanced by the combination of laser scanning confocal microscopy (LSCM) with single-cell voltage-clamp techniques. The confocal elimination of out-of-focus fluorescence combined with high speed laser scanning provides sufficient spatial and temporal resolution of calcium release events to enable the visualization of quantal calcium release in cardiac and smooth muscle preparations (Cheng et al. 1993a; Cannell et al. 1995; Lopez-Lopez et al. 1995; Nelson et al. 1995). The separate and combined use of these methods has greatly advanced the understanding of the molecular processes regulating cellular $[Ca^{2+}]_i$. This review will focus on the role of calcium permeant ion channels in the regulation of cytosolic calcium in smooth muscle.

4.2
Voltage-Dependent Calcium Channels in Smooth Muscle Cells

Smooth muscle   cells contain at least one type of Ca^{2+} channel that is activated by membrane depolarization. Voltage-activated, dihydropyridine-sensitive Ca^{2+} channels appear to be the major calcium entry pathway in a wide variety of types of smooth muscle including blood vessels such as coronary, cerebral, renal, mesenteric, skin, and pulmonary vascular beds (Aaronson et al. 1988; Benham and Tsien 1988; Caffrey et al. 1986; Worley et al. 1986; Nelson et al. 1988; Nelson and Worley 1989; Bean et al. 1986; Ohya and Sperelakis 1989; Ganitkevich and Isenberg 1990). This Ca2+ channel has been referred to as the L-type Ca^{2+} channel to indicate the fact that it inactivates at a relatively slow rate, resulting in a long-lasting current following membrane depolarization. In contrast, currents conducted by low voltage-activated, or T-type, Ca^{2+} channels are transient in nature, inactivate rapidly and completely during a depolarizing stimulus, and are dihydropyridine-insensitive. T-type Ca^{2+} channels are infrequently encountered in vascular smooth muscle cells (Bean et al. 1986; Benham et al. 1987a; Wang et al. 1989; Ganitkevich and Isenberg 1990; Smirnov and Aaronson 1992). The physiological role, if any, of T-type calcium channels in smooth muscle is not clear.

4.2.1 Properties of Ca^{2+} channels in smooth muscle

Ca^{2+} influx through single voltage-dependent Ca^{2+} channels has been measured using physiological levels of Ca^{2+} as a charge carrier (Gollasch et al. 1992) (Rubart et al. 1996). These measurements reveal the extremely high permeation rate of Ca^{2+} ions though single channels. The molecular mechanisms determining the Ca^{2+} permeation rate are not completely understood, but could involve high affinity Ca^{2+} binding to permeation sites in the channel pore or local negative surface charges near or in the external pore mouth (Gollasch and Nelson 1998). In any case, it is likely that the high Ca^{2+} permeation is controlled by the α_1 subunit alone since recordings of CHO cells expressing smooth muscle α_{1Cb} subunits display this large Ca^{2+} permeation rate (Gollasch et al. 1996).

Voltage-dependent Ca^{2+} channels can exist in at least three kinetic states, only one of which is capable of conducting current. The resting and inactivated states are non-conducting conformations of the channel (Hess et al. 1984; Hille 1994)). The open current conducting state can be reached from the resting but not the inactive states. Studies using either 10 mM Ba^{2+} (Nelson et al. 1990; Ganitkevich and Isenberg 1990; Quayle et al. 1993) or 10 and 2 mM Ca^{2+} (Gollasch et al. 1992; Rubart et al. 1996) have demonstrated that there is no discrete threshold for activation of voltage-dependent Ca^{2+} channels. Instead, single Ca^{2+} channels open with low probability in the voltage range of 60 to 40 mV. These openings are crucial in determining steady-state Ca^{2+} influx that underlies smooth muscle tone, and underscore the importance of voltage-dependent inactivation processes in determining the open probability of single Ca^{2+} channels in vascular smooth muscle cells under physiological conditions (see Rubart et al. 1996).

Measurements at the single channel level have allowed estimation of the unitary Ca^{2+} current of ~0.17 pA under relatively physiological conditions (-40mV, 2 mM Ca^{2+}, 1 mM Mg^{2+}) (Gollasch et al. 1992; Rubart et al. 1996)). This current corresponds to a Ca^{2+} influx rate of about 1 million ions per second. As a result of the small volume of a single smooth muscle cell (~1 pl), one open Ca^{2+} channel would conduct a current capable of changing the cytosolic $[Ca^{2+}]$ by 2.3 M/s at 36°C assuming no buffering or extrusion of Ca^{2+}. Assuming a myocyte fast calcium buffering capacity (ratio of total to free Ca^{2+} ions see below) of 100 (Guerrero et al. 1994b; Fleischmann et al. 1996; Kamishima and McCarron 1996), one channel would contribute 23 nM/s. This calculation illustrates the dramatic potential for Ca^{2+} channels to serve as key regulators of $[Ca^{2+}]_i$. Such sustained increases in Ca^{2+} influx markedly influence $[Ca^{2+}]_i$. Fleischmann et al have provided direct evidence that threshold increases in steady I_{Ca} resulting from one or two open Ca^{2+} channels have a significant impact on cytosolic calcium, and that this occurs during sustained depolarizations at negative membrane potentials (Fleischmann et al. 1994). At more positive potentials, Ca^2 channel inactivation results in only transient increases in $[Ca^{2+}]_i$. Thus there is a voltage range, which corresponds to physiological membrane potentials in non-spiking myocytes, over which small increases in Ca^{2+} channel activity produces sustained increases in $[Ca^{2+}]_i$, which has been termed the "calcium window" (Fleischmann et al. 1994). The operation of voltage-dependent Ca^{2+} channels in this low activity mode, where the open-state probability of any single channel is quite low (on the order of 0.01), represents a fundamentally different way in which these channels participate in calcium signaling in non-spiking cells.

4.2.1.1 Molecular Biology of Voltage-Dependent Calcium Channels

Ca^{2+} channels in smooth muscle consist of a central pore-forming α_1-subunit and several auxiliary subunits. At least six different Ca^{2+} channel α_1 genes have been defined (α_{1S}, α_{1B}, α_{1C}, α_{1D}, and α_{1E}), and only one splice variant (α_{1C-b}) is found in smooth muscle (Snutch and Reiner 1992; Hofmann et al. 1994; Birnbaumer et al. 1994). The pore-forming α_{1C} subunit confers the voltage-and dihydropyridine-sensitivity that characterize the Ca^{2+} channel in smooth muscle (Mikami et al. 1989; Biel et al. 1990). It is interesting to note that the cardiac α_1 subunit (α_{1C-a}) and the smooth muscle α_{1C-b} are splice variants of the same gene (α_{1C}). However, the skeletal muscle α_1 subunit is a product of a different gene (α_{1S}), and this is consistent with the fact that the primary physiological function of the Ca^{2+} channel in skeletal muscle is as a voltage-sensor and not as a current-conducting pore.

The basic structure of all known α_1-subunits is well conserved. This subunit consists of four homologous repeating motifs (I-IV), and each motif consists of six membrane-embedded hydrophobic segments (S_1-S_6). Both the amino- and carboxy-terminal domains are intracellular. A positively charged amino acid occupies every third or fourth position in the S_4 domain of each motif. It is suggested that the S_4 domain functions as the voltage sensor in voltage-gated ion channels (Catterall 1995). Ion selectivity is believed to be controlled by a loop between S_5 and S_6 (SS_1-SS_2 region or P loop) (Heinemann et al. 1992). The P loop is predicted to fold into the membrane and form part of the pore region (Guy and Conti 1990). It is also believed that negatively charged glutamate residues in the SS_2 region of each of the four motifs are responsible for the high Ca^{2+} selectivity, and participate in high-affinity binding with divalent cations (Yang et al. 1993; Mikala et al. 1993; Tang et al. 1993; Yatani et al. 1994; Bahinski et al. 1997). The receptor sites for various Ca^{2+} channel antagonists (dihydropyridines, phenylalkylamines, and benzothiazepines) that give these channels their distinctive pharmacological profile have been localized to various regions on the 1 subunit near the pore lining (Catterall and Striessnig 1992; Hofmann et al. 1990; Catterall 1995; Varadi et al. 1995; Mori et al. 1996; Mitterdorfer et al. 1996; Hockerman et al. 1997). All known phosphorylation sites for protein kinase A and many of the phosphorylation sites for protein kinase C have been located to intracellular positions (Varadi et al. 1995). Carboxy-terminal residues in the α_{1C} subunit are responsible for channel inactivation induced by Ca^{2+} influx (Zhou et al. 1997).

Ca^{2+} channels at presynaptic nerve terminals are inhibited by G-protein coupled receptors acting through a membrane delimited pathway (Dolphin et al. 1986; Holz et al. 1989; Herlitze et al. 1996); the molecular basis of this inhibitory modulation has recently been localized to the carboxy terminus of the α_{1A} and α_{1B} subunits (De Waard et al. 1997; Zamponi et al. 1997; Herlitze et al. 1997). $G\beta\gamma$ binds to the consensus binding site QXXER (Chen et al. 1995) found in the $_1$ subunit domain I-II linker, and also found on the GIRK1 potassium channel (Huang et al. 1995). The binding site is contained within the region that binds the channel βsubunits, suggesting that this is a key site of channel modulation. Interestingly, α_{1C} subunits do not contain the QXXER domain (Herlitze et al. 1997), explaining the absence of G protein mediated inhibition of Ca^{2+} currents in smooth muscle.

Functional heterogeneity of Ca^{2+} channels in different tissues may be explained by differential expression of auxiliary subunits β and α_2/δ with the α_1 pore. To date, four different genes encoding β subunits are known ($\beta_1, \beta_2, \beta_3, \beta_4$) and both β_2 and β_3 have been found in smooth muscle (Hullin et al. 1992; Gollasch et al. 1997). The α_2 and δ subunits are post-translationally modified products of the same gene (Jay et al. 1991). Different combinations of β and α_2/δ splice variants can coexist in various tissues (Varadi et al. 1995; Angelotti and Hofmann 1996). These accessory subunits may regulate pore formation and stabilization, gating, channel kinetics, and drug binding (Bosse et al. 1992; Itagaki et al. 1992) (Welling et al. 1993; Neely et al. 1993; Wei et al. 1995; Shistik et al. 1995; Chien et al. 1996; Bangalore et al. 1996; Catterall and Striessnig 1992; Mori et al. 1996; De et al. 1996; Suh-Kim et al. 1996; Hockerman et al. 1997).

4.2.1.2 Pharmacology of Voltage-Dependent Calcium Channels

Three main classes of chemical compounds act as antagonists of L-type Ca^{2+} channels. These are dihydropyridines (e.g. nifedipine), phenylalkylamines (e.g. verapamil), and benzothiazepines (e.g. diltiazem) (Catterall and Striessnig 1992; Hofmann et al. 1990; McDonald et al. 1994; Catterall 1995). The benzothiazepines and phenylalkylamines block L-type Ca^{2+} channels by binding preferentially to the open state (Lee and Tsien 1983; Tung and Morad 1983; Hering et al. 1989; Klockner and Isenberg 1991), whereas dihydropyridines block Ca^{2+} channels by binding to the inactive state (Uehara and Hume 1985; Bean 1984). In contrast to the Ca^{2+} channel antagonists, one dihydropyridine compound in particular has agonistic properties. Thus, the levorotatory enantiomer of BAY K 8644 promotes channel opening at less depolarized potentials by causing a leftward shift in the activation curve (see Worley et al. 1991; Quayle et al. 1993)). Recently, the black mamba venom toxin calciseptine has been shown to inhibit vascular Ca^{2+} channels in a similar way to dihydropyridines (de Weille et al. 1991; Teramoto et al. 1996).

4.2.2 Functional Significance of Ca^{2+} Channels in Vascular Smooth Muscle Cells

The resting membrane potential of vascular smooth muscle cells in arterioles subjected to physiological transmural pressure is in the range of 40 to 60 mV (Brayden and Nelson 1992; Harder 1984; Harder et al. 1987; Nelson et al. 1990). Furthermore, there is a steep dependence of arterial diameter on membrane potential, suggesting that some voltage-dependent processes are fundamental regulators of arterial tone (Brayden and Nelson 1992; Nelson et al. 1990; Knot and Nelson 1995). Membrane potential changes of even a few millivolts can have dramatic effects on arterial diameter (Knot and Nelson 1998a). Ca^{2+} influx via voltage-dependent L-type Ca^{2+} channels is also highly dependent upon membrane potential in the same range as arterial diameter (Rubart et al. 1996; Knot and Nelson 1998a), suggesting the hypothesis that vascular smooth muscle membrane potential regulates arterial tone by altering Ca^{2+} influx through L-type Ca^{2+} channels (Nelson et al. 1990). Thus, any factor that changes smooth muscle membrane potential should have definite effects on arterial diameter. Consistent with this assumption are observations that K^+ channel openers like pinacidil and cromakalim elicit hyperpolarization of smooth muscle and arterial dilation (Nelson and Quayle 1995). Furthermore, inhibition of K^+ channels by drugs like charybdotoxin causes smooth muscle depolarization and arterial constriction (Brayden and Nelson 1992; Nelson and Quayle 1995; Knot et al. 1998b).

The important contribution of Ca^{2+} channels to the regulation of vascular tone becomes very obvious in hypertension, and increases in Ca^{2+} channel activity have been reported in various experimental models of hypertension (Wilde et al. 1994; Ohya et al. 1996). Furthermore, hypertension appears to be correlated with vascular smooth muscle cell depolarization (Hermsmeyer et al. 1982; Harder et al. 1983; Lamb and Webb 1989; Silva et al. 1996; Martens and Gelband 1996), and this would have a dramatic effect on the influx of Ca^{2+} under basal conditions. However, calcium channels may not be directly changed in hypertension. Instead, factors (e.g. membrane potential) that regulate calcium channel open probability may be changed in hypertension.

The signal transduction mechanisms relating to regulation of vascular tone at the single smooth muscle cell level are just now being characterized. It is apparent that vasoconstrictors like norepinephrine, endothelin, angiotensin II, seritonin, and histamine can augment currents conducted by L-type Ca^{2+} channels in vascular smooth muscle cells (Nelson et al. 1988; Hughes et al. 1996; Worley et al. 1991; Bkaily et al. 1988; Inoue et al. 1990). Diacylglycerol produced in response to vasoconstrictor-mediated activation of the phosphoinositide cascade stimulates protein kinase C which may lead to activation of L-type Ca^{2+} channels (see Fish et al. 1988; Vivaudou et al. 1988; Clapp et al. 1987; Lepretre and Mironneau 1994; Lepretre et al. 1994).

Vasodilators such as nitric oxide that act through cGMP/protein kinase G signaling systems may act in part by inhibiting currents conducted by L-type Ca^{2+} channels. Whole-cell currents conducted by voltage-dependent Ca^{2+} channels have been shown to be sensitive to the NO donor sodium nitroprusside in smooth muscle cells from pulmonary artery (Clapp and Gurney 1991), basilar artery (Tewari and Simard 1997), and human coronary artery (Quignard et al. 1997). Single channel studies have revealed that NO reduces Ca^{2+} channel open probability during step depolarizations (Tewari and Simard 1997). This effect was prevented by the protein kinase inhibitor H-8 and mimicked by the membrane-permeable cGMP analogue 8-Br-cGMP, suggesting that NO inhibits L-type Ca^{2+} channels in vascular smooth muscle cells through a pathway involving cGMP and protein kinase G (Tewari and Simard 1997), see also (Quignard et al. 1997). In addition, there is some evidence suggesting that substances acting via cAMP/protein kinase A signaling mechanisms may stimulate L-type Ca^{2+} channels in smooth muscle from coronary artery (Fukumitsu et al. 1990), basilar artery (Tewari and Simard 1994), and portal vein (Ishikawa et al. 1993). However, the effects of protein kinase A on voltage-dependent Ca^{2+} channels may depend upon the specific tissue studied and the types of regulatory subunits expressed with the α_{1C} pore.

4.3 Nonselective Cation Channels

Over the last decade work from several laboratories has demonstrated that acetylcholine, norepinephrine, and ATP, constrictor neurotransmitters released at the smooth muscle neuromuscular synapse, activate nonselective cation channels in smooth muscle cells (Benham et al. 1985; Benham and Tsien 1987; Inoue et al. 1987b; Byrne and Large 1988b; Byrne and Large 1988a; Benham 1989; Inoue and Isenberg 1990a; Vogalis and Sanders 1990; Wang and Large 1991; Loirand et al. 1991; Pacaud and Bolton 1991; Janssen and Sims 1992). Nonselective cation currents are also activated by other spasmogens such as histamine (Komori et al. 1992) neurokinin A and Substance P (Lee et al. 1995), and endothelin (Kume et al. 1995). The expression of calcium permeant ion channels other than voltage-dependent calcium channels was predicted following extensive studies suggesting that neurotransmitters and hormones activated a dihydropyridine resistant calcium influx in smooth muscle (Bolton 1979), and the subsequent measurement of nonselective cation currents in isolated myocytes (Benham et al. 1985; Inoue et al. 1987b). The channels underlying these currents can now be conveniently categorized in terms of their activation mechanisms: ionotropic channels are ligand-gated, whereas metabotropic channels are activated following the binding of a ligand to its G protein -coupled receptor, and the attendant stimulation of a second messenger cascade. While it is clear that at least some of these channels are calcium permeant, resulting in calcium influx under physiological conditions, the degree to which calcium permeates other channels is controversial. This section will summarize current knowledge with respect to the calcium permeability of I_{Cat} channels.

4.3.1 Ionotropic Nonselective Cation Channels

4.3.1.1 P2X Channels
Extracellular ATP binds to ligand-gated ion channels, termed P2X receptors (Burnstock 1990; Bean and Friel 1990; Bean 1992). Currents resulting from the activation of these receptors were first recorded in smooth muscle disaggregated from rabbit ear arteries (Benham and Tsien 1987b); application of ATP or ATPS to a vascular myocyte was shown to activate a nonselective cation current. Single channel recordings in 130 mM Na^+ identified a 20 pS conductance channel, and subsequent measurements by Benham (Benham 1989) established calcium permeation. The successful expression cloning of P2X receptor genes provided molecular confirmation of these early experiments (Valera et al. 1994a; Brake et al. 1994). As shown in **Figure 1** (Surprenant et al. 1995), there is excellent agreement between heterologously expressed P2X receptor currents, and currents activated by ATP in isolated smooth muscle cells.

At least seven members of the P2X gene family of ATP-gated cation nonselective cation channels have now been cloned and re-expressed

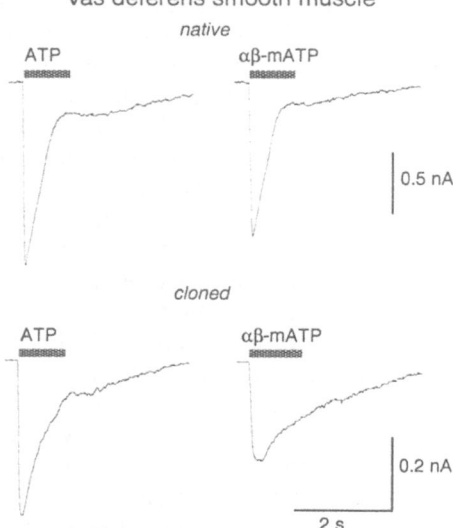

Fig. 1. P2X currents in smooth muscle cells resemble heterologously expressed $P2X_1$ currents. A. Currents evoked by purinergic agonists in single vas deferens smooth muscle cells. B. Currents from HEK293 cells transfected with $P2X_1$ cDNA. Current desensitization and response to the selective agonist ,-me-ATP are similar. From Surprenant et al, 1995 (Figure 3)

(Buell et al. 1996; North 1996). The P2X receptor subtypes that have been reported in smooth muscle include $P2X_1$, $P2X_2$, and $P2X_4$ (Valera et al. 1994a; Brake et al. 1994; Garcia-Guzman et al. 1996). Whereas it has not been possible to unambiguously ascribe ATP activated currents in single smooth muscle cells to a particular P2X subtype, the agonist profile for ATP-activated currents and functional responses suggests that $P2X_1$ channels underlie a major component of these currents in some smooth muscles. Thus α,β methylene ATP, 2-methlythio ATP, and β,γ me-L-ATP evoke currents in isolated myocytes (Khakh et al. 1995; Evans and Kennedy 1994) and contractile responses in smooth muscle (Trezise et al. 1995; Corr and Burnstock 1994; Ziyal et al. 1997)}(Hartley and Kozlowski 1997){Najbar, Li, et al. 1996 ID: 1960}. These agents are potent agonists for $P2X_1$ channels {Valera S, Hussy, N. et al; Surprenant, Buell, et al. 1995 ID: 2047}, whereas $P2X_2$ and $P2X_4$ are not strongly activated by α,β meATP (Brake et al. 1994; Soto et al. 1996)}. It should also be noted, however, that some smooth muscles with functional purinergic responses are poorly activated by α,β meATP, suggesting either that other P2X subtypes (such as $P2X_4$) or G protein coupled P2Y receptors, underlie these functional responses (Abbracchio and Burnstock 1994)

As described above, P2X receptors are nonselective cation channels with clearly demonstrated calcium permeability. Benhams original description of P2X currents in arterial smooth muscle indicated a P_{Ca}/P_{Na} ratio of approximately 3 (Benham and Tsien 1987b). Subsequent calcium fluorescence measurements suggested that approximately 10% of the ATP-gated nonselective cation current is carried by calcium ions (Benham 1989). Quite similar properties have been reported for heterologously expressed P2X channels; using fluorescent measurements in HEK 293 cells, Garcia-Guzman et al (1996) estimated that 8.2% of the $hP2X_4$ current was carried by calcium in physiological solutions. Calculations of P_{Ca}/P_{Na} from the Goldman-Hodgkin-Katz equation have ranged from 4 to 4.8 (Garcia-Guzman et al. 1996; Valera et al. 1994b; Lewis et al. 1995). Thus the expression of ionotropic P2X channels provides a paradigm for neurotransmitter evoked depolarization and calcium influx. The degree to which specific channel subtypes contribute to physiological excitation-contraction coupling in smooth muscle remains to be determined.

4.3.2 Metabotropic Cation Channels

4.3.2.1 Metabotropic Cation Channels Activated by Muscarinic Receptors

4.3.2.1.1 Mechanism of Current Activation

Acetylcholine activates a depolarizing, non-selective cation current observed in isolated vascular and nonvascular smooth muscle cells (Benham et al. 1985; Inoue et al. 1987a; Inoue and Isenberg 1990a; Vogalis and Sanders 1990; Loirand et al. 1991; Komori et al. 1992; Janssen and Sims 1992; Fleischmann et al. 1997). Muscarinic stimulation of single myocytes results in a characteristic, biphasic $[Ca^{2+}]_i$ response that is not observed with simple depolarizations to activate voltage-dependent ion channels, or when calcium release is stimulated with caffeine. Similarly, muscarinic stimulation activates a biphasic inward current, consisting of a brief, inactivating calcium-activated chloride current, followed by a sustained, noisy current of low magnitude, which has been shown to be a nonselective cation current. Activation of this current accounts for the slow EPSPs observed following parasympathetic nerve stimulation of smooth muscle preparations (Byrne and Large 1987). Unlike the ligand-gated P2X receptor currents, muscarinic currents are activated following stimulation of a receptor that is a member of the seven transmembrane spanning, G protein linked receptor family. The metabolic events associated with I_{Cat} current activation have been partially determined. The muscarinic I_{Cat} is not activated by the release of intracellular calcium, as shown by the differential current response to caffeine and methacholine in the same myocyte (**Figure 2**). Studies have also established that intracellular dialysis or exposure to pertussis toxin blocks acetylcholine induced I_{Cat} currents, and that the currents are activated by GTPγS in the absence of acetylcholine (Inoue and Isenberg 1990a; Komori and Bolton 1990; Komori et al. 1992). Similarly, semi-selective M_2 receptor antagonists and dialyzed antibodies directed against the α subunit of G_i or G_o proteins blocks current activation (Wang et al. 1997) While these studies identify the general upstream signaling elements associated with I_{Cat} activation, the studies do not indicate whether α or βγ proteins activate the channels or intermediate proteins (since antibodies directed against subunits would be expected to disrupt signaling by both elements) or identify subsequent transduction steps in channel activation. It should also be noted that Reports have also suggested that either protein kinase C or tyrosine kinase mediates muscarinic I_{Cat} channel opening (Oike et al. 1993; Minami et al. 1994), although these findings are hampered by the specificity of the kinase inhibitors employed. An additional level of complexity in the

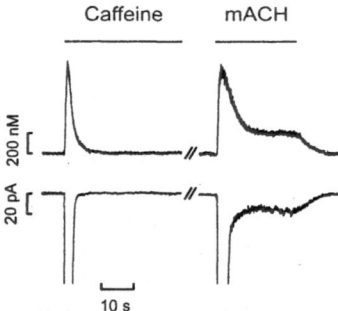

Fig. 2. The nonselective cation current (and a sustained elevation of $[Ca^{2+}]_i$) is activated by methacholine, but not by caffeine. A nondialyzed trachealis myocyte loaded with fura2 was sequentially exposed to caffeine (8 mM) and then to methacholine (50 M). The rapidly inactivating calcium-activated chloride current has been attenuated to better illustrate I_{Cat}. Figure modified from experiment shown in Wang and Kotlikoff, 1997 (Figure 6).

transduction pathway leading to current activation was the finding that the current is strongly facilitated by an increase in intracellular calcium. Inoue and Isenberg (1990c) first reported that I_{Cat} was larger in guinea pig ileum myocytes when exposure to acetylcholine was immediately preceded by activation of the voltage-dependent calcium current, and that in cells dialyzed with EGTA/Ca^{2+}, the amplitude of I_{Cat} activated by acetylcholine increased at higher $[Ca^{2+}]_i$. These findings were extended by Pacaud and Bolton who showed that elimination of calcium release in guinea pig jejunal myocytes by heparin dialysis markedly decreased the muscarinic I_{Cat} (Pacaud and Bolton 1991). Taken together with experiments indicating M_2/G_i or G_o coupling, the finding of calcium facilitation of I_{Cat} suggests that physiological current activation involves the simultaneous activation of M_2 and M_3 receptors, where M_2 agonism activates second messenger pathways responsible for channel gating, and M_3 agonism results in calcium release sufficient to achieve a substantial current. The level of $[Ca^{2+}]_i$ required to facilitate cation channel opening may vary, since in some myocytes the current cannot be evoked in the absence of a rise in $[Ca^{2+}]_i$ above basal levels (Wang et al. 1997). Further evidence of the requirement for dual stimulation in I_{Cat} activation is shown in **Figure 3**. In the experiments shown, I_{Cat} could not be activated by methacholine in the presence of the partially selective M_3 antagonist hexahydro-sila-difenidol at a concentration predicted to block 97% of M_3 receptors and 28% of M_2 receptors. However, simultaneous application of caffeine to evoke calcium release reconstitutes muscarinic coupling to I_{Cat} in the presence of the M_3 antagonist.

The single channel properties of nonselective cation channels underlying the muscarinic I_{Cat} have been reported. Values range from 20-25 pS in guinea pig ileum (Inoue et al. 1987b), 25 pS in rabbit portal vein (Inoue and Kuriyama 1993), and 30 pS in canine pyloric circular myocytes (Vogalis and Sanders 1990). A much larger, calcium-activated conductance has been reported to be activated by muscarinic and adrenergic stimulation (Loirand et al. 1991), although the relevance of such channels is unclear given the results of experiments indicating that a rise in calcium alone is not sufficient to activate I_{Cat}, and given that noise analysis of the muscarinic current does not indicate activation of such a large conductance channel.

4.3.2.1.2   Calcium Permeability
The association of the I_{Cat} current with $InsP_3$-mediated release has made the unequivocal determination of the physiological calcium permeability of this current difficult, since any contribution to $[Ca^{2+}]_i$ made by calcium

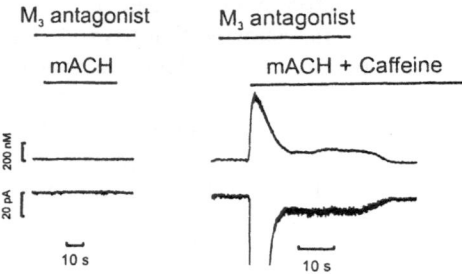

Fig. 3. Activation of I_{Cat} requires M_2 receptor binding and an increase in $[Ca^{2+}]_i$. The M_3 receptor antagonist hexahydro-sil-difenidol (0.4 M) completely inhibits the response to methacholine (50 M). However, if caffeine (8 mM) is used to release intracellular calcium, I_{Cat} is activated and the $[Ca^{2+}]_i$ response is sustained. Figure modified from experiment shown in Wang et al., 1997 (Figure 3)

permeating the channel is obscured by the intracellular calcium release. Early studies indicated that the nonselective current was calcium permeant, since it was observed in external solution containing isotonic calcium chloride (Bolton and Kitamura 1983). However, subsequent reports by Pacaud et al (Pacaud and Bolton 1991) raised doubts about the degree of calcium permeability of the muscarinic receptor -activated I_{Cat} current under physiological conditions. This conclusion was based on the observations that: 1) voltage-dependent calcium channel blockers completely abolish muscarinic contractions in this tissue, and 2) exposure of jejunal smooth muscle cells to caffeine as well as carbachol results in increased membrane permeability to calcium during voltage-clamp steps, suggesting that the permeability was not due to I_{Cat} since caffeine does not activate the current, but due to an increase in calcium permeability related to store depletion.

These findings may not relate to all smooth muscle tissues. Numerous smooth muscle tissues demonstrate substantial dihydropyridine insensitivity, particularly with respect to muscarinic contractions. Moreover, in other smooth muscle cells such as tracheal myocytes, muscarinic stimulation but not caffeine exposure results in an increased membrane calcium permeability (Fleischmann et al. 1997). Several lines of evidence suggest that calcium permeates I_{Cat} currents under physiological conditions. First, muscarinic stimulation is associated with a sustained increase in $[Ca^{2+}]_i$, whereas exposure to caffeine or dialysis of InsP3 results in a transient rise in $[Ca^{2+}]_i$ that promptly falls to baseline (see Figure 2,3). Second, rapid switching to nominally calcium-free solution during (but not before) I_{Cat} activation immediately decreases I_{Cat}. In these experiments I_{Cat} increases in magnitude slightly, suggesting a degree of divalent block by calcium ions under physiologic conditions (Fleischmann et al. 1997). Third, a switch to high Ca^{2+} solution increases $[Ca^{2+}]_i$ during (but not before) I_{Cat} activation. Fourth, rapid application of Ni^{2+} (10 mM) blocks I_{Cat} and reduces $[Ca^{2+}]_i$. Fifth, as in the experiments of Pacaud and Bolton (Pacaud and Bolton 1991), hyperpolarizing voltage steps increased I_{Cat} and $[Ca^{2+}]_i$, while depolarizing voltage steps evoke an outward current and a net drop of $[Ca^{2+}]_i$. Sixth, mACh activates inward currents in monovalent free extracellular solution containing 110 mM Ca^{2+}; the reversal potential is shifted to 12 mV, consistent with a relative Ca^{2+}/Cs^+ permeability of 3.6 (Fatt and Ginsborg 1958).

To obtain a more physiological estimate of the fraction of current carried by Ca^{2+} ions under physiological conditions, I_{Cat} current and $[Ca^{2+}]_i$ were simultaneously measured at high bandwidth following hyperpolarizing voltage-clamp steps, similar to methods reported by Guerrero et al (Guerrero et al. 1994b). This method relies on the calculation of the calcium buffering capacity of the cell, which is the ratio of the total calcium ions added to those appearing free in the cytosol (see below). Assuming that instantaneous calcium buffering is not altered during muscarinic stimulation, the relationship between the buffering capacity calculated from a step depolarization to activate the voltage-dependent calcium current, and that observed during small hyperpolarizing voltage-clamp steps to enhance I_{Cat}, indicate that approximately 14% of the I_{Cat} current is carried by calcium ions (Fleischmann et al. 1997). To the extent that buffering decreases following release of SR calcium stores, this is an overestimate, and probably should be considered as a maximum. However, since the estimate of calcium permeability is derived from the ratio of the calcium buffering capacities determined following rises in $[Ca^{2+}]_i$ associated with the voltage-dependent calcium current and the muscarinic cation current, it is not sensitive to the accuracy of determination of the cell calcium buffering capacity. The latter is influenced by the degree of exogenous calcium buffer (fura 2) in the cell, the estimate of cell volume, and the accuracy of the determination of $[Ca^{2+}]_i$, whereas the fractional calcium permeability derives from the ratios of the integrated calcium and cation currents, which can be determined with substantial accuracy.

4.3.2.2 Metabotropic Cation Channels Activated by Adrenergic R eceptors
The activation of nonselective cation channels by norepinephrine was first
reported by Byrne and Large (Byrne and Large 1988b), but further experiments were limited by the variable responses observed in dialyzed cells
(Amedee et al. 1990). The cation currents were subsequently isolated in
rabbit portal vein cells using perforated patch-clamp methods. In this
preparation norepinephrine activates a nonselective cation current in the
presence of caffeine, which was used to eliminate the calcium-activated
chloride current (Wang and Large 1991). The adrenergic cation current is
not observed in all vascular preparations; norepinephrine does not activate the current in rabbit ear artery cells (Wang et al. 1993). Similar to the
muscarinic I_{Cat}, the adrenergic current is permeable to divalent cations
and may account for sustained calcium influx during adrenergic stimulation (Byrne and Large 1988b) (Wang and Large 1991; Inoue and Kuriyama
1993). Simultaneous measurements of current and calcium in rat portal
vein myocytes indicate a similar relationship between activation of the
current and a sustained increase in $[Ca^{2+}]_i$ (Pacaud et al. 1992). That is, as
with the muscarinic current, hyperpolarization or an increase in extracellular calcium augments the sustained rise in $[Ca^{2+}]_i$ observed during activation of the current.

The adrenergic current is not activated by calcium release, since exposure to caffeine or ionomycin do not activate the current (Wang and Large
1991). Unlike the muscarinic current, however, release of intracellular
calcium does not augment the current (Wang and Large 1991). Thus the
prominent calcium facilitation observed with the muscarinic current does
not appear to be a feature of these channels. Rather, the current is modified by extracellular calcium in a complex manner; I_{Cat} is augmented both
by a decrease in extracellular calcium from physiological levels and by an
increase in $[Ca^{2+}]_o$ when the current is activated in solutions with less than
50 M calcium.

One major question relates to whether the nonselective cation channels
activated by norepinephrine and acetylcholine are the same protein.
Inoue and Kuriyama (Inoue and Kuriyama 1993) studied the activation of
both currents in the rabbit portal vein, one of the few vascular preparations in which excitatory muscarinic responses are observed. Based on
lack of an additional current response once I_{Cat} was fully activated by one
agonist, and similar current I-V relationships, current activation and deac-
tivation kinetics, single channel conductance, and block by Cd^{2+}, they
concluded that adrenergic and muscarinic stimulation activates a common nonselective cation channel. On the other hand, substantial differ-

ences have been reported between the portal vein adrenergic current and the muscarinic current recorded in other preparations. First, as discussed above, unlike the muscarinic current, the adrenergic current is not facilitated by a rise in intracellular calcium. Second, the I-V relationships of the two currents are somewhat distinct. The adrenergic current has a linear current-voltage relationship at negative potentials and is inwardly rectifying at positive potentials (Wang and Large 1991), whereas the muscarinic current is markedly suppressed at negative potentials (Benham et al. 1985; Inoue and Isenberg 1990b). Third, while both currents can be activated by GTPS (Inoue and Isenberg 1990a; Komori and Bolton 1990; Helliwell and Large 1997), the adrenergic current appears to be linked to phospholipase C activation (Helliwell and Large 1997), whereas the muscarinic current is activated by M_2 receptor/$G_{i \text{ or }} G_o$ mechanism (see above). Thus the findings of Inoue and Kuriyama (1993) may indicate that in tissues such as the portal vein that are under predominant adrenergic control, the adrenergic nonselective cation channel is expressed and can be activated after muscarinic phospholipase C stimulation, but these findings may not relate to the principal cation channel that is the activated following M_2 receptor activation in non-vascular smooth muscles. In this regard, tachykinins have been reported to activate a nonselective cation current with similar properties to the adrenergic current (Lee et al. 1995). The current was activated by neurokinin A or Substance P, was not facilitated by the release of intracellular calcium, and had an inwardly rectifying I-V relationship. These findings probably suggest the expression of a cation channel activated following phospholipase C activation in some, but not all, smooth muscle tissues.

4.3.2.3 Calcium Permeant Cation Channels Activated by Other Stimuli

In addition to those described above, cation channels that are activated by mechanical stretch (Kirber et al. 1988; Wellner and Isenberg 1993), by calcium (Loirand et al. 1991), and by caffeine (Guerrero et al. 1994a; Guerrero et al. 1994b) have been reported in smooth muscle. Several of these reports indicate substantial calcium permeability (Kirber et al. 1988; Loirand et al. 1991; Guerrero et al. 1994a; Wellner and Isenberg 1994) to the underlying channels. Stretch-activated cation channels are particularly interesting given the high shear forces encountered by contracted smooth muscle and the well described phenomenon of stretch-induced contraction (Bulbring 1997).

4.4 Calcium Release Channels

Calcium flux into the cytosol from the sarcoplasmic reticulum is mediated by two important calcium permeant channels in myocytes: the inositol trisphospate (InsP₃) receptor and the ryanodine receptor. These channels share substantial sequence homology, protein topology, and likely fourfold symmetry (Marks et al. 1990; Chadwick et al. 1990). They are probably best described as nonselective cation channels, since they have are relatively poorly selectivity for divalent over monovalent cations ($P_{Div/Mon}<10$) (Smith et al. 1988; Liu et al. 1989; Bezprozvanny and Ehrlich 1994). InsP₃ receptors and ryanodine receptors have similar selectivity properties with a rank order of divalent conduction of Ba>Sr>Ca>Mg (Tinker et al. 1993; Bezprozvanny and Ehrlich 1994). The large calcium concentration difference between SR and cytosol (and the roughly equivalent monovalent concentrations) insures that the channels function principally to release intracellular calcium. The activation and modulation of calcium release channel gating by second messengers such as InsP₃ and calcium, phosphorylation, and the degree of SR filling, and the functional interactions between these channels, is currently poorly understood, but an area in which important advances can be anticipated.

4.4.1 Inositol Trisphosphate Mediated Calcium Release

A now well established action of neurotransmitter activated excitation-contraction coupling in smooth muscle is the stimulation of $G_{q/11}$ coupled membrane receptors, activation of phospholipase C, and an attendant rapid rise in InsP₃ concentration within the myocyte (Baron et al. 1984; Somlyo et al. 1985). The role of the InsP₃ receptor in mediating neurotransmitter evoked calcium release has been demonstrated in numerous permeabilized tissue and single cell experiments, in which intracellular dialysis of heparin prevents calcium release following muscarinic, adrenergic, or histaminergic stimulation (Iino 1987; Kobayashi et al. 1989; Iino 1990a; Komori and Bolton 1990; Pacaud and Bolton 1991; Kitamura et al. 1992; Loirand et al. 1992; Ito et al. 1993; Janssen and Sims 1993; Wang and Kotlikoff 1997). Similarly, intracellular application of InsP₃ prevents subsequent neurotransmitter evoked calcium release (Wang and Kotlikoff 1997). These findings and others have clearly established that receptor stimulated phospholipase results in a release of calcium from intracellular stores through InsP₃ receptors in smooth muscle.

4.4.1.1 Structure InsP₃ Receptors

The InsP$_3$ receptor isolated from smooth muscle is a 224 kDa protein that binds InsP$_3$ with a K_D of approximately 2.4 nM (Chadwick et al. 1990). Based on the cloning of cDNAs isolated from various tissues at least three isoforms of InsP$_3$ receptors exist (Furuichi et al. 1989; Mignery et al. 1990a; Parys et al. 1996). All three receptor subtypes are expressed in most tissues, including smooth muscle (De et al. 1994; Morgan et al. 1996). The functional channel is assembled either as a homotetrameric or heterotetrameric complex of the three cloned subtypes (Monkawa et al. 1995; Wojcikiewicz and He 1995; Joseph et al. 1995). The amino termini of the four InsP$_3$ receptor subunits extend into the cytosol and contain InsP$_3$ binding activity (Mignery and Sudhof 1990b; Miyawaki et al. 1991), whereas the carboxy terminus contains six likely transmembrane segments and a putative 5-6 linker that makes up the pore domain, a general structure quite similar to the voltage gated ion channels (Mikoshiba 1997; Joseph 1996). InsP$_3$ receptors also have an extensive regulatory domain that contains binding sites for calcium, calmodulin (types I and II InsP$_3$ receptors), and several consensus protein kinase phosphorylation sequences (Joseph 1996).

Purified InsP$_3$ receptors have been shown to function as InsP$_3$ gated cation channels in liposomes and planar lipid bilayers (Maeda et al. 1991; Ferris et al. 1989), and types I and III InsP$_3$R have been expressed in Sf9 cells using a baculovirus expression system (Yoneshima et al. 1997). Single channel currents are expected to be on the order of 0.1 pA under normal cellular conditions. Since the unitary current and mean open time of the ryanodine receptor are 4 to 5 fold greater than the InsP$_3$ receptor, the calcium flux associated with a single opening of the InsP$_3$ receptor would be more than 20 fold less than for the opening of a ryanodine receptor (see Bezprozvanny 1996). Thus InsP$_3$ calcium release is an intrinsically slower system than CICR, which requires no enzymatic second messenger synthesis, and releases calcium in a more explosive manner, probably indicating the different functional requirements associated with E-C coupling in smooth versus cardiac muscle. The substantially larger calcium flux through ryanodine receptors may also be important for coupling to calcium-sensitive sarcolemmal ion channels (Nelson et al. 1995).

4.4.1.2 Regulation of InsP₃ Receptor Ion Channels

In addition to the regulation of channel open probability by cytosolic InsP$_3$, the receptor/channel is regulated by cytosolic calcium (Iino 1987). The relationship between $[Ca^{2+}]_i$ and channel open state probability is bellshaped in smooth muscle, resulting in a prominent increase in channel opening at a fixed level of InsP$_3$ as calcium rises from resting levels to approximately 300 nM, followed by a reduction in channel gating at higher calcium concentrations (Iino 1990a; Bezprozvanny et al. 1991; Iino and Endo 1992). This prominent regulatory feature suggests that the initial release of calcium may augment InsP$_3$ channel gating, creating a positive feedback loop that insures rapid and sufficient calcium release (Hirose et al. 1993), whereas inhibition of the channel at higher $[Ca^{2+}]_i$ may serve as a classical negative feedback loop (Missiaen et al. 1992a). The biphasic regulation may account for the pulsatile nature of InsP$_3$ receptor mediated signaling, and is also likely essential for regenerative calcium wave propagation (De and Keizer 1992; Atri et al. 1993; Clapham 1995; Parker et al. 1996). In this model, an increase in the cellular concentration of InsP$_3$ results in the simultaneous activation of localized InsP$_3$ receptors; if the local release of calcium is sufficiently great, nearby channels are recruited as calcium diffuses from the initial release site. The calcium induced gating of InsP$_3$ receptors serves to reinforce the propagating calcium, resulting in further InsP$_3$R recruitment. Types I and III InsP$_3$ receptors expressed in Sf9 cells are differentially regulated by calcium (Yoneshima et al. 1997); the affinity of the type I receptor for InsP$_3$ monotonically decreases as calcium is increased, and the type III receptor affinity increases. The extent to which this differential calcium regulation underlies the functional biphasic properties observed in smooth muscle and other tissues is unknown. It has also been proposed that a rise in $[Ca^{2+}]_i$ may initiate InsP$_3$ receptor-mediated calcium release through InsP$_3$ receptor/channels at resting levels of InsP$_3$ (Ehrlich et al. 1994b). If so, it might be predicted that calcium release through ryanodine receptors would activate release through InsP$_3$ receptors. However, in smooth muscle cells dialyzed with and without heparin there is no difference in the caffeine -evoked peak $[Ca^{2+}]_i$, the kinetics of the $[Ca^{2+}]_i$ increase, or the associated calcium-activated chloride current (Wang and Kotlikoff 1997). Similar results were obtained by Komori and Bolton (1991) in jejunal cells, using the maxi-K current to infer calcium release. These experiments suggest that a rise in $[Ca^{2+}]_i$ is not sufficient to activate InsP$_3$ receptors, although it is also possible that modest calcium flux through InsP$_3$ receptors could be obscured when caffeine is used to simultaneously activate ryanodine receptors.

InsP$_3$ receptors are phosphorylated in cells following activation of protein kinase A (PKA) (Yamamoto et al. 1989; Joseph and Ryan 1993), protein kinase G (PKG)(Komalavilas and Lincoln 1994), and tyrosine kinase (Harnick et al. 1995). Interestingly, splice variants of the type-I InsP$_3$ receptors with different phosphorylation characteristics exist; a shorter form of the receptor is found in peripheral tissues such as smooth muscle, whereas the longer form is exclusively expressed in neuronal tissues (Danoff et al. 1991). The short form is a splice variant with 40 amino acids deleted between two consensus PKA phosphorylation sites. Experiments with purified InsP$_3$R from vas deferens, which expresses only the short form, indicated that this form is more readily phosphorylated than the brain form, probably due to a switch in the site at which phosphorylation occurs (Danoff et al. 1991). Controversy exists, however, with respect to the functional consequences of PKA phosphorylation; studies indicate that PKA phosphorylation results in a decrease (Supattapone et al. 1988; Volpe and Alderson-Lang 1990) or an increase (Nakade et al. 1994) in the potency of InsP$_3$ receptor calcium release. Given the multiple functional targets reported for PKA-mediated relaxation of smooth muscle, the importance of InsP$_3$ receptor is not possible to assess currently.

A question of substantial functional importance in tonic smooth muscle tissues relates to the degree to which the activation of InsP$_3$ receptors by phospholipase C linked agonists results in a sustained increase in $[Ca^{2+}]_i$. While it is known that PLC activation and attendant inositol phosphate metabolism is maintained during continued agonist/$G_{q/11}$ -coupled receptor interaction, the extent to which the continued generation of InsP$_3$ underlies the sustained elevation of $[Ca^{2+}]_i$ that is a prominent feature of agonist-evoked calcium responses (Fay et al. 1979; Morgan and Morgan 1982; Himpens et al. 1988), is not known. One indirect mechanism by which continued InsP$_3$ receptor channel activity has been suggested to affect sustained agonist induced increases in $[Ca^{2+}]_i$, is by altering cellular calcium buffering (van Breemen and Saida 1989). Removal of calcium ions from the cytosol by the sarcoplasmic reticulum calcium ATPase (SERCA) occurs during activation. However, since calcium pumped back into the SR is continuously released through InsP$_3$ receptors, this results in an apparent decrease in myocyte calcium buffering capacity. In this manner the continued activation of InsP$_3$ receptors would be essential to the sustained increase in $[Ca^{2+}]_i$ achieved by sustained calcium influx by amplifying small calcium fluxes through sarcolemmal calcium channels.

4.4.2 Ryanodine Receptor Mediated Calcium Release

The open probability of ryanodine-sensitive calcium release channels in cardiac and smooth muscle increases with cytoplasmic calcium, and this property of the ryanodine receptor has been referred to as calcium-induced calcium release" or CICR" (see Meissner 1994; Striggow and Ehrlich 1996). Calcium release through a single or small number of ryanodine receptor channels (unitary CICR) would have little direct effect on cytoplasmic calcium (see section on calcium sparks). However, the activation of a large number of ryanodine receptors by the calcium current could cause global calcium transients, as is the case for cardiac muscle. In cardiac muscle, the local Ca^{2+} influx through a voltage-dependent Ca^{2+} channel activates a nearby ryanodine receptor(s) located in the junctional SR to release Ca^{2+} into the cytoplasm (Figure 4). This close association or fine-tuning of a voltage-dependent (L-type) Ca^{2+} channel and a ryanodine receptor channel is essential for activation of ryanodine receptor channels, since ryanodine receptors have relatively low (μM) Ca^{2+} sensitivity (Lopez-Lopez et al. 1995; Cheng et al. 1996). Any disruption of this close association of L-type Ca^{2+} channels and ryanodine receptors could result in faulty E-C coupling (Gomez et al. 1997). Thus, the function of ryanodine receptor Ca^{2+} channels in cardiac muscle cells is to amplify the Ca^{2+} current across the plasma membrane to induce contraction in the muscle cell.

In smooth muscle, CICR could result in the activation of myosin light chain kinase leading to contraction, but could also be important in activating calcium-sensitive ion channels in the sarcolemma. The local activation of calcium-activated potassium (K_{Ca}) channels would tend to decrease global $[Ca^{2+}]_i$ through membrane potential hyperpolarization. The net effect on contraction would depend on the degree of direct MLCK activation by CICR versus the reduction in global calcium due to the decrease in calcium entry as a result of membrane hyperpolarization. This section will first examine the evidence that ryanodine receptors in smooth muscle can be activated by calcium current to cause global changes in calcium through the spatially and temporally coordinated activation of a number of ryanodine receptors.

4.4.2.1 Structure of Ryanodine Receptors

The solubilized ryanodine receptor from smooth muscle is a high molecular weight protein complex made up of four monomers of approximately 550 kD. Reconstitution of this complex forms a calcium permeant, nonselective cation channel that is inhibited by ruthenium red (Herrmann-Frank et al, 1991; Xu et al. 1994). Ryanodine locks the channel in a subconducting open state, explaining the ability of this compound to release intracellular calcium in many cell types (Hymel et al, 1988). The structure of the three cloned ryanodine receptors (RYR-1, RYR-2, and RYR-3) bears a strong similarity to InsP$_3$ receptors. These two ligand-gated cation chan-

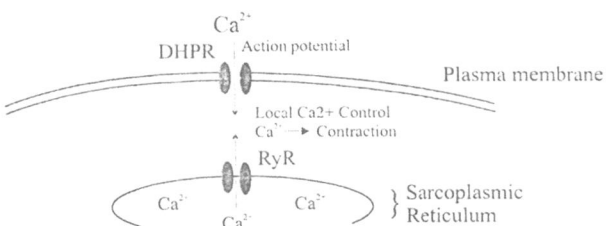

Fig. 4. Many Ca^{2+} sparks, activated in a synchronized manner by Ca^{2+} influx during an action potential, sum to a global Ca^{2+} transient to cause contraction. Ca^{2+} influx through dihydropyridine receptors (DHPR) triggers the release of Ca^{2+} from SR through ryanodine (RYR2) channels, which cause muscle contraction (Berridge 1997).

nels are tetrameric structures with strong homology, particularly in the C terminal region. In smooth muscle ryanodine receptors type 2 and 3 (RYR2 and RYR3), the receptors that act as calcium-activated release channels (rather than dihydropyridine receptor activated channels), are primarily expressed (Giannini et al. 1995), although RYR1 has also been reported in some smooth muscle tissues (Neylon et al. 1995). The existence of more than one isoform raises the possibility of heterotetramer formation. Evidence indicates the formation of heterotetramers in CHO cells, which express all three RYR isoforms (Monkawa et al, 1995). Interestingly, FK506-binding proteins (FKBPs) co-immunoprecipitate with ryanodine receptors (Jayaraman et al, 1992), and also associate with $InsP_3$ receptors (Cameron et al, 1995).

4.4.2.2 Calcium Induced Calcium Release in Smooth Muscle
Calcium-induced calcium release was first indicated in smooth muscle by experiments demonstrating that caffeine, which is known to bind and activate ryanodine receptors, contracts smooth muscle in calcium free solutions (Endo 1977). Subsequently, experiments demonstrated that ryanodine depletes caffeine-sensitive calcium release (Iino et al. 1988; Ito et al. 1991), and CICR was directly measured in permeabilized preparations using a calcium indicator (Iino 1989a). These experiments established that the CICR apparatus exists as a potential calcium release mechanism from smooth muscle sarcoplasmic reticulum, but did not establish evi-

dence of its function during physiological activation. Rather, initial measurements suggested that the calcium requirement for CICR in smooth muscle was too high (greater than 1μ M) to participate in the initiation of contraction, which begins at concentrations just higher than 100 nM (Iino 1989b). Measurements in single cells provided strong evidence that CICR could be triggered in some, but not all, myocytes. If CICR participates in activation of smooth muscle, it could be triggered by: 1) calcium influx associated with the opening of plasmalemmal calcium channels such as voltage-dependent calcium channels or P2X receptors, or 2) the release of calcium from nearby IP_3 receptors. We will summarize evidence for and against the activation of ryanodine receptors by both types of calcium entry into the cytosol.

4.4.2.2.1   CICR Activated by I_{Ca}

Experiments in single myocytes isolated from the guinea-pig urinary bladder (Ganitkevich and Isenberg 1992) and rat portal vein (Gregoire et al. 1993) indicate that activation of the voltage-dependent calcium current (I_{Ca}) triggers calcium release from caffeine sensitive intracellular stores. Ganitkevich and Isenberg (1992) first presented evidence that the $[Ca^{2+}]_i$ transient evoked by depolarization and activation of I_{Ca} was affected by manipulation of SR calcium stores. As shown in **Figure 5** (Ganitkevich and Isenberg 1992, figure 2), a prolonged voltage clamp step to 0 mV produces a phasic and tonic increase in $[Ca^{2+}]_i$, and the phasic component is markedly inhibited following exposure to caffeine. Further evidence that CICR was triggered by I_{Ca} included suppression of the phasic component of the calcium transient by ryanodine and thapsigargin, and augmentation of the $[Ca^{2+}]_i$ transient by increased extracellular calcium and Bay K8644. Subsequent experiments suggested that the low-affinity fluorescent calcium indicator Mag-Indo-1 was a useful reporter for SR calcium release, and that I_{Ca} increased $[Ca^{2+}]_i$ by 16 M, through CICR (Ganitkevich and Hirche 1996).

For other smooth muscle cells the data are conflicting, or suggest that CICR does not contribute measurably to the $[Ca^{2+}]_i$ signal observed following activation of I_{Ca}. Gregoire et al (1993) concluded that Ca^{2+} and Sr^{2+} ions could release calcium from the SR in rat portal vein myocytes, based on the attenuation of the $[Ca^{2+}]_i$ transient following caffeine or ryanodine application, or during neurotransmitter stimulation. Using 1 s depolarization protocols, they determined that CICR resulted in a delayed increase in $[Ca^{2+}]_i$ following the voltage clamp step, rather than affecting the transient $[Ca^{2+}]_i$ increase as observed by Ganitkevich and Isenberg (Ganitkevich and Isenberg 1992). Experiments by Kamishima and McCarron (1996) using the same preparation, however, failed to observe evidence for CICR. They found a similar relationship between I_{Ca} and $[Ca^{2+}]_i$ in the presence and absence of ryanodine, and presented evidence that the continued rise in $[Ca^{2+}]_i$ following termination of a voltage clamp step, observed in some but not all experiments, was associated with calcium influx rather than calcium release, since it was not observed when the cell was stepped to E_{Ca} (Kamishima and McCarron 1996). They concluded that a calcium-activated cation channel, as reported by Loirand et al. (1991), may be responsible for the delayed increase in $[Ca^{2+}]_i$. Kamishima and McCarron (1996) further found that the rise in $[Ca^{2+}]_i$ observed in high bandwidth measurements closely conformed to that predicted from the integrated charge associated with I_{Ca}.

Similar conclusions were reached in equine tracheal myocytes (Fleischmann et al. 1994; Fleischmann et al. 1996); neither ryanodine (Fleischmann et al. 1994) nor caffeine (Fleischmann et al. 1996) alters the relationship between I_{Ca} and $[Ca^{2+}]_i$. Single wavelength measurements of $[Ca^{2+}]_i$ at high bandwidth were used to examine the instantaneous relationship between the calculated increase in $[Ca^{2+}]_i$ due to calcium flux through I_{Ca} (JCa^{2+}/V) and the measured $[Ca^{2+}]_i$ (Fleischmann et al. 1996). As in rat portal vein myocytes, there was an excellent agreement between ion flux through I_{Ca} and the rise in $[Ca^{2+}]_i$. No discontinuity occurred over the full time-course of the current, as would be expected if CICR were triggered by an initial calcium influx, or following an increase in $[Ca^{2+}]_i$ beyond a threshold level, and similar results were observed in 10 mM calcium to augment the calcium current. After repolarization, $[Ca^{2+}]_i$ fell monotonically, without evidence of activation of a regenerative calcium release process. Also, the magnitude of the $[Ca^{2+}]_i$ achieved for a given current integral was small, consistent with calcium entry via I_{Ca} as the only source of calcium ions contributing to the rise in cytosolic calcium (see below). Further, the relationship between I_{Ca} and the peak $[Ca^{2+}]_i$ (or the calcium flux) was equivalent at all voltages, as would be expected if the rise in $[Ca^{2+}]_i$ is due only to the influx of calcium ions through VDCC and if calcium efflux processes are relatively slow. Thus no prominent threshold in the relationship between the integral of the calcium current and the achieved $[Ca^{2+}]_i$ was observed, as might be predicted for (CICR) (Iino 1989a; Iino 1990b).

One obvious feature of the preparation in which CICR has been most clearly demonstrated, guinea pig urinary bladder myocytes, is that there is a prominent sustained component to the biphasic $[Ca^{2+}]_i$ transient associ-

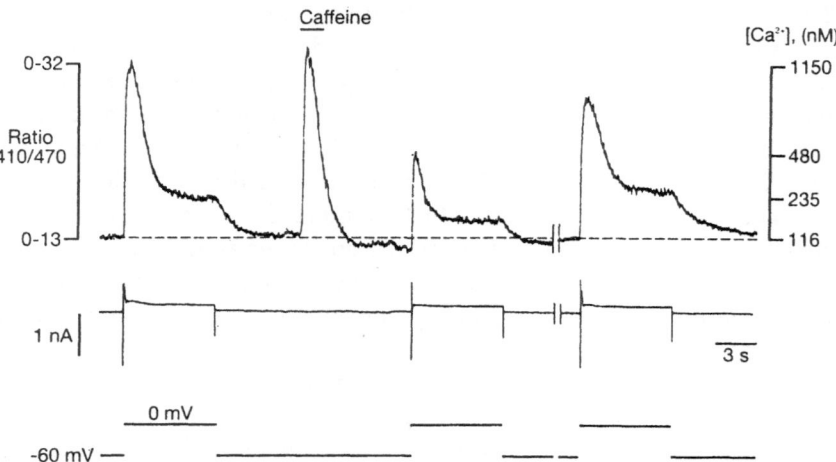

Fig. 5. Calcium-induced cfalcium release in urinary bladder myocytes. Simultaneous recording of current and $[Ca^{2+}]$; during voltage steps to evoke I_{Ca} and exposure to caffeine (10 mM). Depolarization evokes a large $[Ca^{2+}]_i$transient and a sustained increase in $[Ca^{2+}]_i$. Caffeine evokes a $[Ca^{2+}]_i$ transient of similar magnitude; following the release of calcium by caffeine, depolarization transients are markedly smaller, suggesting that CICR amplified the first depolarization calcium transient. From Ganitkevich and Isenberg, 1992

ated with sustained step depolarizations to activate the maximum I_{Ca} (Ganitkevich and Isenberg 1991). A similar sustained increased in $[Ca^{2+}]_i$ associated with sustained depolarizations to potentials more positive than 10 mV is not observed in myocytes from rat portal vein (Kamishima and McCarron 1996) or equine trachealis (Fleischmann et al. 1994; Fleischmann et al. 1996), where sustained increases in $[Ca^{2+}]_i$ associated with sustained depolarizations are observed only at negative potentials at which voltage-dependent channel inactivation is minimized (Fleischmann et al. 1994; Kamishima and McCarron 1996). This obvious difference in depolarization mediated calcium transients may relate to the large calcium current density and incomplete inactivation of I_{Ca} in bladder myocytes (Klockner and Isenberg 1985). Such a non-inactivating current results in a very large net calcium flux (proportional to the integral of the current), when compared to the flux associated with the completely inactivating I_{Ca} of other smooth muscle cells (see above). Bladder myocytes also display much larger increments in $[Ca^{2+}]_i$ associated with a single voltage clamp step (Ganitkevich and Isenberg 1991; Ganitkevich and Isenberg 1992). The $[Ca^{2+}]_i$ transients often reach greater than 1μM and are similar in magnitude to caffeine induced calcium transients, consistent with the increased I_{Ca}, and perhaps an additional source of calcium ions (CICR) contributing to the transient. However, depolarization induced increases in $[Ca^{2+}]_i$ are much more typically observed between 150 and 400 nM in smooth muscle cells (Yagi et al. 1988; Aaronson and Benham, 1989; Becker et al. 1989; Vogalis et al. 1991; Gregoire et al. 1993; Guerrero et al. 1994b; Fleischmann et al. 1994; Fleischmann et al. 1996; Kamishima and McCarron 1996).

A quantitative expression of the relationship between I_{Ca} and the achieved $[Ca^{2+}]_i$ is through the calculation of the cytosolic buffering capacity. The increase in $[Ca^{2+}]_i$ associated with the calculated flux through I_{Ca} provides an estimate of the buffering capacity of the cell, which is the ratio of the total calcium entering through I_{Ca} versus the measured increase in free calcium:

$$B = \sum I_{Ca}dt/(2FV\Delta[Ca^{2+}]_i)$$

where B is the dimensionless calcium buffering capacity, Z is Faradays constant (multiplied by 2 for the charge of the calcium ion), V is the cell volume, and $\Delta[Ca^{2+}]_i$ is the net change in free calcium measured in the cell. B is the instantaneous buffering capacity of the cytosol only to the extent that the removal of calcium occurs slowly and does not affect the peak $[Ca^{2+}]_i$ achieved. In smooth muscle the calcium efflux rate following a calcium transient produced by I_{Ca} is quite slow relative to cardiac muscle. The exponential decay time constant is on the order 6 s, indicating that over the first 300 ms there will be less than a 5% decline in the peak $[Ca^{2+}]_i$ transient. Not surprisingly, Guerrero et al (1994b) showed that for the first 100 ms of the calcium current, the effect of calcium removal was negligible. They obtained a total B (including the buffering of the exogenous calcium indicator) of 82 in *Bufo marinus* cells in the presence of ryanodine, indicating that of 83 calcium ions entering the cytosol, 82 are bound by endogenous buffer and 1 appears as a free ion. To the extent that CICR occurs, the buffering capacity will be artificially lowered, since the released calcium ions contribute to $[Ca^{2+}]_i$ but are not part of the integrated current. Thus in cardiac cells, in which CICR is strongly activate by I_{Ca}, the rise in calcium is substantially greater than that predicted by the integrated cur-

rent (Callewaert et al. 1988). However, in several smooth muscle preparations (Fleischmann et al. 1996; Kamashima and McCarron 1996), the calculated B value is quite similar to that of non-excitable cells (Neher and
Augustine 1992), and to smooth muscle cells dialyzed with ryanodine
(Guerrero et al. 1994b). In other words, in these smooth muscle preparations the increase in $[Ca^{2+}]_i$ for a given current is of a magnitude expected
for the flux of calcium ions associated with the calcium current alone.

Thus it appears that in spiking bladder myocytes with large, incom-
pletely inactivating calcium currents, the $[Ca^{2+}]_i$ threshold required for
CICR (Iino 1989b) is met and calcium is released from the SR through
ryanodine receptors. In many other smooth muscle cells, however, it appears that CICR is not sufficient to cause significant changes in global
calcium, even under optimal conditions in which the full I_{Ca} is achieved.
One would therefore expect that in those cells in which a slow, graded
depolarization occurs during E-C coupling, local CICR causes activation
of K_{Ca} channels (Nelson et al. 1995) and not does contribute directly to
global calcium.   One possible explanation for the lack of global CICR,
even under optimal I_{Ca} conditions, is an insufficient number of voltage-
dependent calcium channels in close proximity to ryanodine receptors to
cause "local" activation.

4.4.2.2.2   CICR Activated by InsP₃ Receptor Calcium Release

Fewer data are available with respect to the role of ryanodine receptor
calcium channels during calcium release mediated by $InsP_3$ receptors.
Ryanodine receptors are expressed on the SR and may occur in close
proximity to $InsP_3$ receptors, although there is evidence of discrete localization in other tissues (Sharp et al. 1993; Martone et al. 1997). If these
receptors are expressed in close proximity, one would expect that calcium
release through neighboring $InsP_3$ receptors would activate ryanodine
receptors, thereby accelerating calcium release. Experiments utilizing ruthenium red, a specific ryanodine receptor antagonist (Ehrlich et al.
1994a), do not support a functional role of RYR receptors during $InsP_3$
mediated calcium release (Wang and Kotlikoff 1997). Following dialysis of
a cell with ruthenium red, caffeine exposure fails to release calcium, indicating functional blockade of ryanodine receptors. Subsequent exposure
to methacholine, however, results in a typical biphasic calcium release and
activation of the calcium-activated chloride current. Conversely, dialysis
of cells with heparin completely blocks muscarinic calcium release,
whereas the caffeine release is unaffected. These findings suggest that
CICR may not play an important role in calcium release triggered by

phospholipase C coupled receptors, although it is quite possible that at low agonist concentrations, CICR provides an important amplifying role in phospholipase C mediated calcium release.

4.4.2.3   RYRs Mediate Elementary Ca^{2+} Release Events

Elementary calcium release events (calcium sparks) have recently been discovered in cardiac, skeletal and smooth muscle, using a laser scanning confocal microscope and the fluorescent calcium indicator, fluo-3 (Cheng et al. 1993b; Nelson et al. 1995; Tsugorka et al. 1995). A calcium spark occurs when multiple, ryanodine-sensitive, calcium release channels (located in the sarcoplasmic reticulum membrane) open, resulting in a transient localized increase in calcium (spark). Therefore, a calcium spark appears to be an elementary CICR event. In vascular smooth muscle, Ca^{2+} sparks activate nearby plasmalemmal K_{Ca} channels (observed as a spontaneous transient outward current or STOC). In cardiac myocytes, the calcium channel is positioned about 20 nm from the ryanodine receptors, and therefore, the opening of a single L-type calcium channels causes a high, local elevation of calcium onto the cytoplasmic surface of the RyR receptor, which leads to a dramatic elevation in calcium spark probability. Thus, this tight structural association is critical for proper E-C coupling in heart, and disturbances can compromise E-C coupling in cases of heart failure (Gomez et al. 1997). Ryanodine receptors appear to be within 20 nm of the surface membrane in smooth muscle (Somlyo 1985). However, the functional coupling of local calcium entry and ryanodine receptors is not known in smooth muscle.

In arterial smooth muscle, calcium sparks can signal myosin light chain kinase for contraction and K_{Ca} channels to drive relaxation, with the latter effect appearing more prominent under the steady-state conditions of arterial tone (Figures 6,7). As indicated above, in urinary bladder smooth muscle, the calcium current can activate a sufficient number of calcium sparks to cause a change in global calcium. Therefore, the relative contributions of the direct effect of calcium sparks to increase global calcium and the indirect effect through K_{Ca} channels to decrease global calcium through membrane hyperpolarization will depend on the spatial and temporal activation of calcium sparks by cytoplasmic calcium.

The frequency and amplitude of calcium sparks are modulated by SR calcium load, protein kinase C (PKC) and cyclic nucleotides/cyclic nucleotide-dependent protein kinases (Bonev et al. 1997; Porter et al. 1997). Calcium spark frequency decreases following activation of PKC (Bonev et al. 1997). This effect appears to be mediated by a direct action of PKC on ryanodine receptor channels rather than a decrease in SR calcium load since the caffeine-induced calcium transient is unaffected by PKC activation (Bonev et al. 1997). Thus, vasoconstrictors that activate PKC may decrease spark frequency, leading to a decrease in K_{Ca} channel activity and membrane depolarization, thereby promoting Ca^{2+} influx through voltage-dependent Ca^{2+} channels. Furthermore, vasodilators that act via cAMP/PKA and cGMP/PKG signaling pathways have been shown to stimulate calcium spark frequency in smooth muscle cells from cerebral and coronary arteries (Porter et al. 1997). cAMP/PKA may increase spark frequency by exerting a direct effect on the ryanodine receptor channel and by increasing SR calcium load, whereas cGMP/PKG may act primarily to increase SR calcium load (Porter et al. 1997; Santana et al. 1998). Stimulation of calcium sparks by vasodilators acting through cyclic nucleotide-dependent processes will result in activation of K_{Ca} channels, membrane hyperpolarization, and decreased Ca^{2+} entry through voltage-dependent Ca^{2+} channels.

Measurements of elementary calcium release events mediated by ryanodine receptors in smooth muscle emphasize the distinction between global and local CICR, and point to a previously unanticipated level of signaling complexity in this tissue. It is likely that local calcium sparks represent an important differential signaling mechanism, whereby low level, localized ryanodine receptor channel activity plays an important role in smooth muscle electrical behavior through coupling to sarcolemmal calcium-activated ion channels. Thus, as with voltage-dependent calcium channels, ryanodine receptor channels appear to act in an efficient, low activity mode; although the average open-state probability of these channels is quite low, openings are efficiently amplified by calcium-dependent conductances.

4.5 Conclusion

We have reviewed the major calcium permeant ion channels in smooth muscle, summarized in **Figure 8**. The obvious complexity and redundancy of these channels almost certainly reflects evolutionary pressures associated with the diversity of functions subserved by smooth muscle. Progress in this area has generally proceeded by the identification and biophysical characterization of these channels using patch-clamp or bilayer techniques, followed by the determination of their molecular structure and regulatory properties. With the notable exception of metabotropic, nonse-

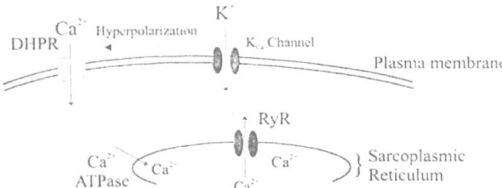

Fig. 6. Individual Ca^{2+} sparks lower global Ca^{2+}. Ca^{2+} sparks activate K_{Ca} (10-100) channels to cause an outward current (STOC) which hyperpolarizes the membrane potential closing voltage-dependent Ca^{2+} channels, and hence decreasing global Ca^{2+}.

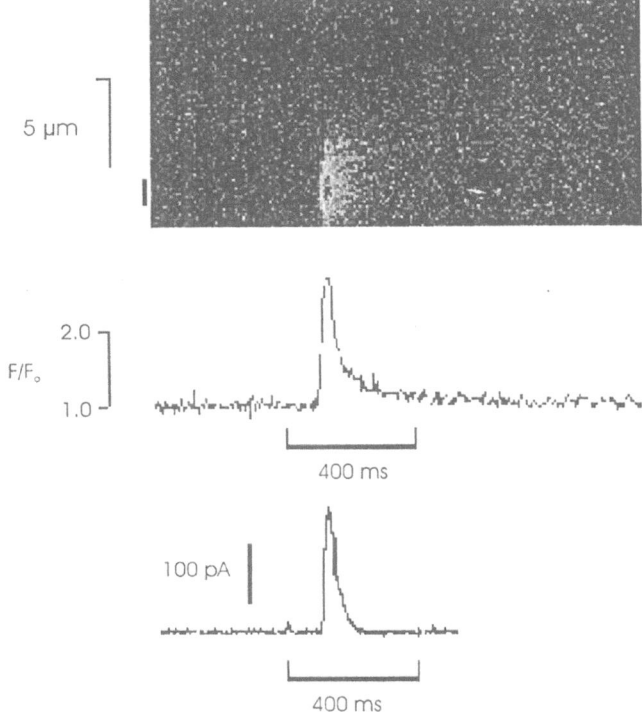

Fig. 7. Line scan image from a vascular smooth muscle cell. Image shows a spark originating at a site towards the edge of the cell. The line scan has had the background subtracted and it has been colorized in order to clearly visualise the spark. In this image time is in the horizontal direction and distance across the cell is in the vertical direction. Beneath the line scan is the averaged fluorescence signal, measured from the section of the line scan indicated by the bar. The figure also shows an example of a STOC to demonstrate the similarity in the time course of these events. (taken from figure 1, Nelson *et al* 1995).

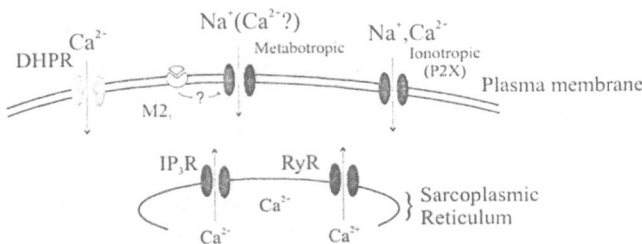

Fig. 8.

lective cation channels, which have not been cloned, substantial structural and regulatory information is now available at the molecular level for all of these channels. It seems likely that the challenge for the next decade will be to develop approaches that allow the determination of the individual function of these proteins in regulating cytosolic calcium, at a level of precision equivalent to that employed in their characterization.

4.6 References

1. Aaronson, P.I. and Benham, C.D. (1989) Alterations in $[Ca^{2+}]_i$ mediated by sodium-calcium exchange in smooth muscle cells isolated from the guinea-pig ureter. Journal of Physiology, 416:1-18.
2. Aaronson, P.I., Bolton, T.B., Lang, R.J., and Mackenzie, I. (1988) Calcium currents in single isolated smooth muscle cells from the rabbit ear artery in normal-calcium and high-barium solutions. J.Physiol.(London), 405:57-75.
3. Abbracchio, M.P. and Burnstock, G. (1994) Purinoceptors: Are there families of P2X and P2Y purinoceptors? Pharmacol.Ther., 64:445-475.
4. Amedee, T., Benham, C.D., Bolton, T.B., Byrne, NG, and Large, W.A. (1990) Potassium, chloride and non-selective cation conductances opened by noradrenaline in rabbit ear artery cells. Journal of Physiology, 423:551-568.
5. Angelotti, T. and Hofmann, F. (1996) Tissue-specific expression of splice variants of the mouse voltage-gated calcium channel alpha2/delta subunit. FEBS Letters, 397:331-337.

6. Atri, A., Amundson, J., Clapham, D., and Sneyd, J. (1993) A single-pool model for intracellular calcium oscillations and waves in the Xenopus laevis oocyte. Biophysical Journal, 65:1727-1739.

7. Bahinski, A., Yatani, A., Mikala, G., Tang, S., Yamamoto, S., and Schwartz, A. (1997) Charged amino acids near the pore entrance influence ion-conduction of a human L-type cardiac calcium channel. Molecular & Cellular Biochemistry, 166:125-134.

8. Bangalore, R., Mehrke, G., Gingrich, K., Hofmann, F., and Kass, R.S. (1996) Influence of L-type Ca channel 2/-subunit on ionic and gating current in transiently transfected HEK 293 cells. American Journal of Physiology, 270:H1521-H1528

9. Baron, C.B., Cunningham, M., Strauss, J.F.3d., and Coburn, R.F. (1984) Pharmacomechanical coupling in smooth muscle may involve phosphatidylinositol metabolism. Proc.Natl.Acad.Sci.USA, 81:6899-6903.

10. Bean, B.P. (1984) Nitrendipine block of cardiac calcium channels: high-affinity binding to the inactivated state. Proceedings of the National Academy of Sciences of the United States of America, 81:6388-6392.

11. Bean, B.P. (1992) Pharmacology and electrophysiology of ATP-activated ion channels. Trends in Pharmacological Sciences, 13:87-90.

12. Bean, B.P. and Friel, D.D. (1990) ATP-activated channels in excitable cells. Ion Channels, 2:169-203.

13. Bean, B.P., Sturek, M., Puga, A., and Hermsmeyer, K. (1986) Calcium channels in muscle cells isolated from rat mesenteric arteries: modulation by dihydropyridine drugs. Circulation Research, 59:229-235.

14. Becker, P.L., Singer, J.J., Walsh, J.V.J., Fay, and FS. (1989) Regulation of calcium concentration in voltage-clamped smooth muscle cells. Science, 244:211-214.

15. Benham, C.D. (1989) ATP-activated channels gate calcium entry in single smooth muscle cells dissociated from rabbit ear artery. Journal of Physiology (London), 419:689-701.

16. Benham, C.D., Bolton, T.B., and Lang, R.J. (1985) Acetylcholine activates an inward current in single mammalian smooth muscle cells. Nature, 316:345-347.

17. Benham, C.D., Hess, P., and Tsien, R.W. (1987) Two types of calcium channels in single smooth muscle cells from rabbit ear artery studied with whole-cell and single-channel recordings. Circulation Research, 61:I10-I16

18. Benham, C.D. and Tsien, R.W. (1987) A novel receptor-operated Ca2+-permeable channel activated by ATP in smooth muscle. Nature, 328:275-278.

19. Benham, C.D. and Tsien, R.W. (1988) Noradrenaline modulation of calcium channels in single smooth muscle cells from rabbit ear artery. Journal of Physiology, 404:767-784.

20. Berridge, M.J. (1997) Elementary and global aspects of calcium signalling. Journal of Experimental Biology, 200:315-319.

21. Bezprozvanny, I. (1996) Inositol (1,4,5)-trisphosphate receptors: functional properties, modulation, and role in calcium wave propagation. Society of General Physiologists Series, 51:75-86.

22. Bezprozvanny, I. and Ehrlich, B.E. (1994) Inositol (1,4,5)-trisphosphate (InsP3)-gated Ca channels from cerebellum: conduction properties for divalent cations and regulation by intraluminal calcium. Journal of General Physiology, 104:821-856.

23. Bezprozvanny, I., Watras, J., and Ehrlich, B.E. (1991) Bell-shaped calcium-response curves of Ins(1,4,5)P3- and calcium- gated channels from endoplasmic reticulum of cerebellum. Nature, 351 :751-754.

24. Biel, M., Ruth, P., Bosse, E., Hullin, R., Stuhmer, W., Flockerzi, V., and Hofmann, F. (1990) Primary structure and functional expression of a high voltage activated calcium channel from rabbit lung. FEBS Letters, 269:409-412.

25. Birnbaumer, L., Campbell, K.P., Catterall, W.A., Harpold, M.M., Hofmann, F., Horne, WA, Mori, Y., Schwartz, A., Snutch, T.P., and Tanabe, T. (1994) The naming of voltage-gated calcium channels. Neuron, 13:505-506.

26. Bkaily, G., Peyrow, M., Sculptoreanu, A., Jacques, D., Chahine, M., Regoli, D., and Sperelakis, N. (1988) Angiotensin II increases Isi and blocks IK in single aortic cell of rabbit. Pflugers Archiv - European Journal of Physiology, 412:448-450.

27. Bolton, T.B. (1979) Mechanisms of action of transmitters and other substances on smooth muscle. Physiological Reviews, 59:606-718.

28. Bolton, T.B. and Kitamura, K. (1983) Evidence that ionic channels associated with the muscarinic receptor of smooth muscle may admit calcium. British Journal of Pharmacology, 78:405-416.

29. Bolton, T.B., Lim, S.P., Salmon, D.M., and Beech, D.J. (1988) Calcium channels and calcium-mediated effects in smooth muscle cells. Journal of Cardiovascular Pharmacology, 12 Suppl 5:S96-S99

30. Bonev, A.D., Jaggar, J.H., Rubart, M., and Nelson, M.T. (1997) Activators of protein kinase C decrease Ca^{2+} spark frequency in smooth muscle cells from cerebral arteries. Am.J.Physiol.(Cell Physiol.), 273:C2090-C2095

31. Bosse, E., Bottlender, R., Kleppisch, T., Hescheler, J., Welling, A., Hofmann, F., and Flockerzi, V. (1992) Stable and functional expression of the calcium channel alpha 1 subunit from smooth muscle in somatic cell lines. EMBO Journal, 11:2033-2038.

32. Brake, A.J., Wagenbach, M.J., and Julius, D. (1994) New structural motif for ligand-gated ion channels defined by an ionotropic ATP receptor. Nature, 371:519-523.

33. Brayden, J.E. and Nelson, M.T. (1992) Regulation of arterial tone by activation of calcium-dependent potassium channels. Science, 256:532-535.

34. Buell, G., Collo, G., and Rassendren, F. (1996) P2X receptors: an emerging channel family. Eur.J.Neurosci., 8 :2221-2228.

35. Bulbring, E. (1997) Correlation between membrane potential, spike discharge and tension in smooth muscle. J.Physiol.(London), 128:200-221.

36. Burnstock, G. (1990) Overview. Purinergic mechanisms. Annals of the New York Academy of Sciences, 603:1-17.

37. Byrne, N.G. and Large, W.A. (1987) Membrane mechanism associated with muscarinic receptor activation in single cells freshly dispersed from the rat anococcygeus muscle. British Journal of Pharmacology, 92:371-379.

38. Byrne, N.G. and Large, W.A. (1988a) Mechanism of action of -adrenoceptor activation in single cells freshly dissociated from the rabbit portal vein. British Journal of Pharmacology, 94:475-482.

39. Byrne, N.G. and Large, W.A. (1988b) Membrane ionic mechanisms activated by noradrenaline in cells isolated from the rabbit portal vein. Journal of Physiology, 404:557-573.

40. Caffrey, J.M., Josephson, I.R., and Brown, A.M. (1986) Calcium channels of amphibian stomach and mammalian aorta smooth muscle cells. Biophysical Journal, 49:1237-1242.

41. Callewaert, G., Cleemann, L., and Morad, M. (1988) Epinephrine enhances Ca^{2+} current-regulated Ca^{2+} release and Ca^{2+} reuptake in rat ventricular myocytes. Proceedings of the National Academy of Sciences of the United States of America, 85:2009-2013.

42. Cameron A.M., Steiner, J.P., Sabatini, D.M., Kaplan, A.I., Walensky, L.D., and Snyder, S.H. (1995) Immunophilin FK506 binding protein associated with inositol 1,4,5-trisphosphate receptor modulates calcium flux. Proceedings of the National Academy of Sciences of the United States of America, 92:1784-1788.

43. Cannell, M.B., Cheng, H., and Lederer, W.J. (1995) The control of calcium release in heart muscle. Science, 268:1045-1049.

44. Catterall, W.A. (1995) Structure and function of voltage-gated ion channels. Annual Review of Biochemistry, 64:493-531.

45. Catterall, W.A. and Striessnig, J. (1992) Receptor sites for Ca^{2+} channel antagonists. Trends in Pharmacological Sciences, 13:256-262.

46. Chadwick, C.C., Saito, A., and Fleischer, S. (1990) Isolation and characterization of the inositol trisphosphate receptor from smooth muscle. Proceedings of the National Academy of Sciences of the United States of America, 87:2132-2136.

47. Chen, J., DeVivo, M., Dingus, J., Harry, A., Li, J., Sui, J., Carty, D.J., Blank, J.L., Exton, J.H., Stoffel, R.H., Inglese, J., Lefkowitz, R.J., Logothetis, D.E., Hildebrandt, J.D., and Iyengar, R. (1995) A region of adenylyl cyclase 2 critical for regulation by G protein subunits. Science, 268:1166-1169.

48. Cheng, H., Lederer, M.R., Lederer, W.J., and Cannell, M.B. (1996) Calcium sparks and $[Ca^{2+}]_i$ waves in cardiac myocytes. American Journal of Physiology, 270:C148-C159

49. Cheng, H., Lederer, W.J., and Cannell, M.B. (1993a) Calcium sparks: elementary events underlying excitation-contraction coupling in heart muscle. Science, 262:740-744.

50. Cheng, H., Lederer, W.J., and Cannell, M.B. (1993b) Calcium sparks: elementary events underlying excitation-contraction coupling in heart muscle. Science, 262:740-744.

51. Chien, A.J., Carr, K.M., Shirokov, R.E., Rios, E., and Hosey, M.M. (1996) Identification of palmitoylation sites within the L-type calcium channel beta2a subunit and effects on channel function. Journal of Biological Chemistry, 271:26465-26468.

52. Clapham, D.E. (1995) Calcium signaling. Cell, 80:259-268.

53. Clapp, L.H. and Gurney, A.M. (1991) Modulation of calcium movements by nitro-prusside in isolated vascular smooth muscle cells. Pflugers Archiv - European Journal of Physiology, *418*:462-470.

54. Clapp, L.H., Vivaudou, M.B., Walsh, J.V.J., and Singer, J.J. (1987) Acetylcholine increases voltage-activated Ca^{2+} current in freshly dissociated smooth muscle cells. Proceedings of the National Academy of Sciences of the United States of America, *84*:2092-2096.

55. Corr, L. and Burnstock, G. (1994) Analysis of P2-purinoceptor subtypes on the smooth muscle and endothelium of rabbit coronary artery. Journal of Cardiovascular Pharmacology, *23*:709-715.

56. Danoff, S.K., Ferris, C.D., Donath, C., Fischer, G.A., Munemitsu, S., Ullrich, A., Snyder, S.H., and Ross, C.A. (1991) Inositol 1,4,5-trisphosphate receptors: distinct neuronal and nonneuronal forms derived by alternative splicing differ in phosphorylation. Proceedings of the National Academy of Sciences of the United States of America, *88*:2951-2955.

57. de Smedt, H., Missiaen, L., Parys, J.B., Bootman, M.D., Mertens, L., Van, D.B., and Casteels, R. (1994) Determination of relative amounts of inositol trisphosphate receptor mRNA isoforms by ratio polymerase chain reaction. Journal of Biological Chemistry, *269*:21691-21698.

58. de Waard, M., Liu, H., Walker, D., Scott, V.E., Gurnett, C.A., and Campbell, K.P. (1997) Direct binding of G-protein betagamma complex to voltage- dependent calcium channels. Nature, *385*:446-450.

59. de Waard, M., Gurnett, C.A., and Campbell, K.P. (1996) Structural and functional diversity of voltage-activated calcium channels. Ion Channels, *4*:41-87.

60. de Weille J., Schweitz, H., Maes, P., Tartar, A., and Lazdunski, M. (1991) Calciseptine, a peptide isolated from black mamba venom, is a specific blocker of the L-type calcium channel. Proceedings of the National Academy of Sciences of the United States of America, *88*:2437-2440.

61. de Young, G. and Keizer, J. (1992) A single-pool inositol 1,4,5-trisphosphate-receptor-based model for agonist-stimulated oscillations in Ca^{2+} concentration. Proceedings of the National Academy of Sciences of the United States of America, *89*:9895-9899.

62. Dolphin, A.C., Forda, S.R., and Scott, R.H. (1986) Calcium-dependent currents in cultured rat dorsal root ganglion neurones are inhibited by an adenosine analogue. Journal of Physiology, *373*:47-61.

63. Ehrlich, B.E., Kaftan, E., Bezprozvannaya, S., and Bezprozvanny, I. (1994a) The pharmacology of intracellular Ca^{2+}-release channels. Trends in Pharmacological Sciences, *15*:145-149.

64. Ehrlich, B.E., Kaftan, E., Bezprozvannaya, S., and Bezprozvanny, I. (1994b) The pharmacology of intracellular Ca^{2+}-release channels. Trends in Pharmacological Sciences, *15*:145-149.

65. Endo, M. (1977) Calcium release from the sarcoplasmic reticulum. Physiol.Rev., *57*:71-108.

66. Evans, R.J. and Kennedy, C. (1994) Characterization of P2-purinoceptors in the smooth muscle of the rat tail artery: a comparison between contractile and electro-physiological responses. British Journal of Pharmacology, 113:853-860.

67. Fatt, P. and Ginsborg, B.L. (1958) The ionic requirements for the production of action potentials in crustacean muscle fibers. J.Physiol.(London), 142:516-543.

68. Fay, F.S., Shlevin, H.H., Granger, W.C.J., and Taylor, S.R. (1979) Aequorin lumines-cence during activation of single isolated smooth muscle cells. Nature, 280:506-508.

69. Ferris, C.D., Huganir, R.L., Supattapone, S., and Snyder, S.H. (1989) Purified inositol 1,4,5-trisphosphate receptor mediates calcium flux in reconstituted lipid vesicles. Nature, 342:87-89.

70. Fish, R.D., Sperti, G., Colucci, W.S., and Clapham, D.E. (1988) Phorbol ester increases the dihydropyridine-sensitive calcium conductance in a vascular smooth muscle cell line. Circulation Research, 62:1049-1054.

71. Fleischmann, B.K., Murray, R.K., and Kotlikoff, M.I. (1994) Voltage window for sustained elevation of cytosolic calcium in smooth muscle cells. Proceedings of the National Academy of Sciences of the United States of America, 91:11914-11918.

72. Fleischmann, B.K., Wang, Y.X., and Kotlikoff, M.I. (1997) Muscarinic activation and calcium permeation of nonselective cation currents in airway myocytes. American Journal of Physiology, 272:C341-C349

73. Fleischmann, B.K., Wang, Y.X., Pring, M., and Kotlikoff, M.I. (1996) Voltage-de-pendent calcium currents and cytosolic calcium in equine airway myocytes. Journal of Physiology (London), 492:347-358.

74. Fukumitsu, T., Hayashi, H., Tokuno, H., and Tomita, T. (1990) Increase in calcium channel current by -adrenoceptor agonists in single smooth muscle cells isolated from porcine coronary artery. British Journal of Pharmacology, 100:593-599.

75. Furuichi, T., Yoshikawa, S., Miyawaki, A., Wada, K., Maeda, N., and Mikoshiba, K. (1989) Primary structure and functional expression of the inositol 1,4,5-trisphos-phate-binding protein P400. Nature, 342:32-38.

76. Ganitkevich, V.Y. and Hirche, H. (1996) High cytoplasmic Ca2+ levels reached during Ca^{2+}-induced Ca^{2+} release in single smooth muscle cell as reported by a low affinity Ca^{2+} indicator Mag-Indo-1. Cell Calcium, 19:391-398.

77. Ganitkevich, V.Y. and Isenberg. (1990) Contribution of two types of calcium channels to membrane conductance of single myocytes from guinea-pig coronary artery. Journal of Physiology (London), 426:19-42.

78. Ganitkevich, V.Y. and Isenberg, G. (1992) Contribution of Ca(2+)-induced Ca2+ release to the $[Ca^{2+}]_i$ transients in myocytes from guinea-pig urinary bladder. Journal of Physiology (London), 458:119-137.

79. Ganitkevich, V.Y. and Isenberg, G. (1991) Depolarization-mediated intracellular calcium transients in isolated smooth muscle cells of guinea-pig urinary bladder. Journal of Physiology (London), 435:187-205.

80. Garcia-Guzman, M., Soto, F., Laube, B., and Stuhmer, W. (1996) Molecular cloning and functional expression of a novel rat heart P2X purinoceptor. FEBS Letters, 388:123-127.

81. Giannini, G., Conti, A., Mammarella, S., Scrobogna, M., and Sorrentino, V. (1995) The ryanodine receptor/calcium channel genes are widely and differentially expressed in murine brain and peripheral tissues. Journal of Cell Biology, *128*:893-904.

82. Gollasch, M., Haase, H., Ried, C., Lindschau, C., Miethke, A., Morano, I., Luft, F.C., and Haller, H. (1997) Expression of L-type calcium channels depends on the differentiated state of vascular smooth muscle cells. FASEB J., (In Press)

83. Gollasch, M., Hescheler, J., Quayle, J.M., Patlak, J.B., Nelson, and MT. (1992) Single calcium channel currents of arterial smooth muscle at physiological calcium concentrations. American Journal of Physiology, *263*:C948-C952

84. Gollasch, M. and Nelson, M.T. (1998) Voltage-dependent Ca^{2+} channels in arterial smooth muscle. Kidney and Blood Pressure Research, *20*:355-371.

85. Gollasch, M., Ried, C., Liebold, M., Haller, H., Hofmann, F., and Luft, F.C. (1996) High permeation of L-type Ca^{2+} channels at physiological [Ca^{2+}]: homogeneity and dependence on the 1-subunit. American Journal of Physiology, *271*:C842-C850

86. Gomez, A.M., Valdivia, H.H., Cheng, H., Lederer, M.R., Santana, L.F., Cannell, M.B., McCune, S.A., Altschuld, R.A., and Lederer, W.J. (1997) Defective excitation-contraction coupling in experimental cardiac hypertrophy and heart failure. Science, *276*:800-806.

87. Gregoire, G., Loirand, G., and Pacaud, P. (1993) Ca2+ and Sr2+ entry induced Ca2+ release from the intracellular Ca2+ store in smooth muscle cells of rat portal vein. Journal of Physiology (London), *472*:483-500.

88. Grynkiewicz, G., Poenie, M., and Tsien, R.Y. (1985) A new generation of Ca^{2+} indicators with greatly improved fluorescence properties. J.Biol.Chem., *260*:3440-3450.

89. Guerrero, A., Fay, F.S., and Singer, J.J. (1994a) Caffeine activates a Ca^{2+}-permeable, nonselective cation channel in smooth muscle cells. Journal of General Physiology, *104*:375-394.

90. Guerrero, A., Singer, J.J., and Fay, F.S. (1994b) Simultaneous measurement of Ca^{2+} release and influx into smooth muscle cells in response to caffeine. A novel approach for calculating the fraction of current carried by calcium. Journal of General Physiology, *104*:395-422.

91. Guy, H.R. and Conti, F. (1990) Pursuing the structure and function of voltage-gated channels. Trends in Neurosciences, *13*:201-206.

92. Hamill, O.P., Marty, A., Neher, E., Sakmann, B., and Sigworth, F.J. (1981) Improved patch-clamp techniques for high-resolution current recording from cells and cell-free membrane patches. Pflugers Arch., *391*:85-100.

93. Harder, D.R. (1984) Pressure-dependent membrane depolarization in cat middle cerebral artery. Circulation Research, *55*:197-202.

94. Harder, D.R., Brann, L., and Halpern, W. (1983) Altered membrane electrical properties of smooth muscle cells from small cerebral arteries of hypertensive rats. Blood Vessels, *20*:154-160.

95. Harder, D.R., Gilbert, R., and Lombard, J.H. (1987) Vascular muscle cell depolarization and activation in renal arteries on elevation of transmural pressure. American Journal of Physiology, 253:F778-F781

96. Harnick, D.J., Jayaraman, T., Ma, Y., Mulieri, P., Go, L.O., Marks, and AR. (1995) The human type 1 inositol 1,4,5-trisphosphate receptor from T lymphocytes. Structure, localization, and tyrosine phosphorylation. Journal of Biological Chemistry, 270:2833-2840.

97. Hartley, S.A. and Kozlowski, R.Z. (1997) Electrophysiological consequences of purinergic receptor stimulation in isolated rat pulmonary arterial myocytes. Circulation Research, 80:170-178.

98. Heinemann, S.H., Terlau, H., Stuhmer, W., Imoto, K., and Numa, S. (1992) Calcium channel characteristics conferred on the sodium channel by single mutations. Nature, 356:441-443.

99. Helliwell, R.M. and Large, W.A. (1997) 1-adrenoceptor activation of a non-selective cation current in rabbit portal vein by 1,2-diacyl-sn-glycerol. Journal of Physiology (London), 499:417-428.

100. Hering, S., Kleppisch, T., Timin, E.N., and Bodewei, R. (1989) Characterization of the calcium channel state transitions induced by the enantiomers of the 1,4-dihydropyridine Sandoz 202 791 in neonatal rat heart cells. A nonmodulated receptor model. Pflugers Archiv - European Journal of Physiology, 414:690-700.

101. Herlitze, S., Garcia, D.E., Mackie, K., Hille, B., Scheuer, T., and Catterall, W.A. (1996) Modulation of Ca^{2+} channels by G-protein beta gamma subunits. Nature, 380:258-262.

102. Herlitze, S., Hockerman, G.H., Scheuer, T., and Catterall, W.A. (1997) Molecular determinants of inactivation and G protein modulation in the intracellular loop connecting domains I and II of the calcium channel $1A$ subunit. Proceedings of the National Academy of Sciences of the United States of America, 94:1512-1516.

103. Hermsmeyer, K., Abel, P.W., and Trapani, A.J. (1982) Norepinephrine sensitivity and membrane potentials of caudal arterial muscle in DOCA-salt, Dahl, and SHR hypertension in rat. Hypertension, 4:49-51.

104. Herrmann-Frank, A., Darling, E., and Meissner, G. (1991) Functional characterization of the Ca^{2+}-gated Ca^{2+} release channel of vascular smooth muscle sarcoplasmic reticulum. Pflugers Archiv - European Journal of Physiology, 418:353-359.

105. Hess, P., Lansman, J.B., and Tsien, R.W. (1984) Different modes of Ca channel gating behaviour favoured by dihydropyridine Ca agonists and antagonists. Nature, 311:538-544.

106. Hille, B. (1994) Modulation of ion-channel function by G-protein-coupled receptors. Trends in Neurosciences, 17:531-536.

107. Himpens, B., Matthijs, G., Somlyo, A.V., Butler, T.M., and Somlyo, A.P. (1988) Cytoplasmic free calcium, myosin light chain phosphorylation, and force in phasic and tonic smooth muscle. Journal of General Physiology, 92:713-729.

108. Hirose, K., Iino, M., and Endo, M. (1993) Caffeine inhibits Ca(2+)-mediated potentiation of inositol 1,4,5-trisphosphate-induced Ca^{2+} release in permeabilized vascular smooth muscle cells. Biochemical & Biophysical Research Communications, 194:726-732.

109. Hockerman, G.H., Peterson, B.Z., Johnson, B.D., and Catterall, W.A. (1997) Molecular determinants of drug binding and action on L-type calcium channels. Annual Review of Pharmacology & Toxicology, 37:361-396.

110. Hofmann, F., Biel, M., and Flockerzi, V. (1994) Molecular basis for Ca^{2+} channel diversity. Annual Review of Neuroscience, 17:399-418.

111. Hofmann, F., Flockerzi, V., Nastainczyk, W., Ruth, P., and Schneider, T. (1990) The molecular structure and regulation of muscular calcium channels. Current Topics in Cellular Regulation, 31:223-239.

112. Holz, G.G., Kream, R.M., Spiegel, A., and Dunlap. (1989) G proteins couple alpha-adrenergic and GABAb receptors to inhibition of peptide secretion from peripheral sensory neurons. Journal of Neuroscience, 9:657-666.

113. Horn, R. and Marty, A. (1988) Muscarinic activation of ionic currents measured by a new whole-cell recording method. Journal of General Physiology, 92:145-159.

114. Huang, C.-L., Slesinger, P.A., Casey, P.J., Jan, Y.N., and Jan, L.Y. (1995) Evidence that direct binding of G to the GIRK1 G protein-gated inwardly rectifying K+ channel is important for channel activation. Neuron, 15:1133-43X.

115. Hughes, A.D., Parkinson, N.A., and Wijetunge, S. (1996) alpha2-Adrenoceptor activation increases calcium channel currents in single vascular smooth muscle cells isolated from human omental resistance arteries. Journal of Vascular Research, 33:25-31.

116. Hullin, R., Singer-Lahat, D., Freichel, M., Biel, M., Dascal, N., Hofmann, F., and Flockerzi, V. (1992) Calcium channel beta subunit heterogeneity: functional expression of cloned cDNA from heart, aorta and brain. EMBO Journal, 11:885-890.

117.Hymel, L., Inui, M., Fleischer, S., and Schindler, H. (1988) Purified ryanodine receptor of skeletal muscle sarcoplasmic reticulum forms Ca^{2+}-activated oligormeric Ca^{2+} channels in planar bilayers. Proceedings of the National Academy of Sciences of the United States of America, 85:441-445.

118. Iino, M. (1987) Calcium dependent inositol trisphosphate-induced calcium release in the guinea-pig taenia caeci. Biochemical & Biophysical Research Communications, 142:47-52.

119. Iino, M. (1989a) Calcium-induced calcium release mechanism in guinea pig taenia caeci. Journal of General Physiology, 94:363-383.

120. Iino, M. (1989b) Calcium-induced calcium release mechanism in guinea pig taenia coli. Journal of Physiology (London), 94 :363-383.

121. Iino, M. (1990a) Biphasic Ca^{2+} dependence of inositol 1,4,5-trisphosphate-induced Ca release in smooth muscle cells of the guinea pig taenia caeci. Journal of General Physiology, 95:1103-1122.

121. Iino, M. (1990b) Calcium release mechanisms in smooth muscle. Japanese Journal of Pharmacology, 54:345-354.

122. Iino, M. and Endo, M. (1992) Calcium-dependent immediate feedback control of inositol 1,4,5-triphosphate-induced Ca^{2+} release. Nature, 360:76-78.

123. Iino, M., Kobayashi, T., and Endo, M. (1988) Use of ryanodine for functional removal of the calcium store in smooth muscle cells of the guinea-pig. Biochemical & Biophysical Research Communications, 152:417-422.

124. Inoue, R. and Isenberg, G. (1990a) Acetylcholine activates nonselective cation channels in guinea pig ileum through a G protein. Am.J.Physiol.(Cell Physiol.), 258:C1173-C1178

125. Inoue, R. and Isenberg, G. (1990b) Effect of membrane potential on acetylcholine-induced inward current in guinea-pig ileum. Journal of Physiology (London), 424:57-71.

126. Inoue, R. and Isenberg, G. (1990c) Intracellular calcium ions modulate acetylcholine-induced inward current in guinea-pig ileum. Journal of Physiology (London), 424:73-92.

127. Inoue, R., Kitamura, K., and Kuriyama, H. (1987a) Acetylcholine activates single sodium channels in smooth muscle cells. Pflugers Arch., 410:69-74.

128. Inoue, R., Kitamura, K., and Kuriyama, H. (1987b) Acetylcholine activates single sodium channels in smooth muscle cells. Pflugers Archiv - European Journal of Physiology, 410:69-74.

129. Inoue, R. and Kuriyama, H. (1993) Dual regulation of cation-selective channels by muscarinic and alpha 1-adrenergic receptors in the rabbit portal vein. Journal of Physiology, 465:427-448.

130. Inoue, Y., Oike, M., Nakao, K., Kitamura, K., and Kuriyama, H. (1990) Endothelin augments unitary calcium channel currents on the smooth muscle cell membrane of guinea-pig portal vein. Journal of Physiology, 423:171-191.

131. Ishikawa, T., Hume, J.R., and Keef, K.D. (1993) Regulation of Ca^{2+} channels by cAMP and cGMP in vascular smooth muscle cells. Circulation Research, 73:1128-1137.

132. Itagaki, K., Koch, W.J., Bodi, I., Klockner, U., Slish, D.F., and Schwartz, A. (1992) Native-type DHP-sensitive calcium channel currents are produced by cloned rat aortic smooth muscle and cardiac 1 subunits expressed in Xenopus laevis oocytes and are regulated by 2- and -subunits. FEBS Letters, 297:221-225.

133. Ito, K., Ikemoto, T., and Takakura, S. (1991) Involvement of Ca^{2+} influx-induced Ca^{2+} release in contractions of intact vascular smooth muscles. American Journal of Physiology, 261:H1464-H1470

134. Ito, S., Ohta, T., and Nakazato, Y. (1993) Inward current activated by carbachol in rat intestinal smooth muscle cells. Journal of Physiology (London), 470:395-409.

135. Janssen, L.J. and Sims, S.M. (1992) Acetylcholine activates non-selective cation and chloride conductances in canine and guinea-pig tracheal smooth muscle cells. Journal of Physiology (London), 453:197-218.

136. Janssen, L.J. and Sims, S.M. (1993) Histamine activates Cl^- and K^+ currents in guinea-pig tracheal myocytes: convergence with muscarinic signalling pathway. Journal of Physiology (London), 465:661-677.

137. Jay, S.D., Sharp, A.H., Kahl, S.D., Vedvick, T.S., Harpold, M.M., and Campbell, K.P. (1991) Structural characterization of the dihydropyridine-sensitive calcium channel 2-subunit and the associated peptides. Journal of Biological Chemistry, 266:3287-3293.

138. Jayaraman, T., Brillantes, A.M., Timerman, A.P., Fleischer, S., Erdjumentbromage, H., Tempst, P., and Marks, A.R. (1992) FK506 binding protein associated with the calcium release channel (ryanodine receptor). Journal of Biological Chemistry, 267:9474-9477.

139. Jiang, H. and Stephens, N.L. (1994) Calcium and smooth muscle contraction. Molecular & Cellular Biochemistry, 135:1-9.

140. Joseph, S.K. (1996) The inositol triphosphate receptor family. Cellular Signalling, 8:1-7.

141. Joseph, S.K., Lin, C., Pierson, S., Thomas, A.P., and Maranto, A.R. (1995) Heteroligomers of type-I and type-III inositol trisphosphate receptors in WB rat liver epithelial cells. Journal of Biological Chemistry, 270:23310-23316.

142. Joseph, S.K. and Ryan, S.V. (1993) Phosphorylation of the inositol trisphosphate receptor in isolated rat hepatocytes. Journal of Biological Chemistry, 268:23059-23065.

143. Kamishima, T. and McCarron, J.G. (1996) Depolarization-evoked increases in cytosolic calcium concentration in isolated smooth muscle cells of rat portal vein. Journal of Physiology (London), 492:61-74.

144. Karaki, H., Ozaki, H., Hori, M., Mitsui-Saito, M., Amano, K., Harada, K., Miyamoto, Nakazawa, H., Won, K.J., and Sato, K. (1997) Calcium movements, distribution, and functions in smooth muscle. Pharmacological Reviews, 49:157-230.

145. Khakh, B.S., Surprenant, A., and Humphrey, P.P. (1995) A study on P2X purinoceptors mediating the electrophysiological and contractile effects of purine nucleotides in rat vas deferens. British Journal of Pharmacology, 115:177-185.

146. Kirber, M.T., Walsh, J.V.J., and Singer, J.J. (1988) Stretch-activated ion channels in smooth muscle: a mechanism for the initiation of stretch-induced contraction. Pflugers Archiv - European Journal of Physiology, 412:339-345.

147. Kitamura, K., Xiong, Z., Teramoto, N., and Kuriyama, H. (1992) Roles of inositol trisphosphate and protein kinase C in the spontaneous outward current modulated by calcium release in rabbit portal vein. Pflugers Archiv - European Journal of Physiology, 421:539-551.

148. Kitazawa, T., Kobayashi, S., Horiuti, K., Somlyo, A.V., and Somlyo, A.P. (1989) Receptor coupled, permeabilized smooth muscle: role of the phosphatidylinositol cascade, G-proteins and modulation of the contractile response to Ca^{2+}. Journal of Biological Chemistry, 264:5339-5342.

149. Klockner, U. and Isenberg, G. (1985) Calcium currents of cesium loaded isolated smooth muscle cells (urinary bladder of the guinea pig). Pflugers Archiv - European Journal of Physiology, 405:340-348.

150. Klockner, U. and Isenberg, G. (1991) Myocytes isolated from porcine coronary arteries: reduction of currents through L-type Ca-channels by verapamil-type Ca-antagonists. Journal of Physiology & Pharmacology, 42:163-179.

151. Knot, H.J. and Nelson, M.T. (1995) Regulation of membrane potential and diameter by voltage-dependent K^+ channels in rabbit myogenic cerebral arteries. American Journal of Physiology, 269:H348-H355

152. Knot, H.J. and Nelson, M.T. (1998a) Regulation of arterial diameter and calcium in cerebral arteries of rat by membrane potential and intravascular pressure. Journal of Physiology, (In Press)

153. Knot, H.J., Standen, N.B., and Nelson, M.T. (1998b) Ryanodine receptors regulate arterial [Ca^{2+}] and diameter in cerebral arteries of rat via KCa channels. Journal of Physiology (London), (In Press)

154. Kobayashi, S., Kitazawa, T., Somlyo, A.V., and Somlyo, A.P. (1989) Cytosolic heparin inhibits muscarinic and alpha-adrenergic Ca^{2+} release in smooth muscle. Journal of Biological Chemistry, 264:17997-18004.

155. Komalavilas, P. and Lincoln, T.M. (1994) Phosphorylation of the inositol 1,4,5-trisphosphate receptor by cyclic GMP-dependent protein kinase. Journal of Biological Chemistry, 269:8701-8707.

156. Komori, S. and Bolton, T.B. (1990) Role of G-proteins in muscarinic receptor inward and outward currents in rabbit jejunal smooth muscle. Journal of Physiology (London), 427:395-419.

157. Komori, S. and Bolton, T.B. (1991) Calcium release induced by inositol 1,4,5-trisphosphate in single rabbit intestinal smooth muscle cells. Journal of Physiology (London), 433:495-517.

158. Komori, S., Kawai, M., Takewaki, T., and Ohashi, H. (1992) GTP-binding protein involvement in membrane currents evoked by carbachol and histamine in guinea-pig ileal muscle. Journal of Physiology (London), 450:105-126.

159. Korn, S.J. and Horn, R. (1989) Influence of sodium-calcium exchange on calcium current rundown and the duration of calcium-dependent chloride currents in pituitary cells, studied with whole cell and perforated patch recording. Journal of General Physiology, 94:789-812.

160. Kume, H., Mikawa, K., Takagi, K., and Kotlikoff, M.I. (1995) Role of G proteins and K_{Ca} channels in the muscarinic and -adrenergic regulation of tracheal smooth muscle. American Journal of Physiology: Lung Cell.and Mol.Physiol., 268 :L221-L229

161. Lamb, F.S. and Webb, R.C. (1989) Potassium conductance and oscillatory contractions in tail arteries from genetically hypertensive rats. Journal of Hypertension, 7:457-463.

162. Lee, H.K., Shuttleworth, C.W., and Sanders, K.M. (1995) Tachykinins activate nonselective cation currents in canine colonic myocytes. American Journal of Physiology (Cell Physiol.), 269:C1394-401.

163. Lee, K.S. and Tsien, R.W. (1983) Mechanism of calcium channel blockade by verapamil, D600, diltiazem and nitrendipine in single dialysed heart cells. Nature, 302:790-794.

164. Lepretre, N. and Mironneau, J. (1994) 2-adrenoceptors activate dihydropyridine-sensitive calcium channels via G_i-proteins and protein kinase C in rat portal vein myocytes. Pflugers Archiv - European Journal of Physiology, 429:253-261.

165. Lepretre, N., Mironneau, J., and Morel, J.L. (1994b) Both 1A- and 2A-adrenoreceptor subtypes stimulate voltage-operated L-type calcium channels in rat portal vein myocytes. Evidence for two distinct transduction pathways. Journal of Biological Chemistry, 269:29546-29552.

166. Lewis, C., Neidhart, S., Holy, C., North, R.A., Buell, G., and Surprenant, A. (1995) Coexpression of P2X2 and P2X3 receptor subunits can account for ATP-gated currents in sensory neurons. Nature, 377:432-435.

167. Liu, Q.Y., Lai, F.A., Rousseau, E., Jones, R.V., and Meissner, G. (1989) Multiple conductance states of the purified calcium release channel complex from skeletal sarcoplasmic reticulum. Biophysical Journal, 55:415-424.

168. Loirand, G., Faiderve, S., Baron, A., Geffard, M., and Mironneau, J. (1992) Autoan-tiphosphatidylinositide antibodies specifically ihibit noradrenaline effects on Ca^{2+} and Cl^- channels in rat portal vein myocytes. The Journal of Biological Chemistry, 267:4312-4316.

169. Loirand, G., Pacaud, P., Baron, A., Mironneau, C., and Mironneau, J. (1991) Large conductance calcium-activated non-selective cation channel in smooth muscle cells isolated from rat portal vein. Journal of Physiology (London), 437:461-475.

170. Lopez-Lopez, J.R., Shacklock, P.S., Balke, C.W., and Wier, W.G. (1995) Local calcium transients triggered by single L-type calcium channel currents in cardiac cells. Science, 268:1042-1045.

171. Maeda, N., Kawasaki, T., Nakade, S., Yokota, N., Taguchi, T., Kasai, M., and Mikoshiba, K. (1991) Structural and functional characterization of inositol 1,4,5-trisphosphate receptor channel from mouse cerebellum. Journal of Biological Chemistry, 266:1109-1116.

172. Marks, A.R., Fleischer, S., and Tempst, P. (1990) Surface topography analysis of the ryanodine receptor/junctional channel complex based on proteolysis sensitivity mapping. Journal of Biological Chemistry, 265:13143-13149.

173. Martens, J.R. and Gelband, C.H. (1996) Alterations in rat interlobar artery mem-brane potential and K^+ channels in genetic and nongenetic hypertension. Circula-tion Research, 79:295-301.

174. Martone, M.E., Alba, S.A., Edelman, V.M., Airey, J.A., and Ellisman, M.H. (1997) Distribution of inositol-1,4,5-trisphosphate and ryanodine receptors in rat neostriatum. Brain Research, 756:9-21.

175. McDonald, T.F., Pelzer, S., Trautwein, W., and Pelzer, D.J. (1994) Regulation and modulation of calcium channels in cardiac, skeletal, and smooth muscle cells. Physiological Reviews, 74:365-507.

176. Meissner G. (1994) Ryanodine receptor/Ca^{2+} release channels and their regulation by endogenous effectors. Annual Review of Physiology, 56:485-508.

177. Mignery, G.A., Newton, C.L., Archer, B.T., and Sudhof, T.C. (1990a) Structure and expression of the rat inositol 1,4,5-trisphosphate receptor. Journal of Biological Chemistry, 265:12679-12685.

178. Mignery, G.A. and Sudhof, T.C. (1990b) The ligand binding site and transduction mechanism in the inositol-1,4,5-triphosphate receptor. EMBO Journal, 9:3893-3898.

179. Mikala, G., Bahinski, A., Yatani, A., Tang, S., and Schwartz, A. (1993) Differential contribution by conserved glutamate residues to an ion-selectivity site in the L-type Ca^{2+} channel pore. FEBS Letters, 335:265-269.

180. Mikami, A., Imoto, K., Tanabe, T., Niidome, T., Mori, Y., Takeshima, H., Narumiya, and Numa, S. (1989) Primary structure and functional expression of the cardiac dihydropyridine-sensitive calcium channel. Nature, 340:230-233.

181. Mikoshiba, K. (1997) The InsP3 receptor and intracellular Ca^{2+} signaling. Current Opinion in Neurobiology, 7:339-345.

182. Minami, K., Fukuzawa, K., and Inoue, I. (1994) Regulation of a non-selective cation channel of cultured porcine coronary artery smooth muscle cells by tyrosine kinase. Pflugers Archiv - European Journal of Physiology, 426:254-257.

183. Missiaen, L., De, S.H., Droogmans, G., and Casteels, R. (1992a) Luminal Ca^{2+} controls the activation of the inositol 1,4,5-trisphosphate receptor by cytosolic Ca^{2+}. Journal of Biological Chemistry, 267:22961-22966.

184. Missiaen, L., De, S.H., Droogmans, G., Himpens, B., and Casteels. (1992b) Calcium ion homeostasis in smooth muscle. Pharmacology & Therapeutics, 56:191-231.

185. Mitterdorfer, J., Wang, Z., Sinnegger, M.J., Hering, S., Striessnig, J., Grabner, M., and Glossmann, H. (1996) Two amino acid residues in the IIIS5 segment of L-type calcium channels differentially contribute to 1,4-dihydropyridine sensitivity. Journal of Biological Chemistry, 271:30330-30335.

186. Miyawaki, A., Furuichi, T., Ryou, Y., Yoshikawa, S., Nakagawa, T., Saitoh, T., and Mikoshiba, K. (1991) Structure-function relationships of the mouse inositol 1,4,5-trisphosphate receptor. Proceedings of the National Academy of Sciences of the United States of America, 88:4911-4915.

187. Monkawa, T., Miyawaki, A., Sugiyama, T., Yoneshima, H., Yamamoto-Hino, M., Furuichi, T., Saruta, T., Hasegawa, M., and Mikoshiba, K. (1995) Heterotetrameric complex formation of inositol 1,4,5-trisphosphate receptor subunits. Journal of Biological Chemistry, 270:14700-14704.

188. Morgan, J.M., De, S.H., and Gillespie, J.I. (1996) Identification of three isoforms of the InsP3 receptor in human myometrial smooth muscle. Pflugers Archiv - European Journal of Physiology, 431:697-705.

189. Morgan, J.P. and Morgan, K.G. (1982) Vascular smooth muscle: the first recorded Ca^{2+} transients. Pflugers Arch., 395:75-77.

190. Mori, Y., Mikala, G., Varadi, G., Kobayashi, T., Koch, S., Wakamori, M., and Schwartz. (1996) Molecular pharmacology of voltage-dependent calcium channels. Japanese Journal of Pharmacology, 72:83-109.

191. Nakade, S., Rhee, S.K., Hamanaka, H., and Mikoshiba, K. (1994) Cyclic AMP-de-pendent phosphorylation of an immunoaffinity-purified homotetrameric inositol 1,4,5-trisphosphate receptor (type I) increases Ca^{2+} flux in reconstituted lipid vesicles. Journal of Biological Chemistry, 269:6735-6742.

192. Neely, A., Wei, X., Olcese, R., Birnbaumer, L., and Stefani, E. (1993) Potentiation by the beta subunit of the ratio of the ionic current to the charge movement in the cardiac calcium channel. Science, 262:575-578.

193. Neher, E. and Augustine, G.J. (1992) Calcium gradients and buffers in bovine chromaffin cells. Journal of Physiology, 450:273-301.

194. Neher, E. and Sakmann, B. (1976) Single-channel currents recorded from mem-brane of denervated frog muscle fibres. Nature, 260:799-802.

195. Nelson, M.T., Chang, H., Rubart, M., Santana, L.F., Bonev, A.D., Knot, H.J., and Lederer, W.J. (1995) Relaxation of arterial smooth muscle by calcium sparks. Sci-ence, 270:633-637.

196. Nelson, M.T., Patlak, J.B., Worley, J.F., and Standen, N.B. (1990) Calcium channels, potassium channels, and voltage dependence of arterial smooth muscle tone. American Journal of Physiology, 259:C3-18.

197. Nelson, M.T. and Quayle, J.M. (1995) Physiological roles and properties of potas-sium channels in arterial smooth muscle. American Journal of Physiology, 268:C799-C822

198. Nelson, M.T., Standen, N.B., Brayden, J.E., and Worley, J.F. (1988) Noradrenaline contracts arteries by activating voltage-dependent calcium channels. Nature, 336:382-385.

199. Nelson, M.T. and Worley, J.F. (1989) Dihydropyridine inhibition of single calcium channels and contraction in rabbit mesenteric artery depends on voltage. Journal of Physiology (London), 412:65-91.

200. Neylon, C.B., Richards, S.M., Larsen, M.A., Agrotis, A., and Bobik, A. (1995) Multiple types of ryanodine receptor/Ca^{++} release channels are expressed in vascu-lar smooth muscle. Biochem.Biophys.Res.Commun., 215:814-821.

201. North, R.A. (1996) P2X receptors: a third major class of ligand-gated ion channels. Ciba.Foundation.Symposium., 198:91-105.

202. Ohya, Y., Abe, I., Fujii, K., and Fujishima, M. (1996) Membrane channels in smooth muscle cells from hypertensive rats. In: Smooth Muscle Excitation. T.B. Bolton and T. Tomita, eds. Academic Press Limited, London, pp. 1-12.

203. Ohya, Y. and Sperelakis, N. (1989) ATP regulation of the slow calcium channels in vascular smooth muscle cells of guinea pig mesenteric artery. Circulation Research, 64:145-154.

204. Oike, M., Kitamura, K., and Kuriyama, H. (1993) Protein kinase C activates the non-selective cation channel in the rabbit portal vein. Pflugers Archiv - European Journal of Physiology, 424 :159-164.

205. Pacaud, P. and Bolton, T.B. (1991) Relation between muscarinic receptor cationic current and internal calcium in guinea-pig jejunal smooth muscle cells. Journal of Physiology (London), 441:477-499.

206. Pacaud, P., Loirand, G., Bolton, T.B., Mironneau, C., and Mironneau. (1992) Intracellular cations modulate noradrenaline-stimulated calcium entry into smooth muscle cells of rat portal vein. Journal of Physiology (London), 456:541-556.

207. Parker, I., Choi, J., and Yao, Y. (1996) Elementary events of InsP3-induced Ca_{2+} liberation in Xenopus oocytes: hot spots, puffs and blips. Cell Calcium, 20:105-121.

208. Parys, J.B., Missiaen, L., Smedt, H.D., Sienaert, I., and Casteels, R. (1996) Mechanisms responsible for quantal Ca^{2+} release from inositol trisphosphate-sensitive calcium stores. Pflugers Archiv - European Journal of Physiology, 432:359-367.

209. Porter, V.A., Bonev, A., Kleppisch, T., Lederer, W.J., and Nelson, M.T. (1997) cAMP/PKA activates Ca^{2+} sparks and K_{Ca} channels in cerebral artery myocytes. Biophys Journal, 72:A170 (Abstract).

210. Quayle, J.M., McCarron, J.G., Asbury, J.R., and Nelson, M.T. (1993) Single calcium channels in resistance-sized cerebral arteries from rats. American Journal of Physiology, 264:H470-H478

211. Quignard, J.F., Frapier, J.M., Harricane, M.C., Albat, B., Nargeot, J., and Richard, S. (1997) Voltage-gated calcium channel currents in human coronary myocytes. Regulation by cyclic GMP and nitric oxide. Journal of Clinical Investigation, 99:185-193.

212. Rubart, M., Patlak, J.B., and Nelson, M.T. (1996) Ca^{2+} currents in cerebral artery smooth muscle cells of rat at physiological Ca^{2+} concentrations. Journal of General Physiology, 107:459-472.

213. Santana, L.F., Etter, E.F., Tuft, R.A., Fogarty, K., Fay, F.S., and Lederer, W.J. (1997) Imaging of Ca2+ sparks using a super fast ultrasensitive CCD camera. Biophys Journal, 72:A44 (Abstract)

214. Sharp, A.H., McPherson, P.S., Dawson, T.M., Aoki, C., Campbell, K.P., and Snyder, S.H. (1993) Differential immunohistochemical localization of inositol 1,4,5-trisphosphate- and ryanodine-sensitive Ca^{2+} release channels in rat brain. Journal of Neuroscience, 13:3051-3063.

215. Shistik, E., Ivanina, T., Puri, T., Hosey, M., and Dascal, N. (1995) Ca2+ current enhancement by 2/ and subunits in Xenopus oocytes: contribution of changes in channel gating and alpha 1 protein level. Journal of Physiology (London), 489:55-62.

216. Silva, E.G., Vianna, L.M., Okuyama, P., and Paiva, T.B. (1996) Effect of treatment with cholecalciferol on the membrane potential and contractility of aortae from spontaneously hypertensive rats. British Journal of Pharmacology, 118:1367-1370.

217. Smirnov, S.V. and Aaronson, P.I. (1992) Ca^{2+} currents in single myocytes from human mesenteric arteries: evidence for a physiological role of L-type channels. Journal of Physiology (London), 457:455-475.

218. Smith, J.S., Imagawa, T., Ma, J., Fill, M., Campbell, K.P., and Coronado, R. (1988) Purified ryanodine receptor from rabbit skeletal muscle is the calcium-release channel of sarcoplasmic reticulum. Journal of General Physiology, 92:1-26.

219. Snutch, T.P. and Reiner, P.B. (1992) Ca^{2+} channels: diversity of form and function. Current Opinion in Neurobiology, 2:247-253.

220. Somlyo, A.P. (1985) Excitation-contraction coupling and the ultrastructure of smooth muscle. Circ.Res., 57:467-507.

221. Somlyo, A.P. and Himpens, B. (1989) Cell calcium and its regulation in smooth muscle. FASEB J., 3:2266-2276.

222. Somlyo, A.P. and Somlyo, A.V. (1994) Signal transduction and regulation in smooth muscle [published erratum appears in Nature 1994 Dec 22-29;372(6508):812]. [Review] [108 refs]. Nature, 372:231-236.

223. Somlyo, A.V., Bond, M., Somlyo, A.P., and Scarpa, A. (1985) Inositol trisphosphate-induced calcium release and contraction in vascular smooth muscle. Proc.Natl.Acad.Sci.USA, 82:5231-5235.

224. Soto, F., Garcia-Guzman, M., Gomez-Hernandez, J.M., Hollmann, M., Karschin, C., and Stühmer, W. (1996) P2X4: An ATP-activated ionotropic receptor cloned from rat brain. Proc.Natl.Acad.Sci.USA, 93:3684-3688.

225. Striggow F. and Ehrlich, B.E. (1996) Ligand-gated calcium channels inside and out. Current Opinion in Cell Biology. 8:490-495.

226. Suh-Kim, H., Wei, X., and Birnbaumer, L. (1996) Subunit composition is a major determinant in high affinity binding of a Ca^{2+} channel blocker. Molecular Pharmacology, 50:1330-1337.

227. Supattapone, S., Danoff, S.K., Theibert, A., Joseph, S.K., Steiner, and Snyder, S.H. (1988) Cyclic AMP-dependent phosphorylation of a brain inositol trisphosphate receptor decreases its release of calcium. Proceedings of the National Academy of Sciences of the United States of America, 85:8747-8750.

228. Surprenant, A., Buell, G., and North, R.A. (1995) P2X receptors bring new structure to ligand-gated ion channels. Trends in Neurosciences, 18:224-229.

229. Tang, S., Mikala, G., Bahinski, A., Yatani, A., Varadi, G., and Schwartz, A. (1993) Molecular localization of ion selectivity sites within the pore of a human L-type cardiac calcium channel. Journal of Biological Chemistry, 268:13026-13029.

230. Teramoto, N., Ogata, R., Okabe, K., Kameyama, A., Kameyama, M., Watanabe, T.X., Kuriyama, H., and Kitamura, K. (1996) Effects of calciseptine on unitary barium channel currents in guinea-pig portal vein. Pflugers Archiv - European Journal of Physiology, 432:462-470.

231. Tewari, K. and Simard, J.M. (1994) Protein kinase A increases availability of calcium channels in smooth muscle cells from guinea pig basilar artery. Pflugers Archiv - European Journal of Physiology, 428:9-16.

232. Tewari, K. and Simard, J.M. (1997) Sodium nitroprusside and cGMP decrease Ca2+ channel availability in basilar artery smooth muscle cells. Pflugers Archiv - European Journal of Physiology, 433:304-311.

233. Tinker, A., Lindsay, A.R., and Williams, A.J. (1993) Cation conduction in the calcium release channel of the cardiac sarcoplasmic reticulum under physiological and pathophysiological conditions. Cardiovascular Research, 27:1820-1825.

234. Trezise, D.J., Michel, A.D., Grahames, C.B., Khakh, B.S., Surprenant, and Humphrey, P.P. (1995) The selective P2X purinoceptor agonist, beta,gamma-methylene-L-adenosine 5'-triphosphate, discriminates between smooth muscle and neuronal

P2X purinoceptors. Naunyn-Schmiedebergs Archives of Pharmacology, *351*:603-609.

235. Tsien, R.Y., Rink, T.J., and Poenie, M. (1985) Measurement of cytosolic free Ca2+ in individual small cells using fluorescence microscopy with dual excitation wavelengths. Cell Calcium, *6*:145-157.

236. Tsugorka, A., Rios, E., and Blatter, L.A. (1995) Imaging elementary events of calcium release in skeletal muscle cells. Science, *269*:1723-1726.

237. Tung, L. and Morad, M. (1983) Voltage- and frequency-dependent block of diltiazem on the slow inward current and generation of tension in frog ventricular muscle. Pflugers Archiv - European Journal of Physiology, *398*:189-198.

238. Uehara, A. and Hume, J.R. (1985) Interactions of organic calcium channel antagonists with calcium channels in single frog atrial cells. Journal of General Physiology, *85*:621-647.

239. Valera, S., Hussy, N., Evans, R.J., Adami, N., North, A., Surprenant, A., and Buell, G. (1994a) A new class of ligand-gated ion channel defined by P2x receptor for extrcellular ATP. Nature, *371*:516-523.

240. Valera, S., Hussy, N., Evans, R.J., Adami, N., North, R.A., Surprenant, A., and Buell, G. (1994b) A new class of ligand-gated ion channel defined by P2x receptor for extracellular ATP. Nature, *371*:516-519.

241. van Breemen, C. and Saida, K. (1989) Cellular mechanisms regulating $[Ca^{2+}]_i$ in smooth muscle. Annual Review of Physiology, *51*:315-329.

242. Varadi, G., Mori, Y., Mikala, G., and Schwartz, A. (1995) Molecular determinants of Ca^{2+} channel function and drug action. Trends in Pharmacological Sciences, *16*:43-49.

243. Vivaudou, M.B., Clapp, L.H., Walsh, J.V.J., and Singer, J.J. (1988) Regulation of one type of Ca^{2+} current in smooth muscle cells by diacylglycerol and acetylcholine. FASEB Journal, *2*:2497-2504.

244. Vogalis, F., Publicover, N.G., Hume, J.R., and Sanders, K.M. (1991) Relationship between calcium current and cytosolic calcium in canine gastric smooth muscle cells. American Journal of Physiology, *260*:C1012-C1018

245. Vogalis, F. and Sanders, K.M. (1990) Cholinergic stimulation activates a non-selective cation current in canine pyloric circular muscle cells. Journal of Physiology (London), *429*:223-236.

246. Volpe, P. and Alderson-Lang, B.H. (1990) Regulation of inositol 1,4,5-trisphosphate-induced Ca^{2+} release. II. Effect of cAMP-dependent protein kinase. American Journal of Physiology, *258*:C1086-C1091

247. Walsh, M.P., Kargacin, G.J., Kendrick-Jones, J., and Lincoln, T.M. (1995) Intracellular mechanisms involved in the regulation of vascular smooth muscle tone. Canadian Journal of Physiology & Pharmacology, *73*:565-573.

248. Wang, Q., Hogg, R.C., and Large, W.A. (1993) A monovalent ion-selective cation current activated by noradrenaline in smooth muscle cells of rabbit ear artery. Pflugers Archiv - European Journal of Physiology, *423*:28-33.

249. Wang, Q. and Large, W.A. (1991) Noradrenaline-evoked cation conductance recorded with the nystatin whole-cell method in rabbit portal vein cells. Journal of Physiology (London), *435*:21-39.

250. Wang, R., Karpinski, E., and Pang, P.K. (1989) Two types of calcium channels in isolated smooth muscle cells from rat tail artery. American Journal of Physiology, *256*:H1361-H1368

251. Wang, Y.-X. and Kotlikoff, M.I. (1997) Muscarinic signaling pathway for calcium release and calcium-activated chloride current in smooth muscle. American Journal of Physiol, *273*:C509-C519

252. Wang, Y.X., Fleischmann, B.K., and Kotlikoff, M.I. (1997) M2 receptor activation of nonselective cation channels in smooth muscle cells: calcium and Gi/G(o) requirements. American Journal of Physiology, *273*:C500-C508

253. Wei, X., Pan, S., Lang, W., Kim, H., Schneider, T., Perez-Reyes, E., and Birnbaumer, L. (1995) Molecular determinants of cardiac Ca^{2+} channel pharmacology. Subunit requirement for the high affinity and allosteric regulation of dihydropyridine binding. Journal of Biological Chemistry, *270*:27106-27111.

254. Welling, A., Bosse, E., Cavalie, A., Bottlender, R., Ludwig, A., Nastainczyk, W., Flockerzi, V., and Hofmann, F. (1993) Stable co-expression of calcium channel 1, and 2/ subunits in a somatic cell line. Journal of Physiology (London), *471*:749-765.

255. Wellner, M.C. and Isenberg, G. (1993) Properties of stretch-activated channels in myocytes from the guinea-pig urinary bladder. Journal of Physiology (London), *466*:213-227.

256. Wellner, M.C. and Isenberg, G. (1994) Stretch effects on whole-cell currents of guinea-pig urinary bladder myocytes. Journal of Physiology (London), *480*:439-448.

257. Wilde, D.W., Furspan, P.B., and Szocik, J.F. (1994) Calcium current in smooth muscle cells from normotensive and genetically hypertensive rats. Hypertension, *24*:739-746.

258. Wojcikiewicz, R.J. and He, Y. (1995) Type I, II and III inositol 1,4,5-trisphosphate receptor co-immunoprecipitation as evidence for the existence of heterotetrameric receptor complexes. Biochemical & Biophysical Research Communications, *213*:334-341.

259. Worley, J.F., Deitmer, J.W., and Nelson, M.T. (1986) Single nisoldipine-sensitive calcium channels in smooth muscle cells isolated from rabbit mesenteric artery. Proceedings of the National Academy of Sciences of the United States of America, *83*:5746-5750.

260. Worley, J.F., Quayle, J.M., Standen, N.B., and Nelson, M.T. (1991) Regulation of single calcium channels in cerebral arteries by voltage, serotonin, and dihydropyridines. American Journal of Physiology, *261*:H1951-H1960.

261. Xu, L., Lai, F.A., Cohn, A, Etter, E, Guerrero, A., Fay, F.S., Meissner,G. (1994) Evidence for a Ca^{2+}-gated ryanodine-sensitive Ca^{2+} release channel in visceral smooth muscle. Proceedings of the National Academy of Sciences of the United States of America, 91:3294-3298.

262. Yagi, S., Becker, P.L., and Fay, F.S. (1988) Relationship between force and Ca^{2+} concentration in smooth muscle as revealed by measurements on single cells. Proceedings of the National Academy of Sciences of the United States of America, 85:4109-4113.

263. Yamamoto, H., Maeda, N., Niinobe, M., Miyamoto, E., and Mikoshiba, K. (1989) Phosphorylation of P400 protein by cyclic AMP-dependent protein kinase and Ca^{2+}/calmodulin-dependent protein kinase II. Journal of Neurochemistry, 53:917-923.

264. Yang, J., Ellinor, P.T., Sather, W.A., Zhang, J.F., and Tsien, R.W. (1993) Molecular determinants of Ca2+ selectivity and ion permeation in L-type Ca^{2+} channels. Nature, 366:158-161.

265. Yatani, A., Bahinski, A., Mikala, G., Yamamoto, S., and Schwartz, A. (1994) Single amino acid substitutions within the ion permeation pathway alter single-channel conductance of the human L-type cardiac Ca^{2+} channel. Circulation Research, 75:315-323.

266. Yoneshima, H., Miyawaki, A., Michikawa, T., Furuichi, T., and Mikoshiba, K. (1997) Ca^{2+} differentially regulates the ligand-affinity states of type 1 and type 3 inositol 1,4,5-trisphosphate receptors. Biochemical Journal, 322:591-596.

267. Zamponi, G.W., Bourinet, E., Nelson, D., Nargeot, J., and Snutch, T.P. (1997) Crosstalk between G proteins and protein kinase C mediated by the calcium channel alpha1 subunit. Nature, 385:442-446.

268. Zhou, J., Olcese, R., Qin, N., Noceti, F., Birnbaumer, L., and Stefani, E. (1997) Feedback inhibition of Ca^{2+} channels by Ca^{2+} depends on a short sequence of the C terminus that does not include the Ca^{2+}-binding function of a motif with similarity to Ca^{2+}-binding domains. Proceedings of the National Academy of Sciences of the United States of America, 94:2301-2305.

269. Ziyal, R., Ziganshin, A.U., Nickel, P., Ardanuy, U., Mutschler, E., Lambrecht, G., and Burnstock, G. (1997) Vasoconstrictor responses via P2X-receptors are selectively antagonized by NF023 in rabbit isolated aorta and saphenous artery. British Journal of Pharmacology, 120:954-960.

Pharmacomechanical coupling: the role of calcium, G-proteins, kinases and phosphatases

Andrew P. Somlyo, Xuqiong Wu, Lori A. Walker and Avril V. Somlyo

Department of Molecular Physiology and Biological Physics, University of Virginia, P.O. Box 10011, Charlottesville, Virginia 22906-0011

Contents

1
Introduction

The concept of pharmacomechanical coupling, introduced to account for membrane potential-independent contractile regulation of smooth muscle (Somlyo and Somlyo 1968a,b), was originally focused on agonist-induced changes in cytosolic $[Ca^{2+}]$. It was also recognized, however, that "some stimuli may also release a potentiator which augments the contractile force at any given free Ca^{2+} level" (Somlyo and Somlyo 1968b). With the subsequent demonstration that Ca^{2+} activates smooth muscle by promoting phosphorylation of the Ser 19 residue of the regulatory myosin light chain (MLC_{20}) by Ca-calmodulin (CaM)-dependent myosin light chain kinase (MLCK; rev. in Hartshorne 1987; Somlyo and Somlyo 1994; Gallagher et al. 1997), it became apparent that pharmacomechanical coupling could operate by both modulating $[Ca^{2+}]_i$ and regulating, independently of $[Ca^{2+}]_i$, the enzymes involved in MLC_{20} phosphorylation and dephosphorylation. Thin filament-associated proteins (i.e., calponin, caldesmon) have also been suggested to contribute to contractile regulation (Walsh 1991; Dabrowska 1994; Marston 1995; Malmqvist et al. 1997), and, at least in tonic smooth muscle containing "slow" myosin isoforms, the high affinity of crossbridges for MgADP may also contribute to force maintenance at low levels of $[Ca^{2+}]_i$ and MLC_{20} phosphorylation (Nishiye et al. 1993; Fuglsang et al. 1993; Somlyo 1993; Murphy 1994; Khromov et al. 1995, 1996; Murphy et al. 1997). We will limit this review to the well-established regulatory mechanisms of MLC_{20} phosphorylation/dephosphorylation.

Smooth muscle, like other cells, can regulate $[Ca^{2+}]_i$ by altering Ca^{2+} flux through two membrane systems: the plasma membrane and the membranes of intracellular organelles. The sarcoplasmic reticulum (SR), which forms a continuous extensive network throughout the smooth muscle cell (Fig. 1B; Nixon et al. 1995) is by far the most important organelle for regulating $[Ca^{2+}]_i$ in muscle (smooth, cardiac and skeletal), whereas in non-muscle cells the related, endoplasmic reticulum (ER) plays a similar role (rev. A.P. Somlyo 1984; A.V. Somlyo 1980; Pozzan et al. 1994; Laporte and Laher 1997). The presence of a functional SR in intact smooth muscle was first established through electron microscopy showing that Sr^{2+}, an electron-opaque "analogue" of Ca^{2+}, was accumulated by the SR (Somlyo and Somlyo 1971). Electron probe analysis further verified that the SR in smooth muscle was a physiological Ca^{2+} store from which Ca^{2+} was re-

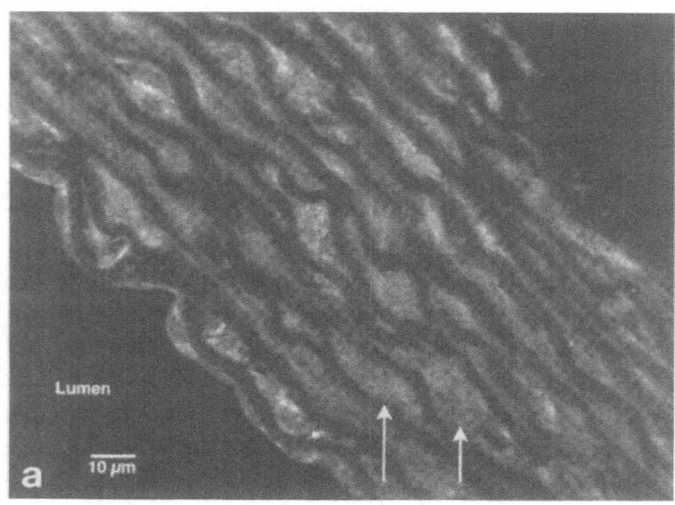

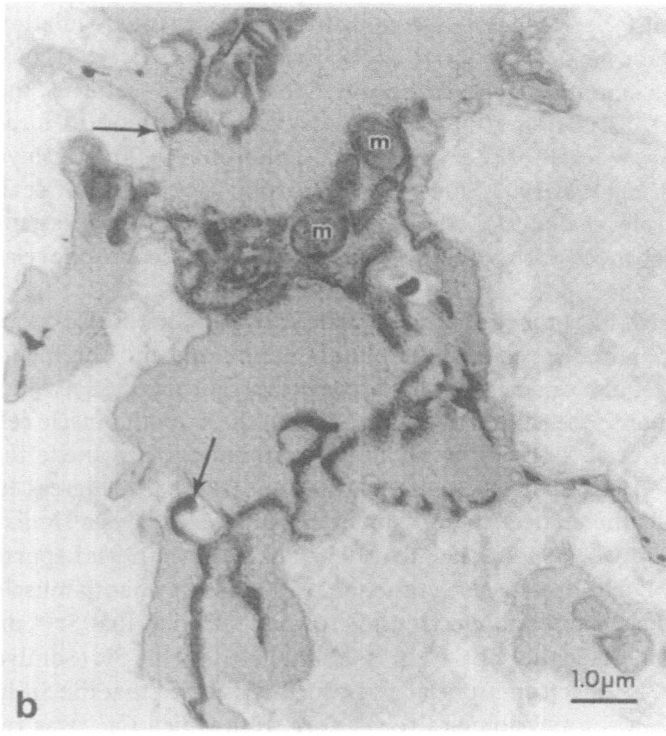

◄──

Fig. 1A. Confocal photomicrograph of guinea pig aorta labeled with a ryanodine receptor antibody and then with a TRITC-conjugated secondary antibody. Cellular non-homogeneous labeling of both endothelial and aortic smooth muscle cells and the mesh-like (arrows) staining pattern is consistent with the distribution of the sarcoplasmic reticulum network (from Lesh et al. 1998). **Fig. 1B.** Electron micrograph of a 70 nm thick section of guinea pig aorta treated with osmium ferrocyanide to selectively stain the sarcoplasmic and endoplasmic reticula showing an extensive network of reticulum throughout the cytoplasm. In stereoscopic views of thick sections (not shown) the densely stained reticulum is continuous and extends throughout the cytoplasm to the plasma membrane, forming surface couplings with the plasma membrane (arrows). Mitochondria (m) are frequently surrounded by closely apposed elements of reticulum. Note that immunolabeling of ryanodine receptors associated with this dense complex of reticulum, such as would be found in the 1μm thick cryostat sections used for immunolabeling (Fig. 1A) could give rise the the patchy, mesh-like staining seen in the confocal fluorescence images (from Lesh et al. 1998).

leased upon stimulation by excitatory agonists (Bond et al. 1984; Kowarski et al. 1985), and numerous subsequent studies of SR isolated from smooth muscle confirmed its Ca^{2+}-ATPase pump activity (rev. in Raeymakers and Wuytack 1996; Karaki et al. 1997; Laporte and Laher 1997).

The discovery that the water-soluble product of the phosphatidylinositol cascade, inositol 1,4,5-trisphosphate ($InsP_3$), released intracellular Ca^{2+} led to the identification of the ER in non-muscle cells (rev. in Berridge 1988) and the SR in smooth muscle (Somlyo et al. 1985, 1992; Walker et al. 1987; Chilvers et al. 1989; LaBelle and Murray 1990; Abdel-Latif 1991; Pijuan et al. 1993; Somlyo and Somlyo 1994) as the source of Ca^{2+} released by $InsP_3$. The physiological importance of this pharmacomechanical coupling pathway is now well established: $InsP_3$-receptors are present on the SR of smooth muscle (Marks et al. 1990; Villa et al. 1993; Nixon et al. 1995; Mackrill et al. 1997), the quantity and kinetics of $InsP_3$-induced Ca^{++}-release are appropriate for activating contraction, both agonist-induced and $InsP_3$-induced Ca^{2+} -release can be inhibited by heparin, an inhibitor acting on the $InsP_3$ receptor, and agonist-induced Ca^{2+}-release is inhibited by phospholipase C (PLC) inhibitors (rev. in Somlyo et al. 1988; Somlyo and Somlyo 1994).

The release of $InsP_3$ from phosphatidylinositol 4,5 bisphosphate (PIP_2) is initiated by the binding of excitatory agonists to heptameric serpentine receptors coupled to trimeric G-proteins ($G\alpha_q$, $G\alpha_{11}$) that activate PLC_β (smooth muscle, see LaBelle and Polyák 1996) or through (tyrosine) phosphorylation of PLC_γ. $InsP_3$-induced Ca^{2+}-release requires adenosine nucleotide (Smith et al. 1985; Somlyo et al. 1992) and can be modulated by $[Ca^{2+}]_i$ (Iino 1987; Iino and Endo 1992).

In addition to $InsP_3$ receptors, the SR of smooth muscle also contains ryanodine receptors (RyRs, Fig. 1; Lesh et al. 1998; Herrmann-Frank et al. 1991). However, the role of RyRs in pharmacomechanical or, for that matter, electromechanical coupling, has not been firmly established. It is often assumed that they respond, as in cardiac muscle, to elevations in Ca^{2+} by Ca^{2+}-induced Ca^{2+}-release (CICR), but in only a few instances has this mechanism been demonstrated in smooth muscle (Ganitkevich and Isenberg 1993a, 1995; Ganitkevich and Hirche 1996; Kohda et al. 1997). It is possible that CICR operates predominantly in the sub-sarcolemmal, junctional SR (Fig. 1B; Somlyo et al. 1971; Devine et al. 1972; Somlyo and Franzini-Armstrong, 1985) where the vicinity of plasma membrane Ca^{2+}-channels (Kargacin 1994) and Na^{2+}/Ca^{2+} exchangers (Ganitkevich and Isenberg 1993a; Juhaszova and Blaustein 1997) may permit large, focal increases in $[Ca^{2+}]_i$. Given the relatively modest physiological increases in bulk $[Ca^{2+}]_i$ in smooth muscle (rev. in Somlyo and Himpens 1989), such large, rapid transients are less likely to occur in the vicinity of the central SR that also contains RyRs (Lesh et al. 1993, 1998). It is also possible that, if the RyRs of central SR are indeed functional, then their "missing effector" is still to be discovered. Cyclic ADP ribose does not appear to have such an effect in some smooth muscles (Nixon et al. 1995), although in other cells it releases Ca^{2+} from caffeine- and ryanodine-insensitive stores (Kannan et al. 1996).

As originally proposed, the concept of pharmacomechanical coupling implied a mechanism that was independent of the changes in membrane potential that mediate electromechanical coupling. More recent studies, however, revealed cross-talk between these two mechanisms of excitation-contraction coupling (EC-coupling): the plasma membrane potential can modulate the extent of PIP_2 hydrolysis and $InsP_3$ release by agonists (Itoh et al. 1992; Ganitkevich and Isenberg 1993b), whereas arachidonic acid, an autacoid released by agonists (Gong et al. 1995), modulates (inhibits) flux through voltage-gated, L-type Ca^{2+}-channels (Shimada and Somlyo 1992; rev. in Somlyo and Somlyo 1994).

2
Ca^{2+}-sensitization: RhoA and Rho-associated proteins

The anticipation of mechanisms that can regulate smooth muscle contraction by changing Ca^{2+}-sensitivity (Somlyo and Somlyo 1968b) received firm experimental support with the arrival of reliable cytosolic Ca^{2+} indicators and methods for controlling $[Ca^{2+}]_i$ at fixed levels in permeabilized smooth muscle. Aequorin, although suitable for the detection of relatively large focal Ca^{2+} transients (van Riper et al. 1996) was the first indicator used to show that a rise in $[Ca^{2+}]_i$ preceded contraction of smooth muscle (Fay et al. 1979), but it is not sufficiently sensitive for the detection of small increases in $[Ca^{2+}]_i$, nor is it a reliable reporter of prolonged, global eleva-tions of $[Ca^{2+}]$ that consume aequorin (rev. in Somlyo and Himpens 1989; van Riper et al. 1996). The use of fura-2, a more reliable indicator of bulk cytosolic $[Ca^{2+}]_i$, verified that the force/$[Ca^{2+}]_i$ ratio can vary significantly in intact smooth muscle depending on the agonist used for activation (Himpens and Somlyo, 1988; Himpens et al. 1990; Rembold 1990), and is generally higher in response to agonists than to depolarization with high K$^+$.

The mechanism(s) through which agonists can increase force without necessarily increasing [Ca^{2+}]$_i$ is referred to as "Ca^{2+}-sensitization." It is of interest that, in the same smooth muscle, the maximal Ca^{2+}-sensitizing efficacies of different agonists are unequal (Himpens et al. 1990), suggesting that the unequal magnitude of pharmacomechanical coupling by different agonists (Somlyo and Somlyo 1968b) may be the result of both unequal changes in [Ca^{2+}]$_i$ and unequal efficiencies of Ca^{2+}-sensitization. An important corollary of the additivity of maximal Ca^{2+}-sensitization by different agonists is that the G-proteins (see below) that mediate it are either not part of a shared pool recruitable by each of the Ca^{2+}-sensitizing receptors or that the efficacy of these receptors in activating the relevant G-protein(s) is saturable.

The Ca^{2+}-sensitizing effect of activation of heptameric serpentine receptors (e.g., muscarinic, α_1-adrenergic, endothelin, thromboxane A$_2$, etc.) coupled by G-proteins to the phosphatidylinositol cascade is mimicked, in permeabilized smooth muscles, by GTPγS. These findings indicated the role of one or more G-proteins in Ca^{2+}-sensitization (Fujiwara et al. 1989; Himpens et al. 1990; Kitazawa et al. 1991; Nishimura et al. 1992; Crichton et al. 1993; Yoshida et al. 1994; Fujihara et al. 1997; Iizuka et al. 1997; rev. in Somlyo and Somlyo 1994), and raised the question of the identity of the G-proteins involved and of the downstream Ca^{2+}-sensitizing mechanism.

RhoA and Rho-associated proteins

RhoA is a member of the Rho family of Ras superfamily low molecular weight (~20-25 kD) monomeric GTP-binding proteins. The crystal structure of RhoA (Fig. 2) contains the common Ras-family fold, but the position of the Switch I region in Rho-GDP diverges significantly from that in Ras-GDP (Wei et al. 1997). Furthermore, like other Rho family proteins, RhoA contains an insert loop (residues 124-136) that is not present in Ras (Wei et al. 1997). Like other G-proteins, RhoA is active when bound to GTP and inactive in the GDP-bound form. The C-terminus of RhoA is prenylated, making the protein hydrophobic and permitting its association with the plasma membranes through the geranyl-geranyl anchor. However, in resting (relaxed) smooth muscle, as in other cells, most of RhoA is cytosolic (Figs. 2, 3); its association with the hydrophobic pocket of guanine nucleotide dissociation inhibitor (GDI; Keep et al. 1997; Gosser et al. 1997) keeps it in solution (Gong et al. 1997a; Fujihara et al. 1997). A small proportion of inactive Rho (presumably RhoA·GDP) is also associated with the plasma membrane in resting smooth muscle (Gong et al. 1997a).

The constitutively active mutant Val14 RhoA·GTP Ca^{2+}-sensitizes permeabilized smooth muscle (Gong et al. 1996), and the Ca^{2+}-sensitizing effect of agonists and GTPγS is inhibited by bacterial exoenzymes (C3, EDIN) that inactivate RhoA or Val14 RhoA by ADP-ribosylating its Asn 41 residue (Gong et al. 1996). It should be noted that inhibition of the maximal effect of GTPγS is only partial, due to either incomplete ADP-ribosylation of endogenous RhoA or to operation of another significant Ca^{2+}-sensitizing pathway (Gong et al. 1996, 1997a).

The Ca^{2+}-sensitizing action of RhoA requires a relatively intact plasma membrane and/or a diffusible co-factor(s). Recombinant RhoA that lacks the hydrophobic, prenylated C-terminus has little or no Ca^{2+}-sensitizing activity and prenylated RhoA does not Ca^{2+}-sensitize smooth muscles that are extensively permeabilized with Triton X-100 (Gong et al. 1996). Furthermore, in smooth muscles stimulated with agonists or GTPγS, cytosolic RhoA dissociates from GDI and translocates to the plasma membrane. This translocation of RhoA temporally and quantitatively correlates with Ca^{2+}-sensitization of force (Gong et al. 1997a,b). The physiological significance of RhoA-mediated Ca^{2+}-sensitization was first verified by the demonstration that inhibition of endogenous RhoA in intact (not permeabilized) smooth muscle also inhibits agonist-induced Ca^{2+}-sensitization. The specific inhibitor of RhoA, C3, can be introduced into intact smooth muscle through a chimeric construct, DC3B, of C3 with the B-subunit of diphtheria toxin that mediates its penetration through the cell membrane (Aullo et al. 1993) and ADP-ribosylation of endogenous RhoA in intact vascular smooth muscle with DC3B inhibits agonist- and, following subsequent permeabilization, GTPγS-induced Ca^{2+}-sensitization and translocation of RhoA to the membrane (Figs. 4, 5; Fujihara et al. 1997). *Clostridium difficile* toxin B also enters cells and inhibits RhoA by monoglucosylating its Thr 37 residue; this toxin inhibits muscarinic Ca^{2+}-sensitization in intact smooth muscle (Otto et al. 1996), but its activity is less specific, as it also inactivates two other Rho family proteins, Cdc 42 and Rac.

In intact smooth muscles, both DC3B (Fig. 4; Fujihara et al. 1997) and the *Clostridium difficile* toxin B (Otto et al. 1996) inhibit the tonic component of agonist (muscarinic and α-adrenergic) -induced contractions, without significantly affecting their initial, phasic components. This suggests that, contrary to earlier belief, the tonic components of agonist-induced contractions are the result of not only, or even largely, Ca^{2+}-influx,

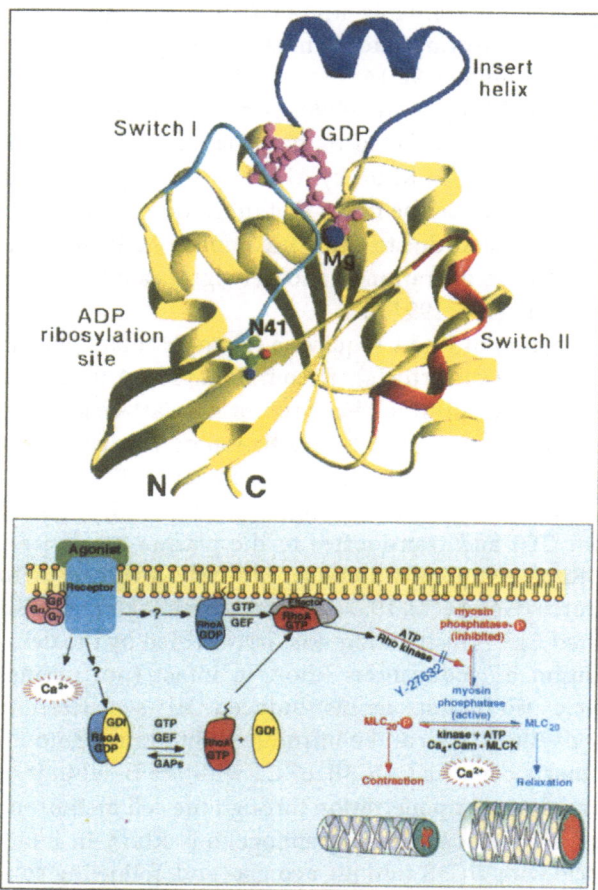

Fig. 2. Upper panel: the three-dimensional structure (main chain only) of human RhoA. Shown in various colors (and identified in the figure) are the biologically relevant features of the structure. Figure generated with RIBBONS (M. Carson). (From Wei et al. 1997.) Lower panel: scheme of suggested RhoA signaling pathways participating in Ca^{2+}-sensitization of smooth muscle. Activation of receptors coupled to certain guanine-nucleotide-binding (G) proteins releases intracellular Ca^{2+} that binds to calmodulin (Cam), and this complex activates myosin light chain kinase (MLCK). By phosphorylating the regulatory light chain of myosin in smooth muscle, MLCK causes vascular smooth muscle to contract and the lumen of blood vessels to narrow. Many of the same receptors also activate RhoA and, with the help of guanine-nucleotide exchange factors (GEFs), dissociate cytosolic RhoA-GDP from guanine-nucleotide dissociation inhibitor (GDI); this allows the exchange of GTP for GDP on RhoA. The active RhoA-GTP activates Rho-associated kinase that phosphorylates – and so inhibits

but result from the combination of Ca^{2+}-influx and the Ca^{2+}-sensitizing effects of these agents.

The abolition of Ca^{2+}-sensitization in Triton-permeabilized smooth muscle (Gong et al. 1996) and the translocation of RhoA to the plasma membrane by Ca^{2+}-sensitizing agonists (Gong et al. 1997a; Fujihara et al. 1997) strongly suggested that Ca^{2+}-sensitization involved activation of a downstream effector (presumably Rho-kinase; see below) through its association with RhoA at the plasma membrane.

Several serine/threonine kinases that have Rho-binding domains and are activated by Rho have recently been identified (Leung et al. 1995; Ishizaki et al. 1996; Fujisawa et al. 1996; Matsui et al. 1996; Leung et al. 1996). Some of these Rho-kinases can mimic the Ca^{2+}-sensitizing effects of agonists, GTPγS and Rho, and are prime candidates as effectors of RhoA-mediated Ca^{2+}-sensitization (Amano et al. 1996; Kimura et al. 1996; Kureishi et al. 1997; Uehata et al. 1997). This conclusion is very strongly supported by the recent demonstration that a relatively specific inhibitor of Rho-kinase inhibits Ca^{2+}-sensitization (Fig. 2; Uehata et al. 1997; rev. in Somlyo, 1997). The most likely downstream target of these kinases is the heterotrimeric smooth muscle myosin phosphatase (SMPP-1M) of which the large subunit (M_{110}) can be phosphorylated by Rho-kinase.

2.1
Protein kinase C, MAP kinase and tyrosine kinases

Prior to the discovery of the role of RhoA and Rho-associated kinases, several other kinases, including protein kinase Cs (PKCs), tyrosine kinases and MAP kinase, have been implicated in Ca^{2+}-sensitization, based largely on the effects of agents that activate or inhibit them.

Phorbol esters that activate some PKCs can cause Ca^{2+}-sensitization (Chatterjee and Tejada 1986; rev. in Walsh et al. 1994, Lee and Severson

---

– myosin phosphatase. Myosin phosphatase dephosphorylates smooth-muscle myosin, causing the smooth muscle to relax and blood vessels to dilate. Y-27632 inhibits Rho-associated kinases, thereby blocking the inhibition of smooth-muscle myosin phosphatase and Ca^{2+}-sensitization (Uehata et al. 1997). So, although Ca^{2+} is the main activator of smooth muscle contraction (through MLCK), the level of force can be modulated independently of it. (From Somlyo 1997).

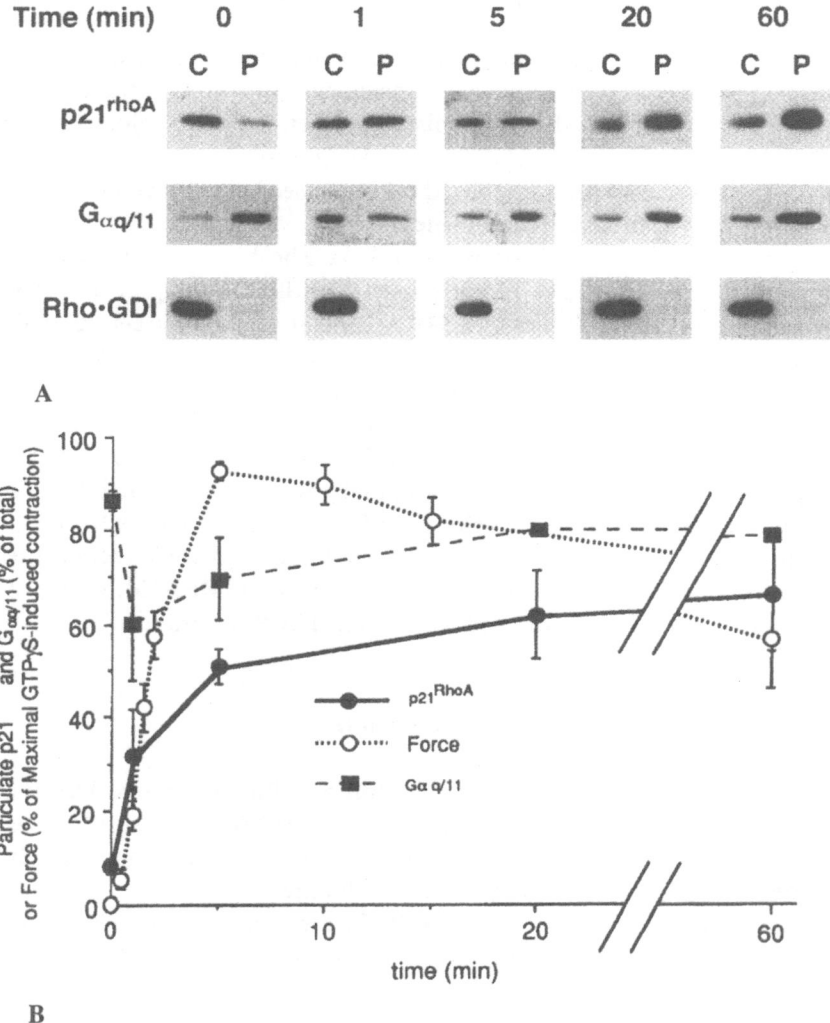

Fig. 3. Time course of translocation of p21rhoA and Gαq/11 and of Ca²⁺-sensitization of force induced by GTPγS. A: representative Western blots shows that translocation of RhoA from the cytosol (C) to the particulate fraction (P) is already detectable at 1 min, the earliest time point checked. The P fraction is the supernatant obtained from the Triton-treated pellet, and thus represents the extracted membrane proteins. Note also that the GTPγS (50 μM)-induced translocation of Gαq/11 is transient, whereas translocation of p21rhoA to the membrane is not reversed during the 1 hr period of

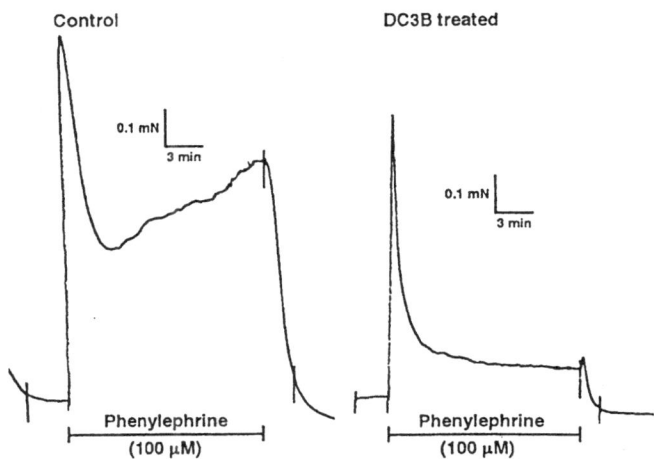

Fig. 4. Phenylephrine-induced contraction in intact portal vein smooth muscle showing the effect of 48 hr treatment with DC3B, the chimeric toxin that ADP-ribosylates and inhibits the activity of endogenous RhoA. Note that the contractile response of untreated smooth muscle was biphasic, consisting of a phasic transient followed by a tonic phase. DC3B treatment inhibited the tonic phase of contraction with little effect on the initial, transient phase (P< 0.0001; from Fujihara et al. 1997).

1994). This effect, reproduced in numerous laboratories, showed that the phorbol ester-sensitive, conventional (c) and novel (n) PKC isoforms that are activated by the physiological product of the phosphatidylinositol cascade, diacylglycerol, can increase force at constant $[Ca^{2+}]_i$. However, recent studies, as well as the known properties of diacylglycerol metabo-

◄————————————————————————————————————

observation. In control strips incubated in pCa 6.5 solution, both p21rhoA and Gαq/11 localization remained constant, and Rho-GDI remained in the cytosol at all time points checked. B: summary of results shown in A (n=3-10 for each point). Force is normalized to maximal contraction (100%) induced by GTPγS (n=12 for each point; from Gong et al. 1997a).

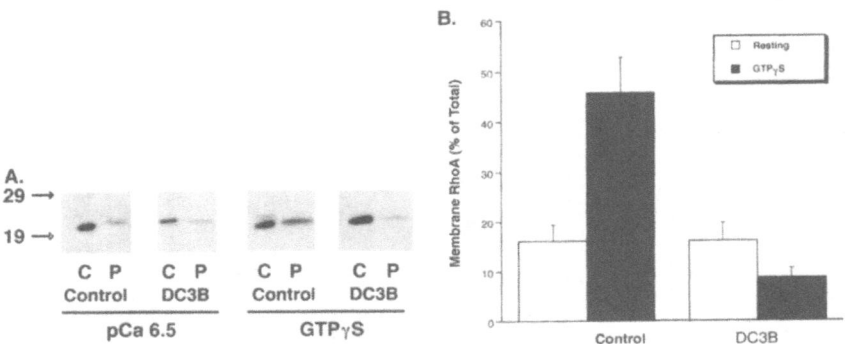

Fig. 5. ADP-ribosylation of RhoA by DC3B inhibits GTPγS-induced RhoA translocation from the cytosolic to the membrane fraction. After incubation with or without the
chimeric toxin DC3B (48 hr), α-toxin-permeabilized tissues were stimulated with
GTPγS (50 μM) for 20 min and homogenized, fractionated and separated into cytosolic
(C) and Triton-extracted particulate fraction (P); translocation of RhoA to the membrane fraction was inhibited by DC3B treatment (48 hr). A: Representative Western
blots of RhoA visualized by enhanced chemiluminescence. Arrows indicate position
of mol wt markers; remainder of gels did not show other bands and are not shown. B:
Summary of the effect of DC3B on translocation of RhoA by GTPγS (50 μM) from the
cytosolic to the membrane fraction. DC3B significantly inhibited translocation of RhoA
(from Fujihara et al. 1997).

lism argue against a major, physiological role of cPKCs and nPKCs in
Ca^{2+}-sensitization. Agonist-induced production of PIP_2-derived diacylglycerol is often transient and even undetectable in smooth muscle (Rembold 1990; Abdel-Latif 1991, but cf. Gong et al. 1995), and inhibition of
agonist-induced Ca^{2+}-sensitization by conventional and novel PKC-spe-
cific inhibitor peptides is inconsistent or absent (Fujita et al. 1995; Gailly et
al. 1997). The strongest evidence arguing against cPKCs and nPKCs being
necessary effectors of physiological, G-protein-coupled Ca^{2+}-sensitization
is that downregulation of these PKCs does not significantly inhibit
agonist- or GTPγS-induced Ca^{2+}-sensitization, although it abolishes the
effect of phorbol esters (Hori et al. 1993; Jensen et al. 1996). Conversely,
downregulation of the G-protein-coupled response does not block phorbol ester-induced Ca^{2+}-sensitization (Gong et al. 1997b). In recent studies
Ca^{2+}-sensitization of force by phorbol esters was found to be associated
with increased phosphorylation of the myosin light chain kinase sites
(Ser19 and Thr18) of MLC_{20} (Itoh et al. 1993; Matsuo et al. 1994; Ikebe and

A. Autoradiograph

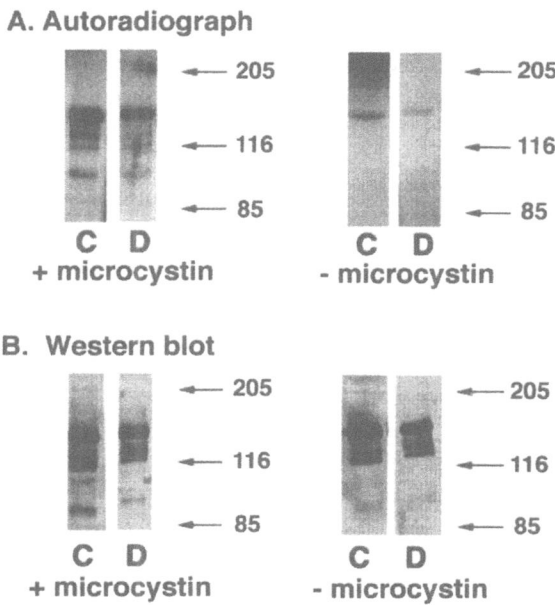

C D
+ microcystin

C D
- microcystin

B. Western blot

C D
+ microcystin

C D
- microcystin

Fig. 6. Myosin light chain kinase (MLCK) is phosphorylated during microcystin-evoked contraction in both control and PDBu downregulated tissues. Control and downregulated tissues were incubated in calcium-free solution containing 10 mM EGTA and then stimulated with 1 µM microcystin for 1 hr in the presence of ^{32}P-ATP, homogenized, and subjected to immunoprecipitation with an antibody against MLCK. A: Autoradiograph of the immunoprecipitate of control (C) and downregulated (D) tissue. B: Western blot of the same membrane for MLCK. Representative of three experiments. Autophosphorylated MLCK is active in solution even in the absence of calcium (Tokui et al. 1995; Andrea and Walsh, personal communication 1997) From Walker et al. 1998.

Brozovich 1996; Gailly et al. 1997). The combination of these results suggests that conventional and/or novel PKCs are on a separate, albeit physiologically not very significant, pathway that converges with the major physiological upstream mechanism of G-protein-coupled Ca^{2+}-sensitization and increases MLC$_{20}$ phosphorylation (see below). There is at present also only limited evidence implicating atypical PKCs that are not activated by phorbol esters in physiological Ca^{2+}-sensitization (Gailly et al. 1997; see below).

Protein kinase C-ε, contrary to previous claims, plays no special role in either G-protein-coupled Ca^{2+}-sensitization or Ca^{2+}-independent contraction of smooth muscle, both of which are retained after complete down-regulation of PCK-ε (Jensen et al. 1996; Walker et al. 1998). Furthermore, the presence or absence of PKC-ε in no way correlated with the ability of smooth muscle to contract in response to phorbol esters, GTPγS and phosphatase inhibitors (e.g., microcystin-LR, calyculin, okadaic acid) in the virtual absence of Ca^{2+} by buffering with high EGTA concentrations (Walker et al. 1998; Gong et al. 1992a). Such contractions, elicited by various inhibitors of smooth muscle myosin phosphatase in the absence of Ca^{2+}, may reflect activation of MLCK through autophosphorylation or through phosphorylation by another kinase (Fig. 6; Tokui et al. 1995; Walker et al. 1998; Andrea and Walsh, personal communication). PDBu-activated PKCs phosphorylate MLCK (Stull et al., 1990), but such phosphorylation is associated with phosphorylation of largely the PKC sites that do not activate the actomyosin ATPase of MLC_{20} (Kamm et al., 1989).

Atypical (a) PKCs are insensitive to phorbol esters, but are activated by arachidonic acid that can also directly inhibit smooth muscle myosin phosphatase (Gong et al. 1992b). A pseudo-substrate peptide inhibitor of aPKCs partially inhibited phenylephrine- and arachidonic acid-induced Ca^{2+}-sensitization (Gailly et al. 1997), raising the possibility that aPKCs, or another kinase associated with smooth muscle myosin phosphatase also activated by arachidonic acid (Ichikawa et al. 1996b), can contribute to Ca^{2+}-sensitization. The strong evidence supporting a major role of Rho kinase (see above) in Ca^{2+}-sensitization indicates that more definitive studies will be required to determine whether aPKCs also contribute to contractile regulation or, alternatively, the PKC pseudopeptide inhibitor also inhibits Rho kinase or other kinases.

The mechanism of Ca^{2+}-sensitization by PKCs may be through activating phosphorylation of a PKC inhibitor or through direct inhibitory phosphorylation of SMPP-1M. Inhibitor I is activated by phosphorylation by cAMP- or cGMP-kinase (Tokui et al. 1996 and refs. therein), but inhibitor I has only modest effect on SMPP-1M holoenzyme (Gong et al. 1992a; Alessi et al. 1992) and, furthermore, cyclic nucleotide activated kinases desensitize, rather than sensitize, smooth muscle (Pfitzer et al. 1986; Wu et al. 1996). The suggestion (Somlyo et al. 1989) that activation (through phosphorylation) of an SMPP-1M inhibitor could be the pathway of PKC-mediated Ca^{2+}-sensitization found experimental support with the isolation, from porcine aorta, of a new protein phosphatase inhibitor that is phosphorylated and activated by PKC (Eto et al. 1995). We are not aware of any

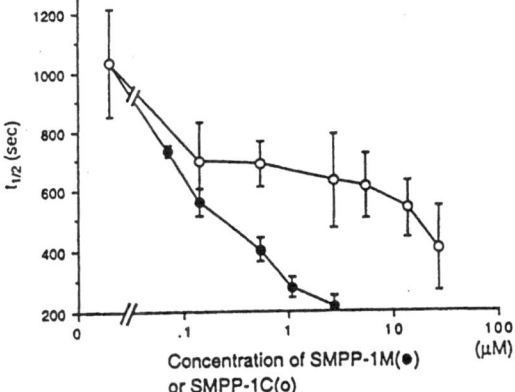

Fig. 7. Concentration-dependent relaxation induced by the purified smooth muscle phosphatase holoenzyme SMPP-1M (187 kDa) and the catalytic subunit PP1C (37 kDa). Rabbit portal vein smooth muscle strips were permeabilized with 0.1% Triton X-100 and the muscles were relaxed in the absence of calcium in the presence of a 1 mM EGTA solution. Subsequent treatment with microcystin-LR (MC, 1 μM), to inhibit endogenous SMPP-1M, induced force development even in relaxing solution (G1, pCa 8). MC was washed away and the purified holoenzyme SMPP-1M or PP1C were added at the indicated concentration and the half-time of relaxation measured. The half-time of relaxation (t1/2) to baseline was used for comparison of the potency of SMPP-1M (solid circles) with that of PP1C (open circles;n=5). (From Shirazi et al. 1994.)

publication to date about the occurrence or effects of phosphorylation of SMPP-1M by cPKCs or nPKCs.

Tyrosine kinases have been implicated in Ca^{2+}-sensitization largely as the result of the inhibitory effects of tyrosine kinase inhibitors, genistein and vanadate, and the correlation between tyrosine phosphorylation and vanadate- or agonist-induced contractions (Di Salvo et al. 1993a,b, 1994, 1997; Steusloff et al. 1995). However, the tyrosine-phosphorylated proteins have yet to be directly linked to a known contractile regulatory mechanism (i.e., MLCK, SMPP-1M) and evaluation of the relationships between tyrosine phosphorylation and modulation of Ca^{2+}-sensitization of contraction in intact muscle is somewhat complicated by concurrent changes in cytosolic Ca^{2+} (Di Salvo et al. 1994, 1997), possibly as the result of activation of PLC-γ (Marrero et al. 1994) or $G\alpha_{q,11}$ (Umemori et al. 1997).

Mitogen-activated protein kinase (MAP kinase) is another kinase that is activated upon stimulation with agonists or depolarization with high K^+ (Adam et al. 1995; Katoch and Moreland, 1995; Gerthoffer et al. 1997).

Caldesmon is one of the substrates of MAP kinase, but exposure of Triton X-100 permeabilized smooth muscles to activated MAP kinase at concentrations sufficient to near-stoichiometrically phosphorylate endogenous caldesmon (Nixon et al. 1995) did not Ca^{2+}-sensitize vascular smooth muscle (Nixon et al. 1995). According to another report, MAP kinase has no effect on contractility of rabbit colonic smooth muscle, but enhances the contraction of canine trachealis (Gerthoffer et al. 1997). The mechanisms responsible for these different findings are not known and, in particular, the effects of MAP kinase on MLC_{20} phosphorylation in the trachealis smooth muscle has not been determined. The very different time courses of, respectively, caldesmon dephosphorylation and relaxation of smooth muscle (Adam et al. 1989) and the failure of MAP kinase to affect contractility of vascular and intestinal smooth muscle argue against a major role of MAP kinase in contractile regulation. The fact that high K^+ and ionomycin (via Ca^{2+}) also activate MAP kinase (ERK1 and ERK2; Abraham et al. 1997; Katoch and Moreland 1995) but does not cause Ca^{2+}-sensitization also argues against a significant role of this kinase in the physiological, G-protein-coupled major mechanism, although *in vitro* MAP kinase (slowly) phosphorylates MLCK and increases its activity (Klemke et al. 1997).

2.2
Smooth muscle myosin phosphatase

The properties of smooth muscle myosin phosphatase (SMPP-1M) and its history have been recently reviewed in depth (Erdödi and Hartshorne 1996; Hartshorne et al. 1998), and here we will only summarize the most salient aspects of the subject. It is generally accepted (rev. in Hartshorne et al. 1998; Somlyo and Somlyo 1994) that the physiological myosin phosphatase dephosphorylates not only MLC_{20}, but also MLC_{20} present in whole myosin. In retrospect, it seems likely that several of the previously isolated enzymes (rev. in Cai et al. 1994; Hartshorne et al. 1998) that have this property contained the 37-38 kDa catalytic subunit, PP1C, variously called $PP1C_\delta$ or $PP1C_\beta$ isoform, with a proteolytic fragment (58-67 kD) of one of the other subunits of the holoenzyme. In addition to PP1C, the holoenzyme consists of a large, regulatory or targeting, subunit (M_{110} or M_{130}) and a 20 kD subunit of unknown function. This composition is characteristic of the enzymes isolated from both chicken gizzard (Alessi et al. 1992; Shimizu et al. 1994) and pig bladder (Shirazi et al. 1994). Dissociation of the catalytic from the $M_{110-130}$ subunit with arachidonic acid increases its glycogen phosphatase and decreases its myosin activity, supporting the notion that the $M_{110-130}$ subunit targets the enzyme to myosin and/or enhances its activity for myosin (Fig. 7; Gong et al. 1992a,b).

Phosphorylation of the $M_{110-130}$ subunit (for detailed structure, see Hartshorne et al. 1998) is thought to be a major physiological mechanism of its (inhibitory) regulation. The region of $M_{110-130}$ involved in regulation and contained within its N-terminus is also involved in its ATP-dependent binding to myosin (Haystead et al. 1995; Ichikawa et al. 1996a; Hartshorne et al. 1998). ATP-dependent binding may be related to regulation, because in the presence of ATP, SMPP-1M binds only to phosphorylated, but not to unphosphorylated, myosin. The $M_{110-130}$ subunit increases the activity of SMPP-1M towards myosin in solution (Alessi et al. 1992; Shirazi et al. 1994) and accelerates dephosphorylation of myosin and relaxation of smooth muscle induced by PP1C in permeabilized smooth muscle (Fig. 7; Shirazi et al. 1994; Gailly et al. 1996). An N-terminal 58 kDa fragment, a common proteolytic product, and even an N-terminal 1-38 peptide can accelerate the relaxation induced by PP1C. The N-terminal portion of $M_{110,130}$ alone can activate the catalytic activity of SMPP-1M, but there is evidence that both the N- and C-terminal regions can bind to myosin (Johnson et al. 1997; Hirano et al. 1997; Hartshorne et al. 1998). Arachidonic acid inhibits both the holoenzyme and the potentiating effect of a 612 N-terminal residue peptide, but not the activity potentiated by the M_{58} fragment, suggesting that the site of inhibition is C-terminal. Phosphorylation of $M_{110-130}$ by a kinase associated with SMPP-1M (Ichikawa et al. 1996b) or by a Rho-associated kinase (Rho-kinase; Kimura et al. 1996) inhibits the activity of the holoenzyme, but the sites phosphorylated by these kinases are different: Thr 654 (130 kD) or Thr 695 (133 kD isoform) is phosphorylated by the phosphatase-associated kinase and residues within 753-1004 by Rho kinase (for further rev., see Hartshorne et al. 1998). Thiophosphorylation of $M_{110-130}$ in permeabilized vascular smooth muscle decreases phosphatase activity and increases Ca^{2+}-sensitivity (Trinkle-Mulcahy et al. 1995), and it is likely that phosphorylation of this subunit plays a physiological role in Ca^{2+}-sensitization by regulating (inhibiting) SMPP-1M.

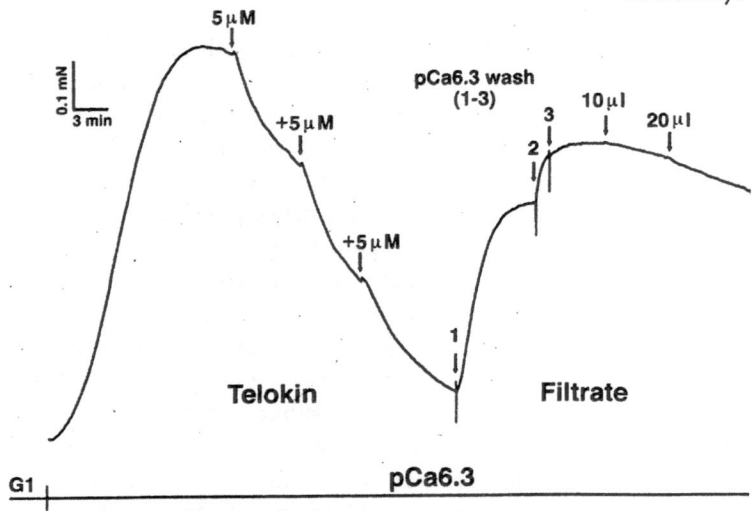

Fig. 8. Relaxation of rabbit ileum smooth muscle by purified endogenous native gizzard telokin. Rabbit ileum smooth muscle strips were depleted of endogenous telokin by permeabilization with Triton X-100 and storage at ~20°C (for up to 4 wks). Following incubation in 0-added Ca, 1 mM EGTA solution (G1), the muscle strips were contracted by exposing them to pCa 6.3 solution with 1 μM calmodulin. Three additions of native telokin of 5 μM each, concentrated by centrifuge filters (Microcon-3K Amicon), induced relaxation. Following 3 washes (1, 2, 3) with pCa 6.3 solution, the relaxant effects of telokin were removed; subsequent additions of 10 and 20 μl of the filtrate separated from the same telokin sample by the centrifuge filter were without effect and served as a control. Representative of n ≈3 experiments. (From Wu et al. 1998.)

3
Desensitization to Ca^{2+}

Desensitization to Ca^{2+} is defined as a decline in MLC_{20} phosphorylation and force in the absence of proportional, or any, decline in $[Ca^{2+}]_i$ (Himpens et al. 1988, 1989; Kitazawa and Somlyo 1990; Somlyo and Somlyo, 1994; Wu et al. 1996, 1998). This time-dependent phenomenon is particularly pronounced in phasic smooth muscles in which, unless the time course of phosphorylation is closely followed, the very transient nature of MLC_{20} phosphorylation (Himpens et al. 1988) can lead to the erroneous conclusion that contraction occurs in the absence of MLC_{20} phosphorylation. Two mechanisms involving, respectively, downregulation of MLCK and upregulation of SMPP-1M are known mechanisms of Ca^{2+}-desensitization. *In vitro* phosphorylation of MLCK within the C-terminal region (site A) of the regulatory domain by cAMP-dependent protein kinase (Conti and Adelstein 1981; Nishikawa et al. 1984), protein kinase C (Nishikawa et al. 1983) and Ca^{2+}/calmodulin-dependent protein kinase II (Hashimoto and Soderling 1990; Ikebe and Reardon 1989) reduces the activity of MLCK by decreasing the affinity (increasing K_{CaM}) for calmodulin approximately tenfold (rev. in Gallagher et al. 1997). Inhibitory phosphorylation *in vivo*, is largely due to CaM-kinase II activity (Tansey et al. 1994; Van Riper et al. 1995; rev. in Gallagher et al. 1997), and represents a negative feedback whereby the rise in $[Ca^{2+}]_i$ first activates MLCK and then, with a slower time course reflecting the lower affinity of CaM-kinase II for Ca^{2+}, inhibits MLCK. A similar negative feedback may also be initiated by GTPγS that also enhances phosphorylation of MLCK and decreases its activity *in situ*, suggesting that SMPP-1M or a related protein phosphatase that is also inhibited by a G-protein-coupled pathway is responsible for dephosphorylation of MLCK (Tang et al. 1993; Word et al. 1994).

The second proposed pathway of Ca^{2+}-desensitization is through enhancement of myosin phosphatase activity (Somlyo et al. 1989; Somlyo and Somlyo 1994). The early studies of Rüegg and co-workers (Pfitzer et al. 1986) and more recent reports (Nishimura and van Breemen 1989; Wu et al. 1996) showed that cyclic nucleotides, through their regulated kinases, could relax permeabilized smooth muscles in the absence of a change in $[Ca^{2+}]_i$. Consequently, cAMP- and cGMP-activated/dependent kinases appeared to be likely upstream activators of smooth muscle phosphatase activity (Somlyo et al. 1989; Somlyo and Somlyo 1994; McDaniel et al. 1994; Kotlikoff and Kamm 1996). Indeed, Wu et al. (1996) demonstrated that 8-br-cGMP accelerated dephosphorylation of MLC_{20} under conditions in which MLCK activity was inhibited, and this finding was confirmed by Lee et al. (1997). Identification of the specific kinase operating in intact smooth muscle, whether cAMP- or cGMP-kinase, is difficult in the presence of accompanying changes in $[Ca^{2+}]_i$ induced by agents that stimulate adenylate cyclase or guanylate cyclase (rev. in Lincoln et al. 1994; Francis and Corbin 1994; McDaniel et al. 1994), and activate the SR Ca^{2+}-ATPase, perhaps by phosphorylating phospholamban (Cornwell et al. 1991). Considerable evidence also indicates cross-stimulation of the cGMP kinase by cAMP (Jiang et al. 1992; rev. in Francis and Corbin 1994; Komalavilas and Lincoln 1996). Although the use of relatively specific cGMP and cAMP inhibitors in permeabilized smooth muscle suggests that cGMP-kinase is largely responsible for Ca^{2+}-desensitization induced by both cAMP and cGMP (Kawada et al. 1997), the possibility remains that both the cAMP- and the cGMP-kinases can mediate this effect *in vivo*.

The mechanism through which cGMP-kinase activates dephosphorylation of smooth muscle may be indirect and involve, as its mediator, telokin. Telokin is a low molecular weight (~17 kD) protein first identified by Hartshorne and co-workers and shown to be identical to the C-terminal sequence of MLCK (Ito et al. 1989). Subsequent determination of the genomic sequence for MLCK indicated that telokin corresponded to the 156 C-terminal segment of MLCK (Gallagher and Herring 1991; Herring and Smith 1996, 1997). Telokin is abundant (70-90 μM) in phasic smooth muscles (Shirinsky et al. 1993), but in tonic smooth muscles only trace amounts are detected by Western blotting (immunoblotting; Gallagher and Herring 1991; Wu et al. 1998). The possibility that telokin may be a mediator of cAMP- and cGMP-mediated activation of smooth muscle dephosphorylation was suggested by our observation that telokin was the most abundant protein phosphorylated when intact and permeabilized smooth muscles were relaxed by, respectively, forskolin and 8-br-cGMP (Wu et al. 1998). Telokin inhibits myosin phosphorylation by MLCK *in vitro*, probably by competing for the MLCK binding site on myosin (Shirinsky et al. 1993; Silver et al. 1997; Sobieszek et al. 1997), but does not inhibit the rate of thiophosphorylation of MLC_{20} *in situ* (Wu et al. 1998). In view of this, and because telokin accelerates dephosphorylation of MLC_{20} and relaxation (Fig. 8) of smooth muscle at constant $[Ca^{2+}]_i$, we consider activation of SMPP-1M by phosphorylated (on cyclic nucleotide kinase sites) telokin to be a physiological Ca^{2+}-desensitizing mechanism. Furthermore, the relationship between telokin and cyclic nucleotide-activated kinase(s) is synergistic: the combination of the two produces significantly greater relaxant effects when added together than the sum of independent additions to smooth muscle at constant $[Ca^{2+}]_i$ (Wu et al. 1998). A proteolytically truncated form of telokin that does not contain the cGMP-kinase phosphorylation site (identical to site B in MLCK) can also relax permeabilized smooth muscle, although without showing synergism with cGMP-kinase. This and the apparent paucity of telokin in tonic smooth muscles indicates that additional studies will be required for a more complete understanding of the mechanism(s) through which cyclic nucleotide-activated kinases activate the SMPP-1M.

4
Summary

The concept of pharmacomechanical coupling, introduced 30 years ago to account for physiological mechanisms that can regulate contraction of smooth muscle independently of the membrane potential, has since been transformed from a definition into what we now recognize as a complex of well-defined, molecular mechanisms. The release of Ca^{2+} from the SR by a chemical messenger, $InsP_3$, is well known to be initiated not by depolarization, but by agonist-receptor interaction. Furthermore, this G-protein-coupled phosphatidylinositol cascade, one of many processes covered by the umbrella of pharmacomechanical coupling, is part of complex and general signal transduction mechanisms also operating in many non-muscle cells of diverse organisms. It is also clear that, although the major contractile regulatory mechanism of smooth muscle, phosphorylation/dephosphorylation of MLC_{20}, is $[Ca^{2+}]$-dependent, the activity of both the kinase and the phosphatase can also be modulated independently of $[Ca^{2+}]_i$. Sensitization to Ca^{2+} is attributed to inhibition of SMPP-1M, a process most likely dominated by activation of the monomeric GTP-binding protein RhoA that, in turn, activates Rho-kinase that phosphorylates the regulatory subunit of SMPP-1M and inhibits its myosin phosphatase activity. It is likely that the tonic phase of contraction activated by a variety of excitatory agonists is, at least in part, mediated by this Ca^{2+}-sensitizing mechanism. Desensitization to Ca^{2+} can occur either through inhibitory phosphorylation of MLCK by other kinases or autophosphorylation and by activation of SMPP-1M by cyclic nucleotide-activated kinases, probably involving phosphorylation of a phosphatase activator. Based on our current understanding of the complexity of the many cross-talking signal transduction mechanisms that operate in cells, it is likely that, in the future, our current concepts will be refined, additional mechanisms of pharmacomechanical coupling will be recognized, and those contributing to the pathogenesis diseases, such as hypertension and asthma, will be identified.

5
Acknowledgements
We wish to thank Ms. Jama Coartney for preparation of the figures and Ms. Barbara Nordin for preparation of the manuscript. This study was supported by NIH grants PO1-HL48807 and PO1-HL19242.

6
References

Abdel-Latif AA (1991) Biochemical and functional interactions between the inositol inositol 1,4,5-trisphosphate-Ca^{2+} and cyclic Amp signalling systems in smooth muscle. Cell Signal 3:371-385

Abraham ST, Benscoter HA, Schworer CM, Singer HA (1997) A role for Ca^{2+}/calmodulin-dependent protein kinase II in the mitogen-activated protein kinase signaling cascade of cultured rat aortic vascular smooth muscle cells. Circ Res 81:575-584

Adam LP, Haeberle JR, Hathaway DR (1989) Phosphorylation of caldesmon in arterial smooth muscle. J Biol Chem 264:7698-7703

Adam LP, Franklin MT, Raff GF, Hathaway DR (1995) Activation of mitogen-activated protein kinase in porcine carotid arteries. Circ Res 76:183-190

Alessi D, MacDougall KL, Sola MM, Ikebe M, Cohen P (1992) The control of protein phosphatase-1 by targeting subunits. Eur J Biochem 210:1023-1035

Amano M, Mukai H, Ono Y, Chihara K, Matsui T, Hamajima Y, Okawa K, Iwamatsu A, Kaibuchi K. (1996) Identification of a putative target for Rho as the serine-threonine kinase protein kinase N. Science 271:648-650

Andrea JE, Walsh MP. (1997) Personal communication

Aullo P, Giry M, Olsnes S, Popoff MR, Kocks C, Boquet P (1993) A chimeric toxin to study the role of the 21 kDa GTP binding protein rho in the control of actin microfilament assembly. EMBO J 12:921-931

Berridge MJ (1988) Inositol lipids and calcium signaling. Proc R Soc (Lond) B 234:359-378

Bond M, Kitazawa T, Somlyo AP and Somlyo AV (1984) Release and recycling of calcium by the sarcoplasmic reticulum in guinea-pig portal vein smooth muscle. J Physiol (London) 355:677-695

Cai S, Nowak G, de Lanerolle P (1994) Myosin dephosphorylation as a mechanism of relaxation of airways smooth muscle. In: Raeburn D, Giembycz MA (eds) Airways smooth muscle biochemical control of contraction and relaxation. Birkhauser Verlag, Basel, pp 233-251

Chatterjee M, Tejada M (1986) Phorbol ester-induced contraction in chemically skinned vascular smooth muscle. Am J Physiol 251:C356-C361

Chilvers ER, Challiss RAJ, Barnes PJ, Nahorski SR. (1989) Mass changes of inositol (1,4,5) trisphosphate in trachealis muscle following agonist stimulation. Am J Physiol 251:C356-C361

Conti MA, Adelstein RS (1981) The relationship between calmodulin binding and phosphorylation of smooth muscle myosin kinase by the catalytic subunit of 3'.5' cAMP-dependent protein kinase. J Biol Chem 256:3178-3181

Cornwell TL, Pryzwansky KB, Wyatt TA, Lincoln TM (1991) Regulation of sarcoplasmic reticulum protein phosphorylation by localized cyclic GMP-dependent protein kinase in vascular smooth muscle cells. Mol Pharmacol 40:923-931

Crichton CA, Templeton AGB, McGrath JC, Smith GL (1993) Thromboxane A₂ ana-logue, U-46619, potentiates calcium-activated force in human umbilical artery. Am J Physiol 264:H1878-H1883

Dabrowska R (1994) Actin and thin-filament-associated proteins in smooth muscle. In: Raeburn D, Giembyzc MA (eds) Airways smooth muscle: biochemical control of contraction and relaxation. Birkhäuser Verlag, Basel, pp 31-59

Devine CE, Somlyo AV, Somlyo, AP (1972) Sarcoplasmic reticulum and excitation-contraction coupling in mammalian smooth muscle. J Cell Biol 52:690-718

Di Salvo J, Pfitzer G, Semenchuk LA (1994) Protein tyrosine phosphorylation, cellular Ca^{2+}, and Ca^{2+}-sensitivity for contraction of smooth muscle. Can J Physiol Phar-macol 72:1434-1439

Di Salvo J, Semenchuk LA, Lauer, J (1993a) Vanadate-induced contraction of smooth muscle and enhanced protein tyrosine phosphorylation. Arch Biochem Biophys 304:386-391

Di Salvo J, Steusloff A, Semenchuk L, Satoh S, Kolquist K, Pfitzer G (1993b) Tyrosine kinase inhibitors suppress agonist-induced contraction in smooth muscle. Biochem Biophys Res Commun 190:968-974

Di Salvo J, Nelson SR, Kaplan N (1997) Protein tyrosine phosphorylation in smooth muscle: a potential coupling mechanism between receptor activation and intracel-lular calcium. Proc Soc Exp Biol Med 214:285-301

Erdödi F, Ito M, Hartshorne DJ. (1996) Myosin light chain phosphatase. In: Bárány M (ed) Biochemistry of smooth muscle contraction. Academic Press, San Diego, pp 131-142

Eto M, Ohmori T, Suzuki M, Furuya K, Morita F (1995) A novel protein phosphatase-1 inhibitory protein potentiated by protein kinase C. Isolation from porcine aorta media and characterization. J Biochem 118:1104-1107

Fay FS, Shlevin HH, Granger WC Jr, Taylor SR (1979) Aequorin luminescence during activation of single isolated smooth muscle cells. Nature 280:506-508

Francis SH, Corbin JD (1994) Progress in understanding the mechanism and function of cyclic GMP-dependent protein kinase. Adv Pharmacol 26:115-170

Fuglsang A, Khromov A, Török K, Somlyo AV, Somlyo AP (1993) Flash photolysis studies of relaxation and cross-bridge detachment: higher sensitivity of tonic than phasic smooth muscle. J Mus Res Cell Motil 14:666-673

Fujihara H, Walker LA, Gong MC, Lemichez E, Bouquet P, Somlyo AV, Somlyo AP. (1997) Inhibition of RhoA translocation and calcium sensitization by *in vivo* ADP-ribosylation with the chimeric toxin DC3B. Mol Biol Cell 8:2437-2447

Fujisawa K, Fujita A, Ishizaki T, Saito Y, Narumiya S (1996) Identification of the Rho-binding domain of p160^{rock}, a Rho-associated coiled-coil containing protein kinase. J Biol Chem 271:23022-23028

Fujita A, Takeuchi T, Nakajima H, Nishio H, Hata F (1995) Involvement of hetero-trimeric GTP-binding protein and *rho* protein, but not protein kinase C, in agonist-induced Ca^{2+} sensitization of skinned muscle of guinea pig vas deferens. J Pharmacol Exper Therap 274:555-561

Fuijiwara T, Itoh T, Kubota Y, Kuriyama H (1989) Effects of guanosine nucleotides on skinned smooth muscle tissue of the rabbit mesenteric artery. J Physiol 408:535-547

Gailly P, Gong MC, Somlyo AV, Somlyo AP (1997) Possible role of atypical protein kinase C activated by arachidonic acid in Ca^{2+} sensitization of rabbit smooth muscle. J Physiol 500:95-109

Gailly P, Wu X, Haystead TAJ, Somlyo AP, Cohen PTW, Cohen P, Somlyo AV (1996) Regions of the 100-kDa regulatory subunit M_{110} required for regulation of myosin-light-chain-phosphatase activity in smooth muscle. Eur J Biochem 239:326-332

Gallagher PJ, Herring BP (1991) The carboxyl terminus of the smooth muscle myosin light chain kinase is expressed as an independent protein, telokin. J Biol Chem 266:23945-23952

Gallagher PJ, Herring BP, Stull JT (1997) Myosin light chain kinases. J Mus Res Cell Motil 18:1-16

Ganitkevich VY, Hirche H (1996) High cytoplasmic Ca^{2+} levels reached during Ca^{2+}-induced Ca^{2+} release in single smooth muscle cells as reported by a low affinity Ca^{2+} indicator Mag-Indo-1. Cell Calc 19:391-398

Ganitkevich VY, Isenberg G. (1993a) Ca^{2+} entry through Na^+-Ca^+ exchange can trigger Ca^{2+} release from Ca^{2+} stores in Na^+-loaded guinea-pig coronary myocytes. J Physiol 468:225-243

Ganitkevich VY, Isenberg G. (1993b) Membrane potential modulates inositol 1,4,5-trisphosphate-mediated Ca^{2+} transients in guinea-pig coronary myocytes. J Physiol 470:35-44

Ganitkevich VY, Isenberg G. (1995) Efficacy of peak Ca^{2+} currents (Ica) as trigger of sarcoplasmic reticulum Ca^{2+} release in myocytes from the guinea-pig coronary artery. J Physiol 484:287-306

Gerthoffer WT, Yamboliev IA, Pohl J, Haynes R, Dang S, McHugh J (1997) Activation of MAP kinases in airway smooth muscle. Am J Physiol 272:L244-L252

Gong MC, Cohen P, Kitazawa T, Ikebe M, Masuo M, Somlyo, AP, Somlyo AV (1992a) Myosin light chain phosphatase activities and the effects of phosphatase inhibitors in tonic and phasic smooth muscle. J Biol Chem 267:14662-14668

Gong MC, Fuglsang A, Alessi D, Kobayashi S, Cohen P, Somlyo AV, Somlyo AP (1992b) Arachidonic acid inhibits myosin light chain phosphatase and sensitizes smooth muscle to calcium. J Biol Chem 267:21492-21498

Gong MC, Fujihara H, Somlyo AV, Somlyo AP (1997a) Translocation of RhoA associated with Ca^{2+}-sensitization of smooth muscle. J Biol Chem 272:10704-10709

Gong MC, Fujihara H, Walker LA, Somlyo AV, Somlyo AP (1997b) Down-regulation of G-protein-mediated Ca^{2+} sensitization in smooth muscle. Mol Biol Cell 8:279-286

Gong MC, Iizuka K, Nixon G, Browne JP, Hall A, Eccleston JF, Sugai M, Kobayashi S, Somlyo AV, Somlyo AP (1996) Role of guanine nucleotide-binding proteins – ras-family or trimeric proteins or both – in Ca^{2+} sensitization of smooth muscle. Proc Natl Acad Sci USA 93:1340-1345

Gong MC, Kinter MT, Somlyo AV, Somlyo AP (1995) Arachidonic acid and diacyl-glycerol release associated with inhibition of myosin light chain dephosphorylation in rabbit smooth muscle. J Physiol (Lond) 486:113-122

Gosser YQ, Namanbhoy, TK, Aghazadeh B, Manor D, Combs C, Cerione RA, Rosen MK (1997) C-terminal binding domain of Rho GDP-dissociation inhibitor directs N-ter-minal inhibitory peptide to GTPases. Nature 387:814-819

Hartshorne DR. (1987) Biochemistry of the contractile process in smooth muscle. In: Johnson LR (ed) Physiology of the Gastrointestinal Tract, ed. 2. Raven Press, New York, pp 423-482

Hartshorne DJ, Ito M, Erdšdi F (1998) Myosin light chain phosphatase. J Mus Res Cell Motil 19:325-341

Hashimoto Y, Soderling TR (1990) Phosphorylation of smooth muscle myosin light chain kinase by Ca^{2+}/calmodulin-dependent protein kinase-II. Comparative study of the phosphorylation sites. Arch Biochem Biophys 278:41-45

Haystead CM, Gailly P, Somlyo AP, Somlyo AV, Haystead TA (1995) Molecular cloning and functional expression of a recombinant 72.5 kDa fragment of the 110 kDa regulatory subunit of smooth muscle protein phosphatase 1M. FEBS Lett 377:123-127

Herring BP, Smith AF (1996) Telokin expression is mediated by a smooth muscle cell-specific promoter. Am J Physiol 270:1656-1665

Herring BP, Smith AF (1997) Telokin expression in A10 smooth muscle cells requires serum response factor. Am J Physiol 272:C1394-C1404

Herrmann-Frank A, Darling E, Meissner G. (1991) Functional characterization of the Ca^{2+}-gated Ca^{2+} release channel of vascular smooth muscle sarcoplasmic reticulum. Pflug Arch 418:353-359

Himpens B, Kitazawa T, Somlyo AP (1990) Agonist dependent modulation of the Ca^{2+} sensitivity in rabbit pulmonary artery smooth muscle. Eur J Physiol 417:21-28

Himpens B, Matthijs G, Somlyo AP (1989) Desensitization to cytoplasmic Ca^{2+} and Ca^{2+} sensitivities of guinea-pig ileum and rabbit pulmonary artery smooth muscle. J Physiol (London) 413:489-503

Himpens BH, Matthijs G, Somlyo AV, Butler TM, Somlyo AP (1988) Cytoplasmic free calcium, myosin light chain phosphorylation, and force in phasic and tonic smooth muscle. J Gen Physiol 92:713-729

Himpens B, Somlyo AP (1988) Free-calcium and force transients during depolarization and pharmacomechanical coupling in guinea-pig smooth muscle. J Physiol (Lond) 395:507-530

Hirano K, Phan BC, Hartshorne DJ (1997) Interactions of the subunits of smooth muscle myosin phosphatase. J Biol Chem 272:3683-3688

Hori M, Sato K, Miyamoto S, Ozaki H, Karaki H (1993) Different pathways of calcium sensitization activated by receptor agonists and phorbol esters in vascular smooth muscle. Br J Pharmacol 110:1527-1531

Ichikawa K, Hirano K, Ito M, Tanaka J, Nakano T, Hartshorne DJ. (1996a) Interactions and properties of smooth muscle myosin phosphatase. Biochemistry 35:6313-6320

Ichikawa K, Ito M, Hartshorne DJ (1996b) Phosphorylation of the large subunit of myosin phosphatase and inhibition of phosphatase activity. J Biol Chem 271:4733-4740

Iino M (1987) Calcium dependent inositol trisphosphate-induced calcium release in the guinea-pig taenia caeci. Biochem Biophys Res Comm 142:47-52

Iino M, Endo M (1992) Calcium-dependent immediate feedback control of inositol 1,4,5-trisphosphate-induced Ca^{2+} release. Nature 360:76-78

Iizuka K, Dobashi K, Yoshii A, Horie T, Suzuki H, Nakazawa T, Mori M. (1997) Receptor-dependent G protein-mediated Ca^{2+} sensitization in canine airway smooth muscle. Cell Calcium 22:21-30

Ikebe M, Brozovich FV (1996) Protein kinase C increases force and slows relaxation in smooth muscle: evidence for regulation of the myosin light chain phosphatase. Biochem Biophys Res Commun 225:370-376

Ikebe M, Reardon S (1989) Location of the inhibitory region of smooth muscle myosin light chain kinase. J. Biol Chem 264:6967-6971

Ishizaki T, Maekawa M, Fujisawa K, Okawa K, Iwamatsu A, Fujita A, Watanabe N, Saito Y, Kakizuka A, Morri N, Narumiya S. (1996) The small GTP-binding protein Rho binds to and activates a 160 kDa Ser/Thr protein kinase homologous to myotonic dystrophy kinase. EMBO J 15:1885-1893

Ito M, Dabrowska M, Guerriero V, Hartshorne DJ (1989) Identification of turkey gizzard of an acidic protein related to the C-terminal portion of smooth muscle myosin light chain kinase. J Biol Chem 264:13971-13974

Itoh H, Shimomura A, Okubo S, Ichikawa K, Ito M, Konishi T, Nakano T (1993) Inhibition of myosin light chain phosphatase during Ca^{2+}-independent vasoconstriction. Am J Physiol 265:C1319-C1324

Itoh T, Seki N, Suzuki S, Ito S, Kajikuri J, Kuriyama H. (1992) Membrane hyperpolarization inhibits agonist-induced synthesis of inositol 1,4,5-trisphosphate in rabbit mesenteric artery. J Physiol 451:307-328

Jensen PE, Gong MC, Somlyo AV, Somlyo AP (1996) Separate upstream and convergent downstream pathways of G-protein and phorbol ester-mediated Ca^{2+}-sensitization of myosin light chain phosphorylation in smooth muscle. Biochemical J 318:469-475

Jiang H, Colbran JL, Francis SH, Corbin JD (1992) Direct evidence for cross-activation of cGMP-dependent protein kinase by cAMP in pig coronary arteries. J Biol Chem 267:1015-1019

Johnson D, Cohen P, Chen MX, Chen YH, Cohen PTW. (1997) Identification of the regions on the M_{110} subunit of protein phosphatase 1M that interact with the M_{21} subunit and myosin. Eur J Biochem 244:931-939

Juhaszova M, Blaustein MP (1997) Na^+ pump low and high ouabain affinity α subunit isoforms are differently distributed in cells. Proc Natl Acad Sci USA 94:1800-1805

Kamm KE, Hsu L-C, Kubota Y, Stull JT (1989) Phosphorylation of smooth muscle myosin heavy and light chains. J Biol Chem 264: 21223-21229

Kannan MS, Fenton AM, Prakash YS, Sieck GC (1996) Cyclic ADP-ribose stimulates sarcoplasmic reticulum calcium release in porcine coronary artery smooth muscle. Am J Physiol 270:H801-H806

Karaki H, Ozaki H, Hori M, Mitsui-Saito M, Amano K-I, Harada K-I, Miyamoto S, Nakazawa H, Wong K-J, Sato K (1997) Calcium movements, distribution, and functions in smooth muscle. Pharmacol. Rev. 49:157-230

Kargacin GJ. (1994) Calcium signaling in restricted diffusion spaces. Biophys J 67:262-272

Katoch SS, Moreland RS (1995) Agonist and membrane depolarization induced activation of MAP kinase in the swine carotid artery. Am J Physiol 269:H222-H229

Kawada T, Toyosato A, Islam MO, Yoshida Y, Imai S (1997) cGMP-kinase mediates cGMP- and cAMP-induced Ca^{2+}-desensitization of skinned rat artery. Eur J Pharmacol 323:75-82

Keep NH, Barnes M, Barsukov I, Badii R, Lian LY, Segal AW, Moody PCE, Roberts GCK (1997) A modulator of Rho family G proteins, rhosGDI, binds these G proteins via an immunoglobulin-like domain and a flexible N-terminal arm. Structure 5:623-633

Khromov A, Somlyo AV, Somlyo AP (1996) Nucleotide binding by actomyosin as a determinant of relaxation kinetics of rabbit phasic and tonic smooth muscle. J Physiol 492:669-673

Khromov A, Somlyo AV, Trentham DR, Zimmermann B, Somlyo AP (1995) The role of MgADP in force maintenance by dephosphorylated cross-bridges in smooth muscle: a flash photolysis study. Biophys J 69:2611-2622

Kimura K, Ito M, Amano M, Chihara K, Fukata Y, Nakafuku M, Yamamori K, Iwamatsu A, Kaibuchi K (1996) Regulation of myosin phosphatase by Rho and Rho-associated kinase (Rho-kinase). Science 273:245-248

Kitazawa T, Gaylinn BD, Denney GH, Somlyo AP (1991) G-protein-mediated Ca^{2+} sensitization of smooth muscle contraction through myosin light chain phosphorylation. J Biol Chem 266:1708-1715

Kitazawa T, Somlyo AP (1990) Desensitization and muscarinic re-sensitization of force and myosin light chain phosphorylation to cytoplasmic Ca^{2+} in smooth muscle. Biochem Biophys Res Comm 172:1291-1297

Klemke RL, Cai S, Giannini AL, Gallagher PJ, de Lanerolle P, Cheresh DA (1997) Regulation of cell motility by mitogen-activated protein kinase. J Cell Biol 137:481-492

Kohda M, Komori S, Unno T, Ohashi H (1997) Characterization of action potential-triggered $[Ca^{2+}]_i$ transients in single smooth muscle cells of guinea-pig ileum. Br J Pharmacol 122:477-486

Komalavilas P, Lincoln TM (1996) Phosphorylation of the inositol inositol 1,4,5-trisphosphate receptor: cyclic GMP-dependent protein kinase mediates cAMP and cGMP dependent phosphorylation in the intact rat aorta. J Biol Chem 271:21933-21938

Kotlikoff MI, Kamm KE (1996) Molecular mechanisms of beta-adrenergic relaxation of airway smooth muscle. Annu Rev Physiol 58:115-141

Kowarski D, Shuman H, Somlyo AP, Somlyo AV (1985) Calcium release by norad-renaline from central sarcoplasmic reticulum in rabbit main pulmonary artery smooth muscle. J Physiol (London) 366:153-175

Kureishi Y, Kobayashi S, Amano M, Kimura K, Kanaide H, Nakano T, Kaibuchi K, Ito M. (1997) Rho-associated kinase directly induces smooth muscle contraction through myosin light chain phosphorylation. J Biol Chem 272:12257-12260

LaBelle EF, Murray BM (1990) Differences in inositol phosphate production in rat tail artery and thoracic artery. J Cell Physiol 144:391-400

LaBelle EF, Polyák E (1996) Phospholipase C β2 in vascular smooth muscle. J Cell Physiol 169:358-363

Laporte R, Laher I (1997) Sarcoplasmic reticulum-sarcolemma interactions and vascu-lar smooth muscle tone. J Vasc Res 34:325-343

Lee MW, Seversen DL (1994) Signal transduction in vascular smooth muscle: diacyl-glycerol second messengers and PKC action. Am J Physiol 267:C659-C678

Lee MW, Ti L, Kitazawa T (1997) Cyclic GMP causes Ca^{2+}-desensitization in vascular smooth muscle by activating the myosin light chain phosphatase. J Biol Chem 272:5063-5068

Lesh RE, Marks AR, Somlyo AV, Fleischer S, Somlyo AP (1993). Anti-ryanodine receptor antibody binding sites in vascular and endocardial endothelium. Circ Res 72:481-488

Lesh RE, Nixon GF, Fleischer S, Airey JA, Somlyo AP, Somlyo, AV (1998) Localization of ryanodine receptors in smooth muscle. Circ Res 82:175-185

Leung T, Manser E, Leung T, Hall C. (1996) Regulation of phosphorylation pathways by p21 GTPases. The p21 ras-related rho subfamily and its role in phosphorylation signalling pathways. Eur J Biochem 242:171-185

Leung T, Manser E, Tan L, Lim L (1995) A novel serine/threonine kinase binding the Ras-related RhoA GTPase which translocates the kinase to peripheral membranes. J Biol Chem 270:29051-29054

Lincoln TM, Komalavilas P, Cornwell TL (1994) Pleiotropic regulation of vascular smooth muscle tone by cyclic GMP-dependent protein kinase. Hypertension 23:1141-1147

Mackrill JJ, Challiss RAJ, O'Connell DA, Lai FA, Nahorski SR (1997) Differential expression and regulation of ryanodine receptor and myo-inositol 1,4,5-trisphos-phate receptor Ca^{2+} release channels in mammalian tissues and cell lines. Biochem J 327:251-258

Malmqvist U, Trybus KM, Yagi S, Carmichael J, Fay FS (1997) Slow cycling of unphos-phorylated myosin is inhibited by calponin, thus keeping smooth muscle relaxed. PNAS 94:7655-7660

Marks AR, Tempst P, Chadwick CC, Riviere L, Fleischer S, Nadal-Ginard B. (1990) Smooth muscle and brain inositol 1,4,5-trisphosphate receptors are structurally and functionally similar. J Biol Chem 265:20719-20722

Marrero MB, Paxton WG, Duff JL, Berk BC, Bernstein KE (1994) Angiotensin II stimulates tyrosine phosphorylation of phospholipase C-γ1 in vascular smooth muscle cells. J Biol Chem 269:10935-10939

Marston S (1995) Ca^{2+}-dependent protein switches in actomyosin based contractile systems. Int J Biochem Cell Biol 27:97-108

Matsui T, Amano M, Yamamoto T, Chihara K, Nakafuku M, Ito M, Nakano T, Okawa K, Iwamatsu A, Kaibuchi K. (1996) Rho-associated kinase, a novel serine/threonine kinase, as a putative target for the small GTP binding protein Rho. EMBO J 15:2208-2216

Matsuo M, Reardon S, Ikebe M, Kitazawa T (1994) A novel mechanism for the Ca^{2+}-sensitizing effect of protein kinase C on vascular smooth muscle: inhibition of myosin light chain phosphatase. J Gen Physiol 104:265-286

McDaniel NL, Rembold CM, Murphy RA (1994) Cyclic nucleotide dependent relaxation in vascular smooth muscle. Can J Physiol Pharmacol 72:1380-1385

Murphy RA (1994) What is special about smooth muscle? The significance of covalent crossbridge regulation. FASEB J 8:311-318

Murphy RA, Walker JS, Strauss JD (1997) Myosin isoforms and functional diversity in vertebrate smooth muscle. Comp Biochem Physiol 117B:51-60

Nishikawa M, de Lanerolle P, Lincoln PM, Adelstein RS (1984) Phosphorylation of mammalian myosin light chain kinases by the catalytic subunit of cyclic AMP-dependent protein kinase and by cyclic GMP-dependent protein kinase. J Biol Chem 259:8429-8436

Nishikawa M, Hidaka H, Adelstein RS (1983) Phosphorylation of smooth muscle heavy meromyosin by calcium-activated, phospholipid-dependent protein kinase. The effect on actin-activated MgATPase activity. J Biol Chem 258:14069-14072

Nishimura J, Moreland S, Ahn HY, Kawase T, Moreland RS, van Breemen C (1992) Endothelin increases myofilament Ca^{2+}-sensitivity in alpha-toxin-permeabilized rabbit mesenteric artery. Circ Res 71:951-959

Nishimura J, van Breemen C (1989) Direct regulation of smooth muscle contractile elements by second messengers. Biochem Biophys Res Comm 163:929-935

Nishiye (1993) The effects of MgADP on cross-bridge kinetics: a laser flash photolysis study of guinea-pig smooth muscle. J Physiol 460:247-271

Nixon GF, Iizuka K, Haystead CMM, Haystead TAJ, Somlyo AP, Somlyo AV (1995) Phosphorylation of caldesmon by mitogen-activated protein kinase with no effect on Ca^{2+} sensitivity in rabbit smooth muscle. J Physiol 487:283-289

Otto B, Steusloff A, Just I, Aktories K, Pfitzer G (1996) Role of Rho proteins in carbachol-induced contractions in intact and permeabilized guinea-pig intestinal smooth muscle. J Physiol 496:317-329

Pfitzer G, Merkel L, Rüegg JC, Hofmann F (1986) Cyclic GMP-dependent protein kinase relaxes skinned fibers from guinea pig taenia coli. J Pflug Arch 407:87-91

Pijuan V, Sukholutskaya I, Kerrick WG, Lam M, van Breemen C, Litosch I. (1993) Rapid stimulation of Ins (1,4,5)P3 production in rat aorta by NE: correlation with contractile state. Am J Physiol 264:H126-H132

Pozzan T, Rizzuto R, Volpe P, Meldolesi J (1994) Molecular and cellular physiology of intracellular calcium stores. Physiol Rev 74:595-636

Raeymakers L, Wuytack F (1996) Calcium pumps. In: Bárány M (ed) Biochemistry of smooth muscle contraction. Academic Press, San Diego, pp 241-253

Rembold CM (1990) Modulation of the $[Ca^{2+}]$ sensitivity of myosin phosphorylation in intact swine arterial smooth muscle. J Physiol 429:77-94

Rembold CM, Weaver BA (1990) $[Ca^{2+}]$, not diacylglycerol, is the primary regulator of sustained swine arterial smooth muscle contraction. Hypertension 15:692-698

Shimada T, Somlyo AP (1992) Modulation of voltage-dependent Ca channel current by arachidonic acid and other long-chain fatty acids in rabbit intestinal smooth muscle. J Gen Physiol 100:27-44

Shimizu H, Ito M, Miyahara M, Ichikawa K, Okubo S, Konishi T, Naka M, Tanaka T, Hirano K, Hartshorne DJ, Nakano T. (1994) Characterization of the myosin-binding subunit of smooth muscle myosin phosphatase. J Biol Chem 269:30407-30411

Shirazi A, Iizuka K, Fadden P, Mosse C, Somlyo AP, Somlyo AV, Haystead TAJ. (1994) Purification and characterization of the mammalian light chain phosphatase holoenzyme. J Biol Chem 269:31598-31606

Shirinsky VP, Vorotnikov AV, Birukov KG, Nanaev AK, Collinge M, Lukas TJ, Sellers JR, Watterson DM. (1993) A kinase-related protein stabilized unphosphorylated smooth muscle myosin minifilaments in the presence of ATP. J Biol Chem 268:16578-16583

Silver DL, Vorotnikov AV, Watterson DM, Shirinsky VP, Sellers JR. (1997) Sites of interaction between kinase-related protein and smooth muscle myosin. J Biol Chem 272:25353-25359

Smith JB, Smith L, Higgins BL (1985) Temperature and nucleotide dependence of calcium release by myo-inositol 1,4,5-trisphosphate in cultured vascular smooth muscle cells. J Biol Chem 260:14413-14416

Sobieszek A, Borkowski J, Babiychuk VS (1997) Purification and characterization of a smooth muscle myosin light chain kinase-phosphatase complex. J Biol Chem 272:7034-7041

Somlyo AP (1984) Cellular site of calcium regulation. Nature 308:516-517

Somlyo AP (1993) Myosin isoforms in smooth muscle: how may they affect function and structure? J Mus Res Cell Motil 14:557-563.

Somlyo AP (1997) Signal transduction: Rhomantic interludes raise blood pressure. Nature (News & Views) 389:908-911

Somlyo AP, Devine, CE, Somlyo, AV, North, SR (1971) Sarcoplasmic reticulum and the temperature-dependent contraction of smooth muscle in calcium-free solutions. J Cell Biol 51:722-741

Somlyo AP, Himpens B (1989) Cell calcium and its regulation in smooth muscle. FASEB J 3:2266-2276

Somlyo AP, Kitazawa T, Himpens B, Matthijs G, Horiuti K, Kobayashi S, Goldman YE, Somlyo AV (1989) Modulation of Ca^{2+}-sensitivity and of the time course of contraction in smooth muscle: a major role of protein phosphatases? Adv Prot Phosphatases 5:181-195

Somlyo AP, Somlyo, AV (1968a) Vascular smooth muscle: I. Normal structure, pathology, biochemistry, and biophysics. Pharmacol Rev 20:197-272

Somlyo AP, Somlyo AV (1994) Signal transduction and regulation in smooth muscle. Nature 372:231-236

Somlyo AP, Walker JW, Goldman YE, Trentham DR, Kobayashi S, Kitazawa T, Somlyo AV (1988) Inositol trisphosphate, calcium and muscle contraction. Phil Trans R Soc Lond B 320:399-414

Somlyo AV (1980) Ultrastructure of vascular smooth muscle. *The Handbook of Physiology: The Cardiovascular System, Vol. II: Vascular Smooth Muscle.* DF Bohr, AP Somlyo, HV Sparks, eds., American Physiological Society, Bethesda, Md., pp. 33-67.

Somlyo AV, Bond, M, Somlyo, AP Scarpa, A (1985) Inositol trisphosphate- induced calcium release and contraction in vascular smooth muscle. Proc Natl Acad Sci USA 82:5231-5235

Somlyo AV, Franzini-Armstrong C (1985) New views of smooth muscle structure using freezing, deep-etching and rotary shadowing. Experientia 41: 841-856

Somlyo AV, Horiuti K, Trentham DR, Kitazawa T, Somlyo AP (1992) Kinetics of Ca^{2+} release and contraction induced by photolysis of caged D-myo-inositol 1,4,5-trisphosphate in smooth muscle: The effects of heparin, procaine and adenine nucleotides J Biol Chem 267:22316-22322

Somlyo AV, Somlyo, AP (1968b) Electromechanical and pharmacomechanical coupling in vascular smooth muscle. J Pharmacol Exp Therap 159:129-145

Somlyo AV, Somlyo AP (1971) Strontium accumulation by sarcoplasmic reticulum and mitochondria in vascular smooth muscle. *Science* 174:955-958.

Steusloff A, Paul E, Semenchuk LA, Di Salvo J, Pfitzer G (1995) Modulation of Ca^{2+} sensitivity in smooth muscle by genistein and protein tyrosine phosphorylation . Arch Biochem Biophys 320:236-242

Stull JT, Hsu L-C, Tansey MG, Kamm KE (1990) Myosin light chain kinase phosphorylation in tracheal smooth muscle. J Biol Chem 265: 16683-16690

Tang D-C, Kubota Y, Kamm KE, Stull JT (1993) GTPγS-induced phosphorylation of myosin light chain kinase in smooth muscle. FEBS Lett 331:272-275

Tansey MG, Luby-Phelps K, Kamm KE, Stull JT (1994) Ca^{2+}-dependent phosphorylation of myosin light chain kinase decreases the Ca^{2+}-sensitivity of light chain phosphorylation within smooth muscle cells. J Biol Chem 269:9912-9920

Tokui T, Ando S, Ikebe M. (1995) Autophosphorylation of smooth muscle myosin light chain kinase at its regulatory domain. Biochemistry 34:5173-5179

Tokui T, Brozovich F, Ando S, Ikebe M (1996) Enhancement of smooth muscle contraction with protein phosphatase inhibitor 1: activation of inhibitor 1 by cGMP-dependent protein kinase. Biochem Biophys Res Comm 220:777-783

Trinkle-Mulcahy L, Ichikawa K, Hartshorne DJ, Siegman MJ, Butler TM (1995) Thiophosphorylation of the 130-kDa subunit is associated with a decreased activity of myosin light chain phosphatase in alpha-toxin-permeabilized smooth muscle. J Biol Chem 270:18191-18194

Uehata M, Ishizaki T, Satoh H, Ono T, Kawahara T, Morishita T, Tamakawa H, Yamagami K, Inui J, Maekawa M, Narumiya S. (1997) Calcium sensitization of smooth muscle mediated by a Rho-associated protein kinase in hypertension. Nature 389:990-994

Umemori H, Inoue T, Kume S, Sekiyama N, Nagao M, Itoh H, Nakanishi S, Mikoshiba K, Yamamoto T (1997) Activation of the G protein Gq/11 through tyrosine phosphorylation of the α subunit. Science 276:1878-1881

Van Riper DA, Chen XL, Gould EM, Rembold CM (1996) Focal increases in $[Ca^{2+}]_i$ may account for apparent low Ca^{2+}-sensitivity in swine carotid artery. Cell Calcium 19:501-508

Van Riper D, Weaver BA, Stull JT, Rembold CM (1995) Myosin light chain kinase phosphorylation in swine carotid artery contraction and relaxation. Am J Physiol 268:H2466-H2475

Villa A, Podini P, Panzeri MC, Sölling HD, Volpe P, Meldolesi (1993) The endoplasmic-sarcoplasmic reticulum of smooth muscle: immunocytochemistry of vas deferens fibers reveals specialized subcompartments differently equipped for the control of Ca^{2+} homeostasis. J Cell Biol 121:1041-1051

Walker JW, Somlyo AV, Goldman YE, Somlyo AP, Trentham DR (1987) Kinetics of smooth and skeletal muscle activated by laser pulse photolysis of caged inositol 1,4,5-trisphosphate. Nature 327:249-251

Walker LA, Gailly P, Jensen P, Somlyo AP, Somlyo AV (1998) The unimportance of being (protein kinase C) epsilon. FASEB J 12:813-821

Walsh M (1991) Calcium-dependent mechanisms of regulation of smooth muscle contraction. Biochem Cell Biol 69:771-800

Walsh MP, Andrea JE, Allen BG, Clement-Chomienne O, Collins EM, Morgan KG (1994) Smooth muscle protein kinase C1. Can J Physiol 72:1392-1399

Wei Y, Zhang Y, Derewenda U, Liu X, Minor W, Nakamoto RK, Somlyo AV, Somlyo AP, Derewenda ZS (1997) Crystal structure of RhoA-GDP and its functional implications. Nature Struct Biol 4:699-703

Word RA, Tang D-C, Kamm KE (1994) Activation properties of myosin light chain kinase during contraction/relaxation cycles of tonic and phasic smooth muscles. J Biol Chem 269:21596-21602

Wu X, Haystead TAJ, Nakamoto RK, Somlyo AV, Somlyo AP. Acceleration of regulatory myosin light chain dephosphorylation and relaxation of smooth muscle by telokin and cyclic nucleotide-activated kinase. J Biol Chem 273:11362-11369

Wu X, Somlyo AV, Somlyo AP (1996) Cyclic GMP-dependent stimulation reverses G-protein-coupled inhibition of smooth muscle myosin light chain phosphatase. Biochem Biophys Res Comm 220: 658-663

Yoshida M, Suzuki A, Itoh T (1994) Mechanisms of vasoconstriction induced by endothelin-1 in smooth muscle of rabbit mesenteric artery. J Physiol 477:253-265

Molecular and Cellular Phenotypes and Their Regulation in Smooth Muscle

Saverio Sartore[1,2], Rafaella Franch[1], Marleen Roelofs[1], and Angela Chiavegato[1]

[1]Department of Biomedical Sciences, University of Padua, I-35121 Padua, Italy
[2]National Research Council Unit for Muscle Biology and Physiopathology, Padua, Italy

[There is a history in all men's lives
Figuring the nature of the times deceas'd;
The which observ'd, a man may prophesy,
With a near aim, of the main chance of things
As yet not come to life, who in their seeds
And weak beginning lie intreasured.]
Henry IV, William Shakespeare

Contents

1
Introduction

In the last decade a huge amount of information has been accumulated pertaining to the structural biology and physiology of smooth muscle (SM) cells (SMC), particularly those of the vascular smooth muscle. This unquestionable success in the field may potentially have a profound impact for an understanding of the basic mechanisms that underline some pathological processes involving SMC. The basic features of some vascular diseases, such as atherosclerosis, hypertension, restenosis after angioplasty and vein-graft, are close to being disclosed by the extraordinary efforts of both cell and molecular biology investigations.

Understanding of the regulatory mechanisms that control growth and differentiation in SMC is still unsatisfactory, especially if compared with the existing levels of knowledge about striated muscle biology. However, the recent experimental achievements obtained with vascular and non-vascular SMC are encouraging and the gap with the skeletal and cardiac muscle is closing. Interestingly, in the course of experimental and spontaneous conditions of regeneration/repair of adult SM [1,2] tissue, SMC undergo a complex phenotypic change which is the expression of the (partial or complete?) recapitulation of the developmental process [1,2]. In this circumstance the phenotypic behavior displayed by SMC is basically similar to the skeletal muscle response-to-injury and cardiac muscle abnormality [3–6]. Thus, all three muscle tissues retrieve an "embryonic-fetal" growth/differentiation program when alteration or loss of muscle tissue occurs, possibly indicating the existence of a common "default"-type phenotypic response expressed in all the muscle cell lineages. The expression of some SM-specific protein markers in skeletal and cardiac muscle tissues in the early stages of development, along with their transient reiteration in some pathological conditions, might be indicative of similarities among the "immature" cell phenotypes of the three muscle cell lineages.

Based on these data, it is reasonable to assume that the structural and functional delineation of the basic features of developing SM tissues is of paramount importance in establishing the pathogenetic mechanisms which govern the processes implicating cellular hyperplasia, hypertrophy and migration in the adult [2]. In addition, the molecular and cellular *in vitro* studies of SMC growth and differentiation have revealed important results about SMC phenotypic stability and heterogeneity that are of particular interest in the proper interpretation of SMC proliferation and mobilization occurring in atherosclerosis and restenosis. The goal of this review is to discuss: (1) the biology of vascular and non-vascular SM tissues; (2) the phenotypic changes that characterize the SMC lineage(s); (3) growth and differentiation of the SMC *in vitro*; (4) non-muscle cells as an alternative source of new SMC; and (5) experimental models involving proliferation/migration of SMC whose characterization is certainly helpful to the understanding of human vascular diseases.

Since the vascular SM tissue/cells have been the most throughly studied by cell biologists in comparison with the other SMs, the discussion will be devoted primarily to this tissue.

2
The morphological organization of the adult and developing smooth muscle

In the normal adult, the different histological, electrophysiological, biochemical, immunochemical and pharmacological properties displayed by the various SM tissues and cells might be tentatively typified in a "SM tree" (Fig. 1; see also text ahead). As far as the arterial SM tissue is concerned, the end-point of this "classification tree" is represented by the two wall layers: the intima and the media [7]. The intima is composed of a single layer of endothelial cells, the subendothelial membrane and the internal elastic

———————————————————————▶

Fig. 1. Flow diagram showing the morphological, electrophysiological, biochemical, immunochemical and pharmacological differences in the levels of SM organization. The five SM tissues depicted (vascular, gastrointestinal, ocular, respiratory and urogenital) are differently connected to the "classification tree", whose complexity progressively diminishes from the bottom to the top where SMC subpopulations or clones are expressed. For sake of clarity, SM types that display "mixed" structural-functional properties (such as the penile *corpus carvernosus*, i.e., blood vessels that behave as sphincters) have been omitted. Pericytes, mioepithelial cells and myofibroblasts are cells which share some structural and functional properties with the SM tissues/cells. Ductus arteriosus has also been included in this diagram for the peculiar structural changes that undergoes around birth (see text). a, Artery; b, branch; lon, longitudinal; circcircular.

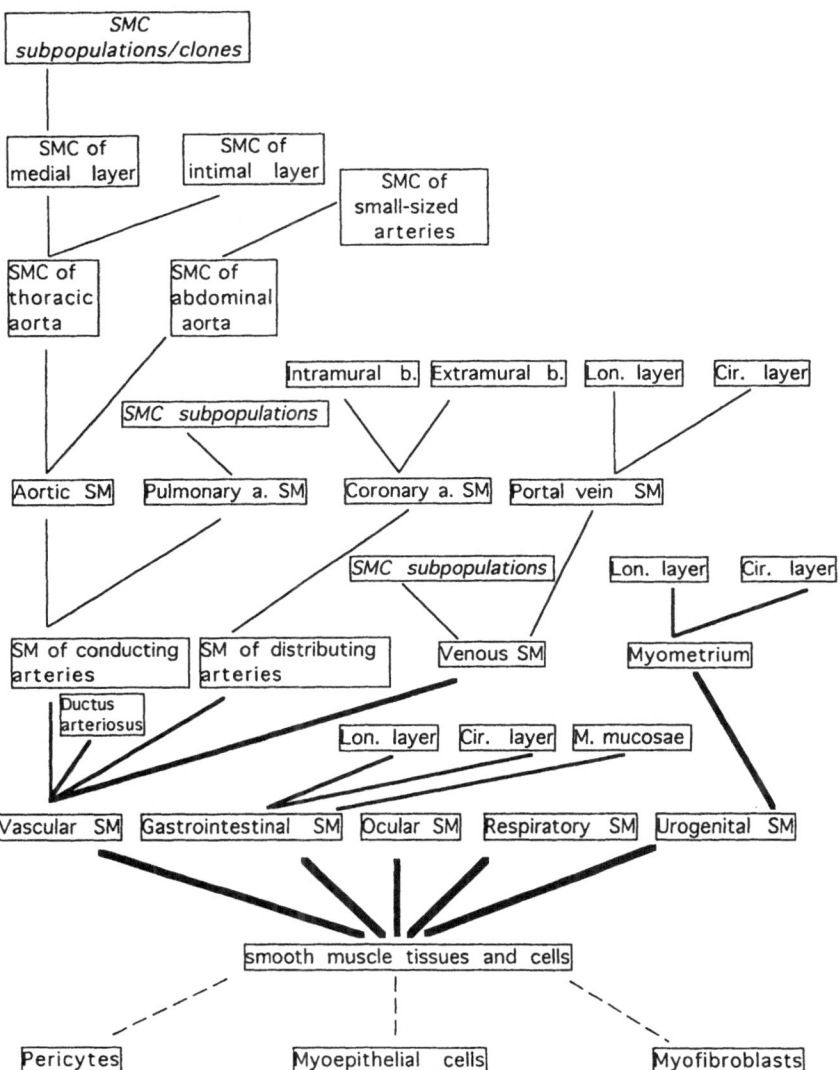

membrane. According to the species, the subendothelial space may contain a variable number of SMC [7]. The media comprises SMC interspersed with elastic fibers, the outermost of which is represented by the external elastic lamina. Three structural-functional interfaces can be identified in the arterial wall: the blood-endothelium, the endothelium-SM tissue and the medial SM-adventitia (Fig. 2). In avian species, however, the organization is more complex inasmuch as fibroblast-like (interlamellar) cells and

SMC (lamellar) cells are co-segregated in the tunica media [8,9]. The former cell type lacks a basement membrane, myofilaments and dense bodies. In addition, bovine aorta and pulmonary artery SMC can have distinct structural profiles as regards shape, location, cell orientation and pattern of elastic *laminae* [10–13].

The epithelium and the serosal layer play an important role in maintaining the structural-functional integrity of visceral SM tissue. Erosion of pseudostratified epithelium of the bladder wall [14] or serosal necrosis [15] have direct consequences on the stability of the differentiation pattern of SMC. Two [16] or possibly three [17] interfaces can also be shown here: the epithelial-SM tissue and the SM tissue-serosa (the submesothelial mesenchymal cells; Fig. 2).

The basic morphology of SMC in the various organs is very similar and related to contractile performance shared by all the SM tissues. In fact, the cytoplasm of fully matured SMC in the adult is abundantly filled with myofilaments, attachment bodies and peripheral vesicles but scarcely equipped with rough endoplasmic reticulum and Golgi complex [18,19]. Despite this common structural profile and the general dislocation of SMC in hollow organs, the functional behavior of SM tissues is heterogeneous, ranging from the phasic response of faster contracting visceral SMs to the tonic response of slow contracting SMs from large arteries [20]. The distinct physiological performance of vascular and visceral SMs develops in a unique structural context. For example, in the media layer of large arteries subjected to large pulsatile pressure fibrocollagenous tissue is present, whereas in the bladder muscle bundles of SMC are arranged in spatially-oriented layers with scarce connective tissue [21]. In the arterial media, single SMC are covered by a basement membrane and embedded in and connected to the extracellular matrix (ECM) meshwork.

Fig. 2. Parietal organization of vascular and bladder wall and the functional interfaces existing between blood/urine *vs.* endothelium/epithelium, endothelium/epithelium *vs.* SM tissue, and SM tissue vs. adventitia/serosa. Additional functional regions might be identified in the intestinal wall between the longitudinal and circular SM layers (presence of a dense innervation) and around the *muscularis mucosae*, whose embryological formation and functional interaction with the epithelium during development is still unclear.

Three continous layers can be shown in the wall of embryonic human aorta. The innermost layer contains undifferentiated SMC, the intermediate layer consists of differentiated cells. In the outermost layer, possibly destinated to become the tunica adventitia, there are fibroblasts/generic mesenchymal cells [22]. At the same time, SMC become linked to the developing elastic *laminae* by bundles of microfibrils [23], giving rise to "contractile-elastic units", i.e., a continous line of structures that link adjacent elastic *laminae*. Contacts between differentiating SMC and endo-

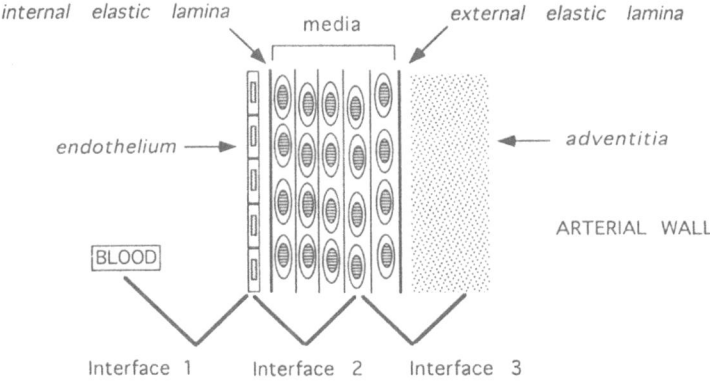

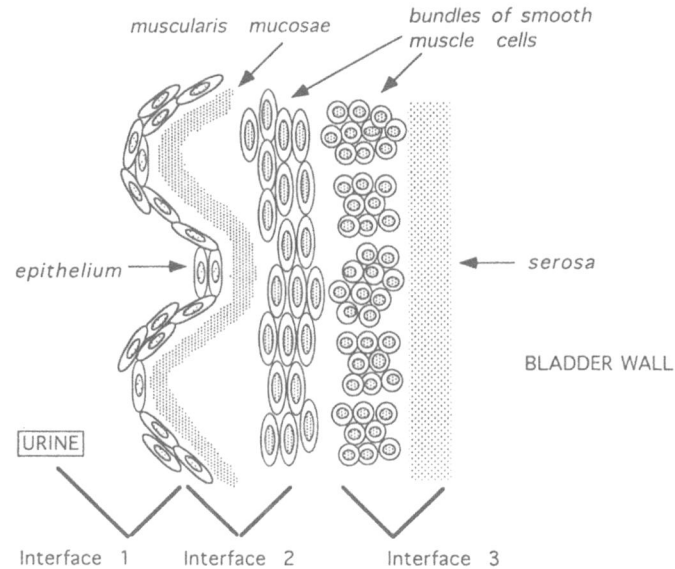

thelial cells ("myoendothelial" contacts) in the form of cytoplasmic projections can also be demonstrated in human fetal aorta [24]. Such contacts might permit the activation of paracrine mechanisms involved in the differentiation process of vascular SMC.

The basic structure of arterial vessels is largely developed at birth [25]. The cytoplasm of medial SMC contains the apparently complete array of filaments and subcellular structures of adult-type SMC [26]. As to the extracellular components, elastin fiber content increases during prenatal and postnatal periods. *In utero*, this occurs particularly with the slowing down of proliferation rate of SMC, concomitantly with the beginning of the fetal phase [27] and postnatally in response to changing hemodynamic conditions and wall stress [28,29]. Elastic fibers continue to develop until complete elastic laminae are formed (about 4 weeks from birth; [30]).

The intima layer in almost all the arterial vessels develops postnatally in some species, such as humans and pig, whereas in other (e.g., rabbit) it is lacking [7,11]. As an exception to this rule, "intimal cushions" spontaneously form before birth in the ductus arteriosus, contributing to its closure [31–33]. As concerns the adventitial layer, it is not clear when and how the directional migration of mesenchymal cells from the periendothelial region to the growing vessel wall, with the inherent morphological cell conversion to SMC [34,35], ceases and a definitive adventitial layer is formed. In particular, it should be established whether "competent" periendothelial mesenchymal cells are completely exausted after the migration/incorporation process or, alternatively, a subpopulation remains even in the adult and becomes active on demand (see Section 8).

In the intestine, at approximately mid-gestation the organogenesis is completed and the individual muscle layers are already detectable [36], though their final size is reached postnatally by a combined hyperplasia and hypertrohy of individual SMC [37]. Developing SMC contain numerous free ribosomes, and abundant rough endoplasmic reticulum, few *caveolae* and scarce dense bodies and myofilaments. In the developing chicken gizzard it has been seen by an ultrastructural analysis that assembly of contractile and cytoskeletal elements is orchestrated temporally according to seven consecutive stages [38]. The final differentiated SMC structure can be visualized one week after hatching [37]. Proliferating SMC can be seen up to day 20 *in ovo* [37,38] and mitotic cells display a high degree of differentiation indicating that the acquisition of a differentiated SMC phenotype does not imply a proliferation blockade [37].

3
Molecular and functional diversity in adult and developing smooth muscle cells

The apparent homogeneity in the morphology of adult normal SM tissue is not confirmed when the distribution of some differentiation markers, particularly those inherent to contractility, is taken into account. The contractile activity of adult SM tissues is accomplished by means of specific sets of proteins whose qualitative and quantitative availability determines the final outcome of the functional performance. To fulfill the criterion of the integrated action in developing this function, "social" interactions among SMC, on the one hand, and between SMC and ECM, on the other, must be established [39]. The cell phenotype that confers the contractile activity to SMC is denominated the "contractile phenotype" [18,40,41]. Changes in the composition of extracellular milieu can have profound consequences on the differentiation profile of SMC. The high responsiveness of SMC to specific environmental cues coupled with the inherent phenotypic instability of SMC give rise to SMC showing the so-called "synthetic" phenotype, i.e., SMC almost deprived of myofilaments and enriched in organelles deputed to synthesis of ECM proteins [18,40,41]. Before discussing this point (see Sections 5 and 6), it is important to identify the "basic molecules" that are the hallmark of the "contractile" phenotype in the adult SMC.

The cytoskeletal and contractile proteins along with proteins that make the SMC responsive to enviromental influences (receptors, signal-transducing molecules, ion channels, etc.) are fully expressed in the "contractile" SMC phenotype (see Fig. 3) by a time-dependent differentiation-maturational process(es) specific to SMC lineage pathways (see Section 5). Though the "range" of phenotypical SMC stability (see Section 7) is quite limited, the expression of these proteins in the "contractile" SMC phenotype is strictly coordinated and regulated. In some experimental models and in pathological conditions, alterations of the growth pattern of adult SM tissues is accompanied by re-expression of embryonic/fetal-type proteins (Fig. 4). Thus, elucidation of the structural organization of genes encoding such proteins will be particularly useful for the understanding of the mechanisms responsible for induction or de-induction of gene expression (see Sections 7.1 and 7.2). In the arterial wall, inactivation or down-regulation of the genes encoding these "basic molecules" might be related

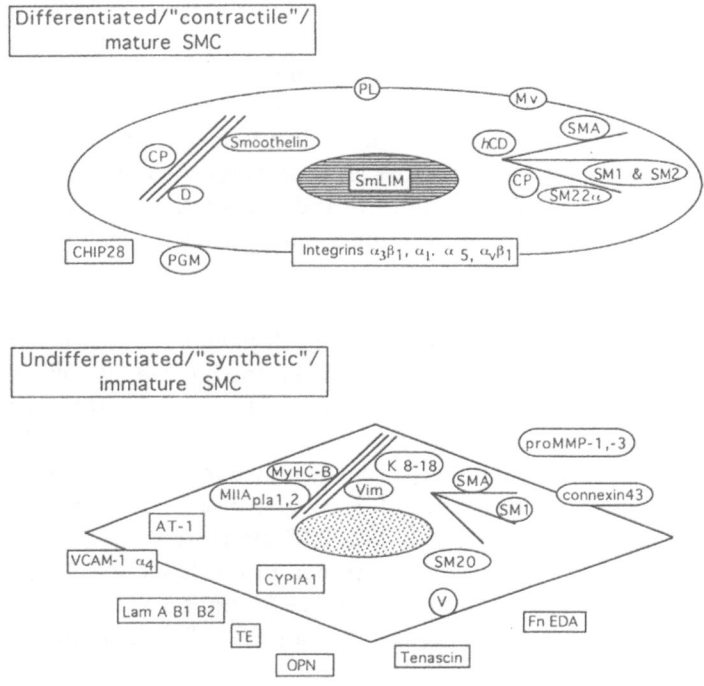

Fig. 3. Key: D, desmin; CP, calponin; h-CD, *h*-caldesmon; FnEDA, fibronectin EIIIA; Lam, laminin; Mv, meta-vinculin; PL, phospholamban; PGM, phosphoglucomutase-related protein; OPN, osteopontin; pro-MMP, pro-metalloproteinase; SMA, SM-type α-actin; SmLIM, SM-type LIM; SM20, 20 kDa SM-type protein; TE, tropoelastin; V, vinculin.

to the acquisition of specific competences for proliferation and/or migration.

Two general phenomena can be identified in the pattern of protein expression in the SMC lineage(s) from the embryonic/fetal to the adult phase of development: the up-regulation of SM - specific (Fig. 5) and the down-regulation of non-muscle (NM) - specific (Fig. 6) proteins. These latter molecules either belong to NM cell lineages (e.g., keratins) or are generically shared by many NM tissues (e.g., the NM myosins).

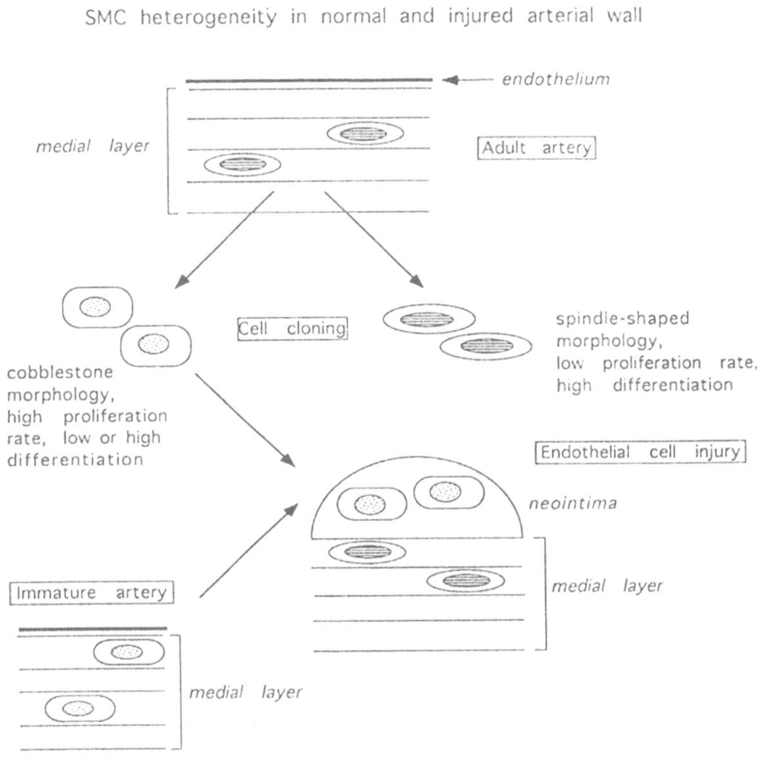

Fig. 4. SMC of immature artery, neointima and cloned SMC from adult artery share some common structural and functional properties.

Only the biochemistry of SMC lineage-specific SM proteins whose expression is up-regulated in the course of development, and that of NM-specific proteins down-regulated during SM development will be discussed here. The reader will find more information about these topics in the excellent review by GK Owens [2].

3.1
Up-regulated expression of smooth muscle-specific differentiation markers

3.1.1
Specific markers of developing-adult smooth muscle tissues

3.1.1.1
Smooth muscle-type myosin isoforms

The myosin molecule consists of six subunits: two heavy (MyHC) and four non covalently associated light (MyLC) chains. The MyHC form a dimer consisting of two globular NH_2-terminal heads and COOH-terminal α-helical coiled-coil tails. The heads contain the actin-activated Mg^{2+}-ATPase activity, whereas the tails are specifically involved in assembly of myosin molecules (for a review see [42, 43]. It is believed that myosin plays a crucial role in SMC contraction and NM cell motility through activation of Ca^{2+}/calmodulin-dependent MyLC kinase and phosphorylation of 20 kDa MyLC [43]. The two heavy chains (about 200 kDa) and the four light chains (17-20 kDa) are expressed as multiple isoforms in SM and NM systems and are differently regulated in tissue- and developmental-specific manners [42,44,45]. The NM-MyHC variants are also present in adult vascular and non-vascular SM tissues, though in very low amounts (see Section 3.2.1). SM-type MyHC are considered to be the only reliable marker of SMC lineage inasmuch as *in vivo* they are expressed exclusively in developing and adult SM tissues [46].

The various SM-MyHC isoforms are generated by distinct alternative splicing processes at 3'-COOH or 5'-NH_2 terminus trascript from a single gene [47–52] mapped to chromosome 16, region q12 according some authors [53] or region p13.13-p13.12 according to Deng et al. [54]. The two isoforms named SM1 and SM2 (204 and 200 kDa, respectively) differ at the COOH-terminal: SM2 contains 9 amino acids encoded by a unique 39 nucleotide exon, whereas SM1 contains a COOH terminal with 43 distinct amino acids [47,48].

The ratio SM1SM2 is variable from tissue to tissue and some differences appear to be species specific [10,11,55–60]. While SM1 is expressed early during development in vascular SM tissue of rat, rabbit and human, SM2 is detectable only postnatally (Figs. 5 and 6). In urinary bladder $SM_2 > SM_1$ before birth and $SM_2 < SM_1$ after birth [61], whereas in myometrium from non-pregnant rabbit $SM_2 < SM_1$ [60] and from non-pregnant rat only SM1 is expressed [62]. In porcine airway SM [63], stochiometry of SM isoforms

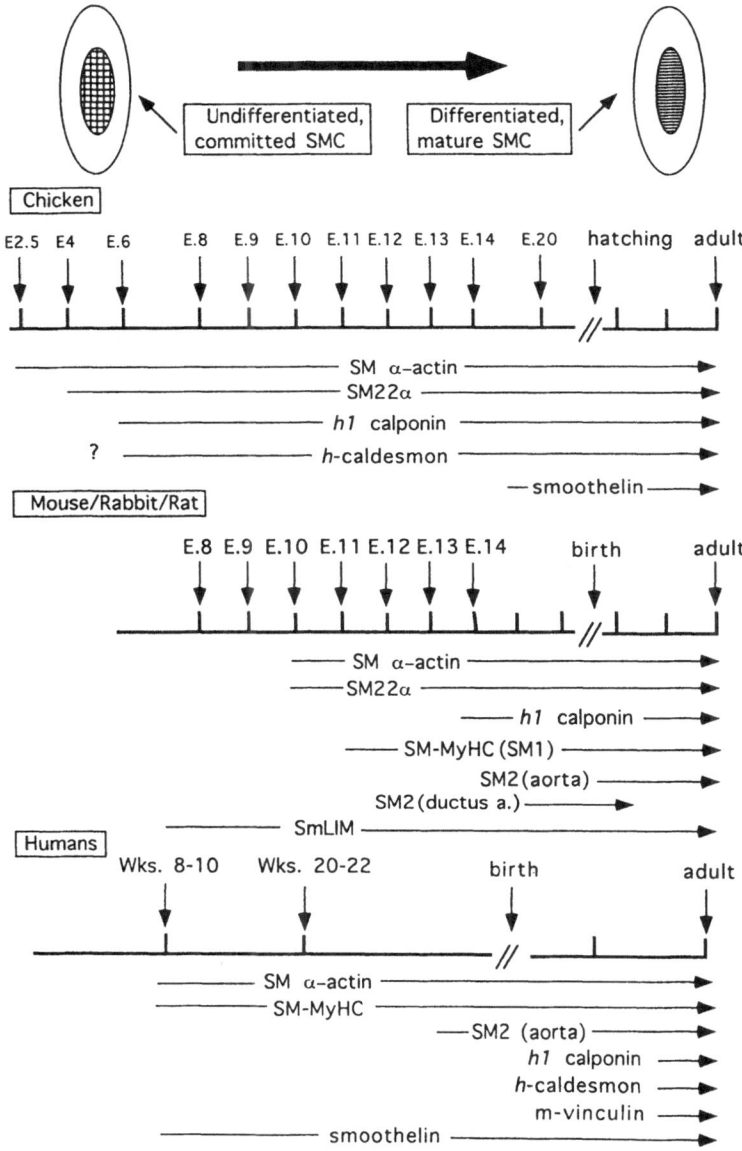

Fig. 5. Time-course of some SM-type protein markers during development in birds and mammals.

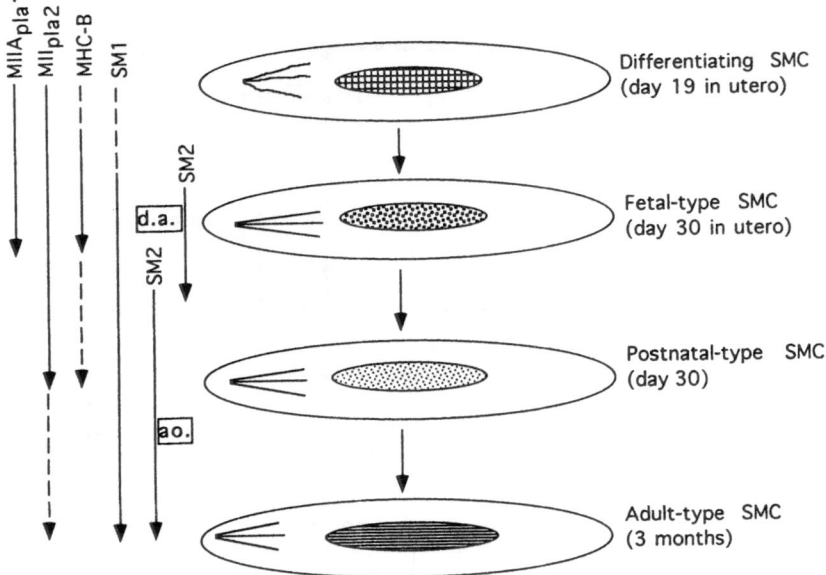

Fig. 6. Myosin isoforms and SMC phenotypes during rabbit development. d.a., ductus arteriosus; ao., aorta.

during development follows the same pattern described for the vascular SM tissue, whereas in human there is no variation from newborn to adult [64]. Interestingly, a differential expression of SM1 and SM2 is exhibited when SM tissue from airways is compared to that from pulmonary artery in rat [65,66].

It has been established that SM-MyHC from vascular and visceral SM tissues differ by an insert of 7 amino acids at the 25-/50-kDa junction of the S1 subfragment of the myosin head [50,51]. It is important to note that the two MyHC isoforms display distinct velocity of movement of actin filaments *in vitro* and actin-activated Mg^{2+}-ATPase activity. These data have been confirmed by Rovner et al. [67] who found that homogeneous populations of heavy meromyosin from tonic or phasic SM tissues incorporating or lacking the 7 amino acid insert near the active site of MyHC display different enzymatic and motility properties not influenced by the essential MyLC. However, other investigators have found a high structural-functional correlation at the level of 17 kDa MyLC exclusively [62,68].

Some information is now available about the transcription factors involved in regulating SM-MyHC in vascular, respiratory and visceral SM tissues and NM cells in different species. A combination of both positive and negative *cis*-acting promoter elements, CArG box elements, multiple E boxes, and CCTCCC sequences are required for SM-specific gene expression [69–73]. In addition, a differential gene regulation can be demonstrated in vascular and respiratory SMC [71] and a SM-specific enhancer exists in vascular and visceral SMC that might be activated by different levels of transcription factor(s) [70]. Some promoter sequences are well conserved in different species such as mouse, rat and rabbit [69-71], whereas others, such as the MEF2-like element is present in rabbit but not in the mouse [72]. Mutation of GC-rich sequences within the species-conserved 227-base pair domain of promoter gives rise to a marked increment of trascriptional activity [73]. A GC-rich sequence within CArG or CArG-like motifs in rat and CCTCCC squences in mouse act as negative *cis*-elements for trascription by members of the Sp1 family of transcription factors [72,73].

3.1.1.2
Smoothelin

The van Eys's group has recently reported that a 59 kDa protein that maps to chromosome 22 named smoothelin is exclusively expressed in the "contractile" SMC phenotype in evolutionarily different species [74–76]. The "synthetic" SMC phenotype expressed at early stages of development in humans [75] and in chicken [74] does not contain this protein (see also Fig. 5). Smoothelin is not expressed in striated muscles or in mixed SMC-NM cell phenotypes such as myofibroblasts, pericytic venules and myoepithelial cells [74,75]. In SMC, smoothelin is not localized in the intermediate filament network but probably with the actin cytoskeletal systems [76]. In the chicken gizzard, smoothelin exists as multiple isoelectric variants of the 59 kDa molecule, whereas in the vascular SM is present as a 95 kDa variant [76].

During development, smoothelin is detectable around day 18 *in ovo*, later than calponin which is expressed at embryonic day 5 (see Fig. 5). In human fetal vessels it appears at week 10–11 of gestation and the total number of arterial SMC declines thereafter becoming about 10 and 30–50% in the media of adult elastic and large muscular vessels, respectively [75]. Since it is largely represented in the media of small muscular arteries, van der Loop et al. [75] have hypothesized that this protein's function is in some way related to the pulsatile contraction of this latter category of blood vessels. Details of smoothelin physiology are to be determined.

3.1.1.3
SmLIM

A member of the LIM family of zinc-finger proteins named SmLIM is expressed in developmentally regulated manner in aortic SM tissue of rat [77]. Its presence is highly represented *in utero* at about 7.5 days *p.c.*, concomitantly with the embryonic-extraembryonic diversification of circulation and then it becomes down-regulated, keeping its mRNA level unchanged throughout later developmental stages (see also Fig. 5). In the adult, this protein is expressed to a lesser extent in other SM tissues (myometrium and intestine), barely detectable in the venous SM tissue and absent in striated muscles [77]. This 194-amino acid protein is localized in the nucleus and maps to chromosome 3 [77]. The LIM domain functions as a modular protein binding interface and for class 2 of this protein, i.e., the protein in which the homeodomain is lacking, it has been hypothetically assigned a regulatory role in SMC-specific gene expression as happens with muscle LIM-domain protein [77].

3.1.2.
Specific markers of adult smooth muscle tissues

αActin, SM22, calponin, and *h*-caldesmon are all protein markers specific for differentiated (adult) SM tissues (see Fig. 3) and present in sarcomeric and non-sarcomeric muscle early during development. This behavior might indicate that a distinct regulatory (negative) mechanism(s) exists which, though activated in a slightly different temporal manner, switches off the respective genes in non-SMC cells during late development. Alternatively or in addition, the SMC-specific gene program is kept switched on in developing SM tissues by a positive regulatory mechanism which might counterbalance the muscle-shared negative mechanism. Recent data, reviewed by Firulli and Olson [78], point out that unique combinations of transcription factors are likely to be responsible for the specific spatiotemporal expression of lineage-related muscle genes and, hence, for the final differentiation pattern in the muscle system.

3.1.2.1 Smooth muscle-type actin isoforms

Six isoforms of actin are expressed in mammalian SM tissues that are generated by distinct genes [79] and show an highly conserved amino acid sequences across species. The expression of actin isoforms is developmentally regulated in a temporal-spatial manner [80–83]. Fully differentiated SMC contain α- (mainly in vascular tissue) and γ-(mainly in the enteric structures) actins and trace amounts of β- and γ-cytoplasmic actin isoforms [79,84–87]. The SM α-isoform is expressed also in striated muscle [88], and the SM γ-isoform in post-meiotic sperm [89], and the increased SM γ-actin content in hypertrophied bladder SM [86,90]. The unique intracellular distribution of actin filamentous networks [91–93] and the decreased SM α-actin content in proliferating vascular SMC ([94]; see Section 6) point to an isoform diversification based on specific functional requirements.

In the adult and in developing animals, SM α-actin is not expressed in SMC solely. It has also been demonstrated in the cells with mixed SM and NM cell characteristics such as myofibroblasts and myoepithelial cells [95,96] as well as in cells of the eye lens, hair follicles, and bone marrow stromal cells [97–99]. During early development, SM α-actin is transiently expressed in cardiac and skeletal muscle tissue and its disappearance is concomitant with the achievement of maturation in the respective tissue [88,100,101].

Northen blot analysis has shown that in rat SM tissues a biphasic pattern of expression occurs for SM α- and γ- actin which peaks at about day 18 *p.c.* and at early postnatal time (day 4–5), respectively [82]. Interestingly, the developmental profile of isoactin expression is SM tissue specific and is basically maintained in the adult [82]. In the very early stages of development, the α-actin expression precedes that of γ-actin in the gastrointestinal, urogenital, respiratory and vascular SM tissues of the mouse [83]. In the chicken, SM-type α-actin has been detected as early as embryonic day 2–3 in the presumptive SMC in close contact with the developing endothelial tube (see Fig. 5 and Section 5; [102]). Similarly, α-actin is precociously expressed in the thick mesenchymal coat around the primitive airways in developing rat lung which presumably furnish the SMC to the branching epithelium of the respiratory system [103]. The α-actin achieves the adult levels in vascular SMC about one month after birth when other fetal SMC markers are markedly down regulated [42].

Work from Owens's laboratory has shed some light about the regulatory mechanisms of SM-type α-actin expression which might be involved in restricting the differentiative options in SM related cells and, perhaps, in the correct spatiotemporal expression of this isoactin in vascular SM tissue during development [104,105]. The 5'-flanking region of rat and chicken α-actin contains CArG elements and E boxes analogous to the SM-MyHC gene [73]. Different *cis* segments of this region, that act as positive or negative elements, control the transcription level in chicken myoblasts and fibroblasts, putatively using distinct mechanisms [104]. Relatively longer constructs containing 547 bp of the promoter display transcription activity in non-SMC which endogeneously express α-actin during some stage of their differentiation. On the contrary, in cells which normally do not express α-actin the construct has no efficacy. Site-directed mutagenesis experiments applied to the CArG motifs of α-actin promoter have different transcriptional activity depending on the type of cells used (SMC *vs.* endothelial cells; [105]). In addition, an electrophoretic mobility assay shows that CArG box-binding proteins (serum response factor (SRF) or SRF-like factor) or a complex of such proteins is involved, possibly via binding to multiple DNA elements, in regulating the endogeneous expression of α-actin in different cell types [105]. These data, along with those reported above and those obtained with SM22α promoter by Li et al. [106], point to the existence of specific transcription mechanism(s) for at least some SM-specific genes based on the interaction of SRF or SRF-like factor with CArG sequence [73].

3.1.2.2

SM22

SM22 is a 22 kDa protein relatively abundant in adult SMs [107] and not, apparently, associated with the cytoskeletal apparatus. Northen blotting analysis has demonstrated that SM22 is well represented in all SM tissues [108,109]. Isoelectric focusing has shown that SM22, which displays a sequence homology with the SM thin filament regulatory protein calponin and the *Drosophila* muscle protein mp20 [110] and NP 25 [111], exists as isoforms of distinct pI, named α, β, γ, and δ [107,112]. Peculiar combinations of SM22 variants, exhibited in tissue- and species-specific manner [107,112] are not the product of different levels of phosphorylation. The available monoclonal antibodies to SM22 show a uniform distribution of immunoreactivity in adult SM tissues [113]. SM22α is encoded by a single copy gene and the primary transcript does not undergo alternative splicing [109]. Conversely, during development in the chicken gizzard the α-isoform is detectable around day 10 in embryo (Fig. 5), whereas the β-isoform is up-regulated after hatching and the final combination α, β, γ is achieved in the adult [112]. SM22α is expressed by all muscle tissues early during development but becomes down-regulated in skeletal and cardiac muscle at later stages of gestation [114] and continously expressed in SM tissue [115]. Both *in situ* hybridization in the mouse [114] and immuno-histochemistry with E-11 monoclonal antibody to SM22 in the chicken [113] have revealed that the specific mRNA and the relative protein for SM22 are expressed precociously (Fig. 5). The protein is co-distributed with calponin and *h*-caldesmon, and appears later with respect to other markers such as tropomyosin, desmin, filamin, SM α-actin and MyLC kinase which are SM-specific in the adult, but shared with the other muscle or NM cells in embryo [113].

The murine SM22α gene and the 5'-flanking region have been studied by three laboratories. The promoter region contains *cis*-acting regulatory sequences: two CArG/SRF boxes, a CACC box, one potential MEF-2 binding site, and eleven E-boxes [106,109,116]. Analysis of transcriptional activity of SM22α promoter in transgenic mice provided evidence that, contrary to the endogeneous expression, constructs of this promoter are able to induce the reporter expression in arterial SM tissues but not in venous or visceral SM tissues [106,116]. Two-nucleotide mutations in the two SRF sites that eliminate SRF binding also abolish the promoter activity in the arterial SMC [117]. This finding is consistent with the concept that not all the SM tissues are created equal in the hitherto believed homogeneous SM system and support the notion that the development of a SM tissue-specific differentiation program is achieved via selective transcriptional regulation. Such mechanism involves multiple transcriptional factors whose SM-type specific combinatorial action would be responsible for the unique spatiotemporal expression of this protein [78,117,118].

3.1.2.3
Calponin and *h*-caldesmon

Calponin [119] and caldesmon [120] are two thin filament associated proteins that bind to F-actin, tropomyosin and calmodulin. Interaction of 34 kDa calponin with F-actin and tropomyosin takes place in a Ca^{2+}-independent manner, whereas that with calmodulin is regulated in a Ca^{2+}-dependent manner. The key role of calponin and caldesmon in SM is to down-regulate actomyosin ATPase activity *in vitro* [120,121]. Thus, they may participate in regulation of contractile performance. Despite their apparent functional similarity, sequence analysis indicates that calponin and caldesmon are not related proteins. They act by different mechanisms of inhibition and bind to distinct thin filament populations in SMC [120, 122, 123].

Both proteins exist as structural variants: *h1* (equivalent to chicken α-isoform) and *l* calponin (homologous to chicken β-isoform), and *h*- and *l*-caldesmon [124]. A new type of calponin (h2), containing 57 amino acids at the C00H-terminus with a strong acidic domain, is also expressed in adult rat SM and NM tissues/cells [125]. The *h1* calponin and *h*-caldesmon are mainly or exclusively found in SMC [126,127], whilst the *l*-caldesmon is widely expressed in NM tissues/cells. The molecular weight deduced from the respective cDNA for *h*-caldesmon and *l*-caldesmon is 87–93 kDa and 59–63 kDa, respectively. The two isoforms of caldesmon have identical sequences except for the insertion of the central repeating domain in *h*-caldesmon.

Based on the calponin expression pattern and nucleotide sequence differences it has been established that three genes generate the calponin variants [122,125,128]. Miano and Olson [127] reported that *h1* calponin transcription begins at two closely located initiation sites in the promoter. This is probably due to the absence of either a TATAA box or an initiator consensus element, which makes this promoter quite unique compared to those discussed above. In addition, transient transfection assay shows that this promoter is active in cell lines that do not express the endogeneous *h1* calponin gene. The presence of an upstream polypurine sequence in the promoter might potentially attenuate the level of transcription and, thus, hamper or exclude *h1* calponin in non-SMC cells [127].

Genomic structure analysis has revealed that caldesmon isoforms are encoded by a single gene mapped to 7q33-q34 locus [129], whose primary transcript is alternatively spliced. Using the same procedure it has been established that the expression of *h*- or *l*-caldesmon depends on a unique selection of two 5'-splice sites with exon 3 (the so-called gizzard-type and brain-type promoter; [130]). The gizzard-type promoter activity in "contractile" SMC is higher than in "synthetic" SMC and this correlates with the amount of *h*-caldesmon. Cell type specificity of promoter activity is strictly related to CArG1 element, whereas E-boxes are not directly involved. The fact that CArG1 is an important *cis*-regulatory element of caldesmon gene structure is demonstrated by gel shift assay showing interaction with nuclear protein factors [131].

Expression of *h1* calponin and *h*-caldesmon occurs very early during development [112, 113, 127, 132; see Fig. 5]. During mouse embryogenesis, *h1* calponin is transiently expressed in the developing heart (from 8.5 to

13.5 day *p.c.*), before its appearance in gastrointestinal, respiratory and vascular SM tissues (at day 13.5 *p.c.*; [127]). In chicken, analogously to SM22, only the α-isoform of calponin is initially expressed, followed by multiple variants after hatching [112]. In the same species, calponin and SM22 are coexpressed in vascular and gastrointestinal SM tissues beginning at embryonic day 6.5. The appearance of *h*-caldesmon, calponin and SM22 is delayed compared to other SM markers [113]. In humans, calponin and *h*-caldesmon appear relatively late in aortic SM during in utero development and after the SM-type MyHC [132,133]; Fig. 6). Interestingly, these two proteins are already present in visceral SM tissue at the time (10–20 wks of gestation) when they are barely detectable in vascular SM [132].

3.2
Down-regulated expression of non-muscle markers in smooth muscle tissues

While the majority of markers for differentiated SMC are located in the cytoplasm (linked with or associated to contractile filaments), those of "immature" SMC" belong mainly to the ECM or, to a lesser extent, plasma membrane (Fig. 3) and are shared by NM tissues/cells.

3.2.1
Non-muscle-type myosin isoforms

Multiple NM-MyHC isoforms are expressed in developing SM tissues of avian and mammalian species and become down-regulated to a various extent around birth [56,57,59,134–138]. These isoforms can be grouped into two MyHC-A and MyHC-B types, the latter being predominantly expressed in brain and the former essentially in muscle and in NM-non-brain tissues [139–141]. The entire gene sequences of the two major isoforms and their respective chromosomal locations (MyHC-A: chromosome 22q11.2; MyHC-B: chomosome 17p13) are now known [142–144]. The major difference between MyHC-A and MyHC-B relies on the final 20 amino acids of the COOH-terminal tail region of the molecule [142–144]. The two isoforms, analogous to SM-type MyHC (see Section 3.1.1.1), display markedly different actin-activated Mg^{2+}-ATPase activity and velocity of movement of actin filaments [51,145]. Additional isoforms are also produced by alternative splicing in correspondence of the 25/50 kDa (near the ATP binding domain) and the 50/20 kDa (near an actin binding domain) fragment, respectively, of the the so-called proteolytically cleaved subfragment-1 [140,146,147]. These alternatively spliced variants are confined, however, to the nervous system. The existence of MyHC-A and MyHC-B-related isoforms has been established on the basis of the immunoreactivity distribution of peptide-specific polyclonal antibodies [137,140,148] and platelet MyHC-specific monoclonal antibodies [42, 134, 149, 150]. Elzinga and colleagues [141] and our own group [42,150] have established that a platelet-type MyHC (MyHC$_{pla}$) is also expressed in SM tissues of vascular and non-vascular type [61, 66, 149]. Though the molecular physiology of these NM-MyHC isoforms in SM tissues is still unknown, some recent data obtained in NM cells seem to indicate that the A and B isoforms of NM-MyHC play a role in cell proliferation/migration of SMC. For example, antisense oligonucleotides to common NM-MyHC isoform sequences suppress SMC proliferation *in vitro* [151]. Experiments performed in *Xenopus* XTC cells show that alternatively spliced MyHC-B isoforms are phosphorylated at a specific serine residue by cdc2 kinase, whereas MyHC-A remains unphosphorylated [152]. Since this enzyme, along with the regulatory subunit cyclin, is a part of the protein complex (named maturation-promoting factor) controlling the cell cycle [153], it is possible that such an enzymatic process is of importance in the regulation of the proliferative level in SMC. In addition, a Mts-1 protein positively associated with metastatic phenotype in neoplastic cells forms a complex with MyHC-B myosin isoform [154]. These facts, along with the finding that MyHC-A and MyHC-B are differentially localized in human [155] and *Xenopus* [145] cells grown *in vitro* are indicative that the two isoforms have different functions within the NM cells. It must be elucidated what kind of interaction, if any, can be established between NM-MyHC and SM-MyHC isoforms in SMC thus resulting in the two "cytoplasmic domains" of cytoskeletal/cytocontractile apparatuses described by North et al. [93,123].

B- and A$_{pla}$-type NM-MyHC isoforms are down-regulated in arterial SM tissue during development in rabbit, rat and in humans [10,11, 56, 57, 138, 156]. In rabbit aortic media, MyHC-B disappears around birth (Fig. 6; [56]), whereas in the pig [157] and humans [138] it persists even in the adult. The human aortic *vasa vasorum* and intramural branches of coronary artery, however, are devoid of B-type NM-MyHC isoform [138]. Down-regulation of A$_{pla}$-type NM-MyHC has a different time-course with respect to MyHC-B. In rabbit, one isoform named A$_{pla1}$ disappears a few days after birth, whilst the other isoform named A$_{pla2}$ is down-regulated about 30 days from birth (Fig. 6; [42, 149, 150]. Trace amounts of A_{pla2} persist even in adults in some loci known to be prone to the development of proliferation/migration of medial SMC into the subendothelial space [158]. MyHC-B in human myometrium [59] and A_{pla2}-type MyHC in rabbit bladder [159] and myometrium [160] also undergo a down-regulation of during development. Taken together these data indicate that "immature" SMC display a "hybrid" myosin isoform content irrespective of the type of SM tissue. Thus, deinduction of NM myosin expression and induction of adult specific-SM2 isoform take place during the same range of time and represent the "passage" from the differentiation to the maturation step in the SMC cell lineage (see Section 5).

3.2.2
Keratin 8 and 18

It well documented that the intermediate filament proteins of keratin-type are expressed in SMC of adult human normal myometrium [161], umbilical vessels [162,163] and in occasional medial SMC of muscular arteries [164]. Keratin 8, and the timely associated fibrinectin variants A and B, are expressed in the aortic SMC of 10-week-old human fetus and become down-regulated thereafter [133]. On the basis of the heterogeneous presence of mesenchymal-type intermediate filament protein vimentin and of muscle-specific intermediate filament protein desmin in adult vascular SM tissues [84, 165], two categories of SMC can be identified with respect to keratin expression, i.e., those which coexpress the combinations of keratin-vimentin and those expressing keratin-desmin. Interestingly, the innermost layer of human umbilical arteries, where keratin 8 and 18 are present, does not contain desmin [162]. Though the possible developmental regulation of desmin expression has not yet been fully ascertained [2], the possibility might exist that the switching off of keratin gene expression early in development is associated with activation of desmin expression. The prevalence of keratins in fetal vascular SM tissue might be related to the high proliferation index of "immature" SMC, though in atherosclerotic lesions, where a proliferation/migration of medial SMC takes place, such a correlation is not supported by experimental data [166]. It worth noting that bladder SMC also does not evidence keratin expression during development or in experimental conditions characterized by SMC hyperplasia [167].

3.2.3
Integrins and vascular adhesion molecule

Cell-matrix interactions mediated through integrins play an essential role in modulating the differentiation level of SMC [39]. Though the various types of integrin studied so far in the vascular system display a marked tendency to increase their expression at late stages of development [168], two adherence proteins show an opposite regulation. The two adhesion proteins Very Late Antigen-4 (VLA-4; $\alpha_4\beta_1$) and Vascular Cell Adhesion Molecule-1 (VCAM-1), are markedly expressed in 10-week-old human fetal aorta, down-regulated at the 24th week of gestation and practically absent in the adult aortic SM tissue [169]. This finding contrasts with the distribution of α_4 during ontogenesis in mouse aorta, where apparently there is no down-regulation, and VCAM-1 expression is lacking in vascular SMC [170]. The fact that the expression of these two proteins is coordinated and associated with a specific SMC phenotype is supported by *in vitro* data. Blocking of VCAM-1/α_4 interaction inhibits the precocious expression of SM-type MyHC in cultured SMC grown in serum-free medium [170]. Since the VCAM-1/α_4 combination reappears in the intimal thickening, the hypothesis has been put forward that these two proteins are important in a potential re-differentiation process which might occur in the atherosclerosis lesion.

3.2.4
Fibronectins

Fibronectins are high molecular weight glycoproteins involved in different cellular activities, such as adhesion, migration and differentiation [39,171]. Fibronectin variants originate from alternative splicing from the single gene transcript at three sites named EIIIA, EIIIB and V segments [172]. In particular, if the V region is totally included in the transcript the CS-1 variant is produced. While EIIIA variant is always expressed in visceral SM tissues, both in early development and in the adult, EIIIB is almost exclusively present in the fetus, both in vascular and non-vascular SM tissues [173-175]. In aortic media, EIIIA is expressed at high levels only in the fetus, though it is also contained in the SMC of tunica intima of adult humans [174] but not in the media where low levels of IIICS fibronectin are expressed [176].

Experiments performed with rat aortic rings support the notion that the selective induction of EIIIA isoform in cultured SMC, grown in serum-free medium, is associated with a phenotypic shift in vascular SMC to a "synthetic" state [177]. EIIIA and EIIIB activation is also visible in experimental conditions and in vascular disease involving proliferation/migration and SMC hypertrophy in adult animals and humans [176, 178–181]. Here, it represents an early and long-lasting event and seems to be one of the first markers that parallels the appearance of the "synthetic" SMC phenotype" [181].

3.2.5
Tropoelastin

Synthesis of the elastin pro-protein tropoelastin during development is strictly related to the structural-functional modifications of arterial SMC. The synthesis of tropoelastin mRNA in rat aorta peaks at late fetal and early postnatal life, i.e., just after the burst of initial SMC replication and continuing after the *postpartum* drop of SMC proliferation [27, 182]. Sixty days after birth tropoelastin expression becomes undetectable in the tunica media [27]. The time course for tropoelastin expression in pulmonary artery follows the same trend [182,183]. Tropoelastin mRNA distribution in aorta and pulmonary artery forms complex patterns along both radial and longitudinal axes of the blood vessel in the sense that the elastogenic activity follows an orderly, sequential, proximal-to-distal deployment from the heart towards the vascular periphery. After birth, in the outer medial layer of rat pulmonary artery, elastogenesis declines more rapidly compared to the inner layer [184]. Results of Durmowicz et al. [185] in developing bovine pulmonary artery confirm the regional distribution reported above and point to existence of a biphasic pattern of tropoelastin mRNA expression. In early gestation expression of tropoelastin is present throughout the vessel wall, whereas in midgestation expression is decreased in the outer layer and maintained in the inner layer. In late gestation, expression is extended to the whole vessel wall though it gives a heterogeneous pattern of distribution. It has been suggested, by comparison with the distribution of SM-specific markers, that this pattern reflects the phenotypic changes of differentiating SMC and the hemodynamic forces which act perinatally in these vascular regions [185].

4
Gene expression and cell heterogeneity in adult and developing smooth muscle cells

As discussed above, the activation of a time-dependent gene program in SM tissues entails the coordinating up-/down-regulation of a number of molecular isoforms and largely contributes to the physiological remodeling of various organs [186]. This process is accomplished by a series of sequential changes of sets of molecules that identify discrete cell phenotypes (cell isoforms). Such a mechanism would permit the progressive adaptation of the (molecular) structure to the variable physiological conditions imposed by development. It is reasonable to assume that each step during this adaptive process is dictated by the combined effect of an endogeneous SMC commitment (see Section 5) and the responsiveness of SMC genes to enviromental cues (see Section 7).

The study of distribution of SM and NM proteins in SM tissues has allowed the identification of distinct SMC populations and their spatiotemporal dynamics during development. Though the existence of these SMC populations does not imply that the respective structural (molecular) differences are stable and heritable from generation to generation of cells, the information obtained is helpful to trace the differentation pathway that the SMC follow in some experimental models and in some pathological settings.

The first evidence for vascular SMC heterogeneity was obtained studying the distribution of the intermediate filament proteins desmin vs. vimentin in the rat [187]. Though these two proteins are not SMC specific, the data indicate that an extensive cellular heterogeneity exists and is dependent upon the vascular level considered. More recently, Johansson et al. [164] have shown that the desmin/vimentin distribution in human vessels is related to the specific anatomical location as also occurs in rabbit [188]. These authors have also established that a desmin-positive SMC population of aorta but not of femoral artery undergoes a phenotypic change with development.

Heterogeneity of SM-type α-actin isoform distribution, as determined by a sequence specific monoclonal antibody [189], in vascular SM tissue is especially evident in newborn rats where a population of SMC is negative for this marker [80,190]. Apparently, a comparison of the spatiotemporal distribution of vimentin/desmin/actin during development and adult is still lacking.

Using monoclonal antibodies to MyHC$_{pla}$ and to MyHC-A, we and others have shown that bovine aortic media is heterogeneous as regards SM- and NM-type MyHC isoform distribution [10,137]. In agreement with data from Stenmark's group [13,29], we also found that such a composition is development time-dependent and spatially-specific arranged with respect to the fibro-elastic structure of arterial wall [10]. Major details about time-dependent changes in SMC populations with development have been obtained in rabbit thoracic aorta [149] and canine carotid artery [191]. As also shown in Fig. 7, in the rabbit two distinct SMC populations exist in the medial layer of adult aorta, namely the "adult" (SM-MyHC isoform only), and "postnatal" (characterized by the expression of SM + NM-MyHC$_{pla2}$), SMC phenotypes [150]. During early stages of development, the aortic SMC are all homogeneous for SM, NM-MyHC$_{pla1}$ and MyHC$_{pla2}$ composition. Immediately after birth, MyHC$_{pla1}$ becomes down-regulated, but all SMC continues to express SM myosin + MyHC$_{pla2}$ up to about 30 days when the adult-type SMC population appears (Fig. 6). In the various adult

SMC populations in rabbit aorta during development

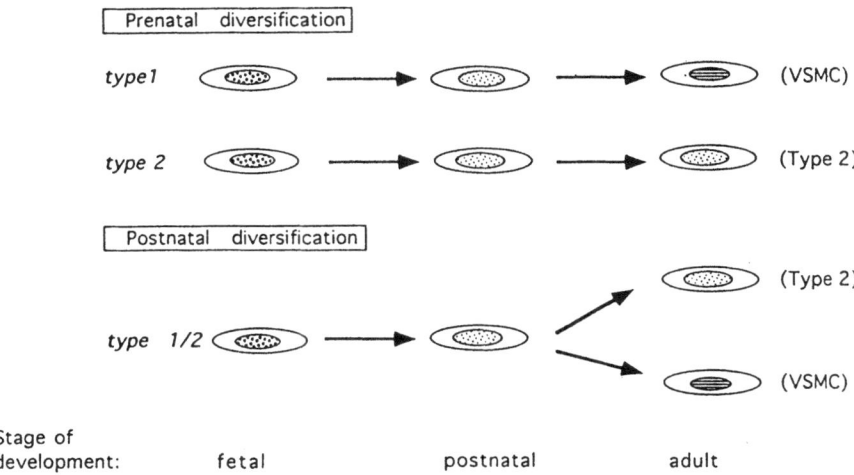

Fig. 7. Diversification of SMC populations in rabbit aorta based on selective SM- and NM-MyHC$_{pla}$ isoform expression. In brackets is reported the nomenclature proposed by Holifield et al. [191] and Seidel et al. [193], possibly corresponding to the SMC phenotypes identified by Giuriato et al. [149].

vessels, the "postnatal"-type SMC are uniquely deployed and such specific localization is of importance in the development of the atherosclerotic lesion [158]. It is not clear whether these two SMC phenotypes also display a corresponding "fetal"- and "adult"-type SM1 to SM2 ratio [192], i.e., up- or down-regulation of myosin isoform expression occurrs concomitantly at the level of single cell. In the dog, arterial and venous SMs also contain two similar cell populations that are distinguishable on the basis of ab-sence (Type 2) or presence (VSMC) of SM-type α-actin and MyHC ([191]; Fig. 7). The Type 2 SMC, similarly to the "postnatal"-type SMC [158], are believed to be involved in neointimal formation, possibly on the basis of a unique migratory ability (Fig. 8; [193]).

Based on cell location and immunohistochemical criteria, multiple dis-tinct SMC populations have also appeared in bovine pulmonary artery [12]. Here, cytoskeletal and ECM proteins are distinctly distributed in the lumenal *vs.* adventitial layer of the tunica media, similarly to the canine carotid artery [191]. Each of four identified SMC populations appears to

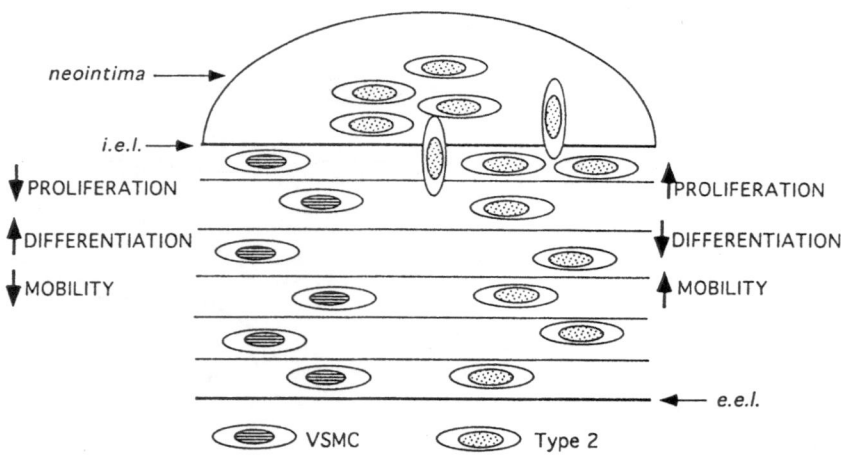

Fig. 8. SMC subpopulations.

progress along distinct developmental pathways, though in the fetal period only one SMC type has been recognized [12,29]. It is worthy of mention that in the presence of a hypoxic stimulus, neonatal hypertensive calves increase the meta-vinculin negative SMC population without any involvement of the other cell populations [194]. This specific SMC subtype is observed in the proliferative response occurring in the outer media of this artery. Also in this model, proliferation and differentiation of vascular SMC appear to be strictly associated.

While all these examples of *in vivo* cell heterogeneity develop in an apparently homogeneous SMC context, in the chicken system morphological and immunophenotypic diversification are closely linked. For example, Yablonka-Reuveni et al. [195] have shown that neural crest-derived lamellar and interlamellar aortic SMC are distinguishable by morphological and immunophenotypic (desmin and SM-type α-actin distribution) criteria. It is not known whether the chicken SMC lacking desmin and SM-type α-actin correspond to the small-sized cell population of bovine aorta [10] and canine carotid artery [191] lacking SM-type MyHC or to the SM-type α-actin negative SMC population expressed in human atherosclerotic plaque [196].

While in the large conduit vessels developing SMC undergo the abovementioned molecular transitions, in the periphery the arteriolar vessels seem less susceptible to phenotypic modifications. For example, Pauletto et al. [197] and Giuriato et al. [158] have shown that in different microvascular regions the majority of peripheral vessels show the "fetal"-type SMC composition, though a minority does not express SM-MyHC. During postnatal development, in agreement with data from Price et al. [198], the size of SM negative microvessels is reduced [158].

Taken together these data suggest that vascular SMC heterogeneity might be a more favorable and flexible condition for the development of an adaptive response compared to a monocellular system.

5
The smooth muscle cell lineage pathways

In the previous sections we have highlighted the fundamental morphological and molecular features of adult and developing SM tissues with the aim of furnishing a conceptual support for the interpretation of the behavioral characteristics of vascular and non-vascular SMC in pathology. It turns out, however, that this issue when approached by a study restricted to the fetal stage of development is not adequate to face the challenge. For example, the induction, after endothelial lesion or angioplastic intervention, of an abrupt rise in proliferation of quiescent medial SMC accompanied by migration from this compartment to the subendothelial space, cannot be completely explained as a mere recapitulation of a "fetal behavior" by adult SMC [199]. Fetal SMC have already acquired some "basic muscle properties" that might differ from those expressed during neointima formation. The temporal sequence of SM-specific gene activation at the early stages of development suggests that the commitment to SMC lineage occurs well before the relatively late expression of SMC markers. Thus, to substantiate the working hypothesis proposed in this review a comparison with the very early stages of SMC development is needed. In this section, we will review briefly the molecular morphogenetic process of blood vessel formation, giving more emphasis to those aspects that are of potential interest in interpreting the mechanism of neointima formation.

5.1
The initial step

At the very early stages of development, after uncommitted cells have achieved a developmental imprinting through a pattern formation process, mesodermal cells are either recruited to the nascent endothelium (vascular system) or epithelium (non-vascular systems; [200]. New vessels are formed through two distinct morphogenetic processes: vasculogenesis (*de novo* emergence in the early embryo) and angiogenesis (new vessels arise from preexisting ones; [201-203]). In adult, new vessels are formed by sprouting from preexisting ones on tissue demand [1,204]. In embryos during vasculogenesis, the endothelial cells organized as small "cell clusters" represent the primitive recruiting structure that, in association with differentiating SMC, will ultimately give rise to the complete and mature blood vessels by a multistep process. The intial step of SMC pathway is preceded by the angioblastic differentiation from the surrounding mesoderm or from angioblasts at the site of blood vessel origin (Fig. 9; [203]. During ontogeny, vasculogenesis or angiogenesis seems to occur depending on whether the mesoderm is associated with endoderma or ectoderma [205].

There are no specific precursors in the cellular milieu which surrounds the respective developing endothelial and epithelial tubes. In contrast to skeletal muscle [206], that are derived from unique populations of mesodermal precursor cells, SMC arise throughout the embryo from diverse precursor cell types (Figs. 9, 10). Cells destined to become the major constituent of the developing blood vessels have three distinct origins: from lateral mesoderm-derived mesenchyme (tentatively named Mes SMC; [207]; see Figs. 10, 11) and from cardiac neural crest (i.e., of neuroectodermic origin [208,209], tentatively named Ect SMC; [207]; see also Figs. 10,11). As to mesodermal-derived SMC, it is important to ascertain when and how these cells are committed to become SMC. In addition, it is not known if migrating cardiac neural crest cells have completely acquired SMC lineage commitment at time of colonization-incorporation in developing blood vessels ([210]; see also Fig. 9).

A neural crest-derived population, however, contributes to the SM tissue formation in the vessel segments proximal to the heart [209]. Endothelial-mesenchymal [2,203] and epithelial-mesenchymal [211-214] cell interactions are of fundamental importance to the establishment of the proper environmental conditions that allow the incorporation and acquisition of SM-type characteristics by the mesenchymal cells during organogenesis.

The morphogenesis of coronary artery system has been elucidate by the use of retroviral markers [215] and chicken-quail chimeras [216]. Coronary vasculature does not stem, as previously hypothesized, from an outgrowth process from aorta, but it seems derived from locally migrating subepicardial endothelial and SMC precursors [215], possibly by vasculogenesis [217]. It is only after the connection with aorta and the right atrium that a vascular wall organization with the formation of a tunica media and an adventitial layer takes place [217]. Cardiac neural crest cells, though not present in immature coronary arteries seem, however, able to influence the survival, in this region, of the definitive vascular structure [218,219].

Among the factors involved in the endothelium-mesenchymal cell interaction there are: transforming growth factor-β (TGFβ), platelet-derived growth factor (PDGF; [1,2]), vascular endothelial growth factor (VFGF; [220]) and the recently discovered angiopoietin-1 [221]. According to a model proposed by Folkman and D'Amore [222], the directional migration-differentiation of mesenchymal cells that takes place around the tube containing proliferating endothelial cells is a two-phase process. In the

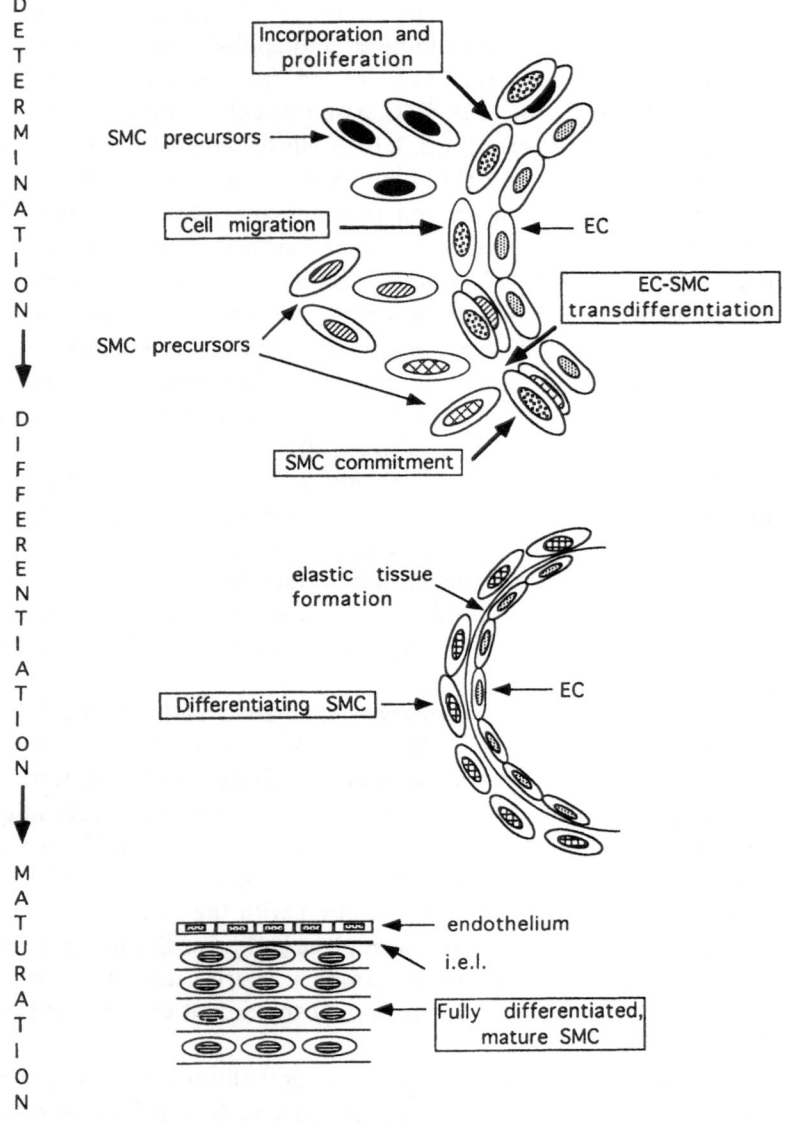

Fig. 9. The arterial SMC lineage pathway during determination, differentiation and maturation phases of development. Note that SMC precursors from different origin are incorporated in the developing arterial wall. EC, endothelial cells; i.e.l., internal elastic membrane.

SMC lineages in avian embryo

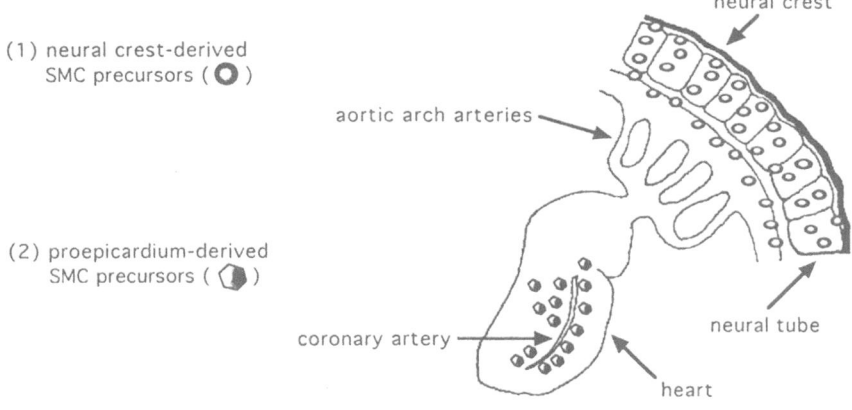

(1) neural crest-derived
    SMC precursors (O)

(2) proepicardium-derived
    SMC precursors (◗)

(3) locally recruited mesodermal
    SMC precursors (◕)

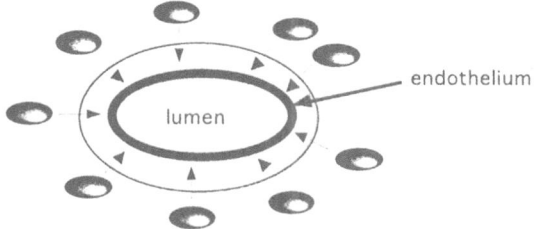

Fig. 10. The three major sources of arterial SMC precursors.

first phase angiopoietin-1 produced by mesenchymal cells [221] activates the TI2 receptors [222] on endothelial cells that in turn release the chemotactic/recruiting factors PDGF-BB or, perhaps, PDGF-BB and HB-EGF (heparin-binding epidermal growth factor). In the second phase, once the recruited mesenchymal cells have been approached and an endothelial-mesenchymal cell contact has been established, TGFβ is locally activated and a block of proliferation and induction of differentiation takes place in endothelial and "activated" mesenchymal cells, respectively. It is worth noting that this hypothetical mechanism might explain the SM tissue formation from mesodermal cells but not the one from neuroectodermal

SMC lineage diversity in avian embryo

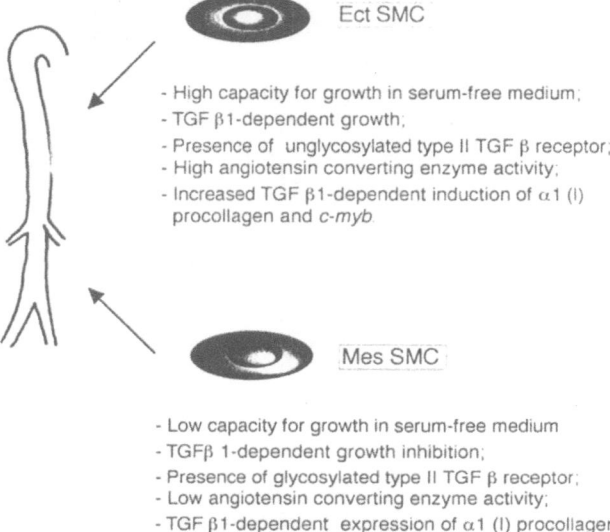

Ect SMC

- High capacity for growth in serum-free medium;
- TGF β1-dependent growth;
- Presence of unglycosylated type II TGF β receptor;
- High angiotensin converting enzyme activity;
- Increased TGF β1-dependent induction of α1 (I) procollagen and c-myb

Mes SMC

- Low capacity for growth in serum-free medium
- TGFβ 1-dependent growth inhibition;
- Presence of glycosylated type II TGF β receptor;
- Low angiotensin converting enzyme activity;
- TGF β1-dependent expression of α1 (I) procollagen unaffected; c-myb not expressed.

Fig. 11. SMC diversity in chicken embryo.

cells. For example, TGFβ1 increased DNA synthesis markedly when applied to cultured cells from chick embryonic aortic segments enriched with neural crest-derived SMC but not from segments containing mesodermal SMC [207]. Interestingly, in avian embryos both VFGF [223] and *flk-1* tyrosine kinase VFGF receptor [224] are markedly expressed in close proximity of developing endothelium at the level of endothelial cell precursors. Since a diffusible form of VFGF protein induces marked alterations in the vascular organization and architecture [224], it is higly plausible that VFGF plays an important role both in vasculogenesis and vascular patterning.

Newly wall-incorporated cells undergo a differentiation process through which specific elements of structural organization of SMC are progressively acquired during *in utero* or *in ovo* development [1,2]. Some information is now available about the first step in the determination-differentiation-maturation sequence that, like the other muscle pathways [225], is involved in vascular myogenesis. Using monoclonal antibodies to

SM-type α-/γ-actin and other markers, some authors have begun to study the commitment of mesodermal cells to SMC [83,102,226-228]. According to Hungerford et al. [102], the chicken cells that express α-actin and the 1E12 antigen (probably an isotype of SM α-actinin [229]) are destined to become SMC, whereas those containing only α-actin are presumably presumptive myoblasts. Only cells located adjacent to the endothelium show the expression of α-actin and this event seems to take place initially at the ventral surface in the chicken [102] and at the dorsal surface in the mouse [227]. On the contrary, in the gastrointestinal tract of developing mouse such an event does not require a direct epithelium-differentiation SMC contact [83]. This author has also hypothesized that myoblast and immature SMC can be distinguished on the basis of α-/γ-actin content. This view has been recently challegend by DeRuiter et al. [230] who have suggested that embryonic endothelial cells can transdifferentiate into mesenchymal cells showing SM α-actin expression. Along with the "centripetal" direction hypothesis of migration-incorporation of presumptive SMC in blood vessel formation, there is an opposite proposal: migration-incorporation from inward to outward (see also Fig. 9). A similar event occurs associated with the inducible transformation of endothelium into cushion mesenchyma cells during valve morphogenesis [231]. In addition, though the SM-type α-actin expression in mesodermal cells is likely to be a marker of SMC precursor cells, it remains to be determined whether this cell phenotype is really unique to SMC lineage or is shared by the other developing muscle tissues that also express this marker.

5.2
Smooth muscle cell diversification in embryo

The SMC cell lineages are differently distributed in large elastic arteries: Ect SMC are mainly localized in the vessels proximal to the heart (outflow tract and aortic arch arteries), whereas Mes SMC are found in the distal part of the same arteries ([207]; see also Fig. 11). One exception to this rule is represented by the formation of coronary arteries that are locally formed by ingrowth of epicardial layer precursors and subsequently connected to the root of the aorta [215]. This process is apparently carried out without the cellular contribution of cardiac neural crest, but this tissue is able to influence the spatial order of developing coronary arteries [218,219]. Future studies must address the point of comparing the process of recruitment-incorporation of SMC precursors in developing coronary arteries *vs.* large conduit vessels.

The two types of embryonic SMC have been analyzed *in vitro* and found to differ in the proliferative response to TGFβ1. This cytokine, when applied to chicken SMC cultures from the thoracic (mainly Ect SMC) aorta, induces DNA synthesis. When added to cultures from abdominal (mainly Mes SMC) aorta TGFβ1 is growth inhibitory. The two types of cultures also differ in the extent of glycosylation of type II TGFβ1 receptor [207]. Results from Rosenquist's laboratory substantially confirm such differences as regards to TGFβ1 effects on cultured Ect *vs.* Mes SMC and suggest, in addition, that PDGF-BB and c-*myb* are also able to produce distinct effects in the two cultures [232,233]. Tissue homogenates from developing aortic tissue containing Ect SMC and Mes SMC, respectively, are also different as regards the increased level of angiotensin converting enzyme, suggesting again the existence of intrinsically different properties in the two cell types ([234]; see also Fig. 11).

5.3
Smooth muscle cell diversification in adult and newborn

It is generally accepted that when vascular SMC are grown *in vitro* they lose to a various extent their peculiar "contractile" features in favor of the "synthetic" phenotype (see Section 6.2). Recent results of cell cloning experiments on aortic SM tissue from adult and newborn rats point, however, to the existence of multiple and phenotypically stable SMC phenotypes *in vivo* and could furnish a biological justification for SMC heterogeneity found with the differentiation markers. Unlike some SMC lines, these clones are derived from plating at different densities and not *via* mutagenesis, transformation [235,236] or from conditionally immortalized cells [237].

Aortic SMC from adult aorta, when cultivated in the presence of whole-blood serum, give the so-called A (adult) - SMC phenotype, i.e., spindle-shaped morphology, no secretion of PDGF, expression of PDGF-A but not of PDGF-B ([238]; Fig. 12). By contrast, aortic SMC from newborn (P-SMC or "pup" SMC) display a "cobblestone" or "epithelioid" morphology, secrete a PDGF-like mitogenic activity and express both PDGF-A and -B. Cell cloning procedures have shown that two distinct cell types can be obtained from 12-day-old rat aortic SM. The first proliferating in plasma-derived serum, showing the "cobblestone" appearance, and containing

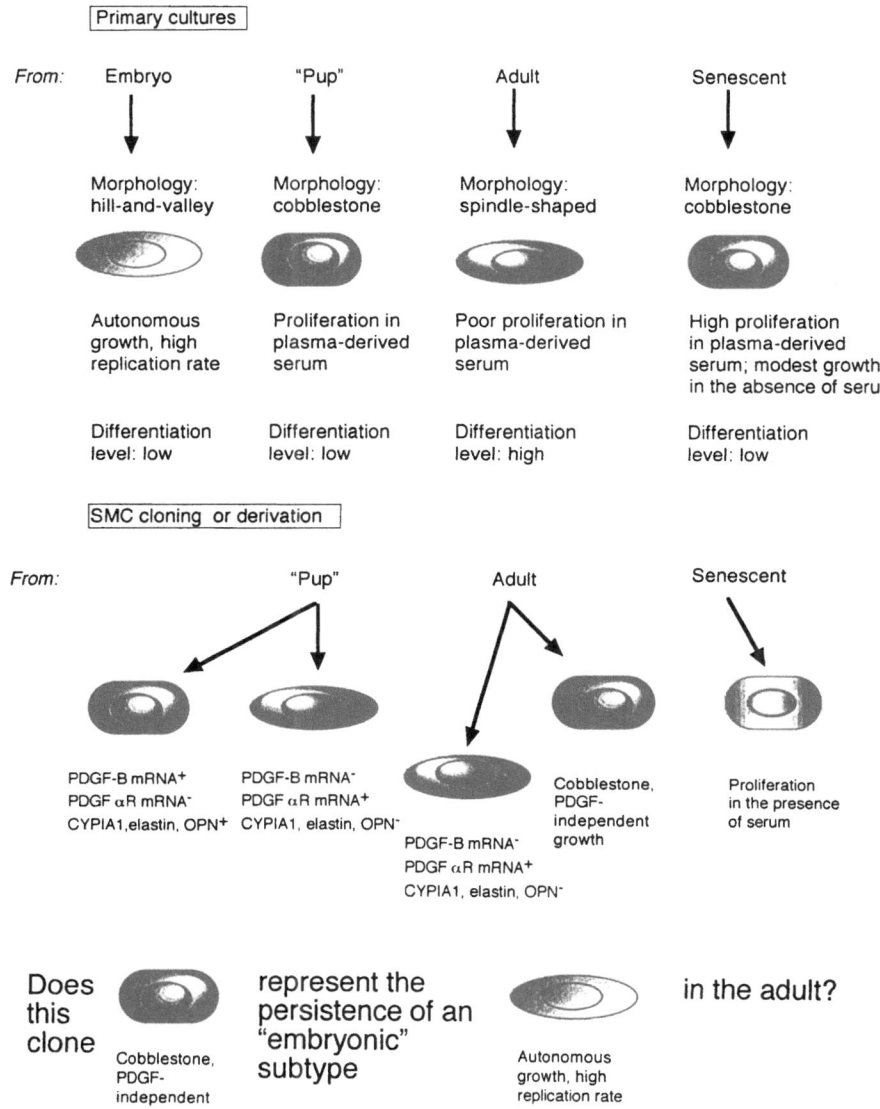

Fig. 12. Morphological and functional properties of primary and cloned/selected vascular SMC. OPN, osteopontin.

PDGF-B mRNA, high levels of CYPIA1, osteopontin and elastin mRNAs but lacking PDGF-α receptor. The other, proliferating poorly in plasma-derived serum, displaying a spindle-shaped morphology, lacking PDGF-B mRNA, containing low levels of CYPIA1, osteopontin and elastin mRNAs but expressing PDGF-αreceptor [238-240]. A-SMC when grown in plasma-derived serum resemble morphologically the P-SMC but do not produce PDGF though they show the presence of PDGF receptors [241].

Bochaton-Piallat et al. [242], have also reported on the differences between arterial SMC from newborn and adult rats grown in presence of serum and the persistence, after cell cloning, of some differentiation markers (SM myosin and α-actin and desmin) in the former. Bochaton-Piallat et al. [243] subsequently confirmed the tendency for clones from newborn or young rats to have a higher degree of differentiation than those from aged rats, the latter being more proliferative than the former (see also Fig. 12). Based on cell morphology, serum-dependent or serum-independent growth, [³H]-thymidine incorporation, migratory ability and differentiation marker distribution, it was shown that the medial layer of adult rat aorta is comprised of four phenotypes: spindle-shaped, cobblestone, thin-elongated and senescent [244].

Taken together these data support the notion that even in adult and developing rat aorta, where medial SMC are apparently identical from a morphological point of view, at least two distinct cell types exist. Their respective functional features are disclosed only in coincidence with wall damage or when they are grown in cultures.

6
Differentiation vs. proliferation in smooth muscle cells in vivo and in vitro

6.1
In vivo

SMC in the various SM tissues are remarkably quiescent [29,37,61,244-246]. In particular, vascular SMC from aorta and pulmonary artery show a replication rate of .06% per day [29,246]. Looking at the SMC lineage pathway during the embryonic/fetal and postnatal phases of development a much different pattern can be observed [27,29,228,247]. In the embryonic rat aorta, the SMC display a very high replication index (75-80%/day).

The index declines to about 40% concomitantly with the developmental transition from the embryonic to fetal stage [27,247]. In newborn rats this index is about 2% [80] and becomes .5% and .06% by 1 and 3-4 months after birth, respectively [246,247]. In the pulmonary artery, however, it is about 30% from day 19 *in utero* up to day 7 after birth [29]. Though it seems that during development proliferation and differentiation are mutually exclusive, recent data from Lee et al. [228] point to different picture. In aortic media from day 2.5-19 chick embryo, the kinetics of replication is temporally distributed in two waves: a rapid (15-17%; from days 4-12) and a slow (less than 5%; from day 16 to hatching) proliferation phase. Interestingly, expression of the precocious muscle lineage marker SM α-actin occurs independetly from the level of proliferation [228].

It must also be kept in mind that the various SMC proliferation steps in the developmental process occuring in the course of physiological vascular remodeling are likely to be counteracted to a various extent by apoptosis [248,249]. The genetic program for cell death in association with mechanisms that control proliferation and regulate differentiation of SMC might contribute to determining the definitive growth status of vascular SM tissues. For example, *c-myc* is involved in controlling proliferation in endothelial-injured neointima formation [250,251] and deregulation of this protooncogene induces apoptosis [252]. Similarly, PDGF, which is known essentially as an inducer of SMC proliferation [253], can partially reverse apoptosis in cultured, passaged SMC from human atherosclerotic plaque thus acting as a survival factor [254].

It is important to emphasize that the "embryonic SMC" when grown *in vitro* display growth properties, such as the ability to proliferate in serum free media (Fig. 12), that are quite distinctive compared to the cultured SMC from newborn rats (the so-called "pup"/neointimal SMC or π-SMC) identified by Schwartz and coworkers ([7]; see also Section 6.2). These "embryonic" SMC share the same growth properties with the SMC subpopulation isolated *in vitro* by Schwartz et al. [241], i.e., an autonomous (serum-independent) growth. It is tempting to speculate that persistence of such "embryonic" SMC in adult may represent a type of "stem-like" SMC population that activates on demand (see also Fig. 12).

Judging from the time-course of proliferation behavior during development, one might argue that the abrupt drop around birth is correlated with the phenotypic changes observed with some differentiation markers such as the disappearance of NM- type MyHC-B or MyHC$_{pla}$ (see Section 3.2.1). Though these two parameters are not as linked as occurs with other muscle systems [5,255], it seems that a relationship between proliferation and differentiation also exists in vascular SMC during *in vivo* differentiation. To support this notion, Belknap et al. [27] found that the highest amount of tropoelastin mRNA was produced in rat aorta after the initial burst of replication, i.e., late fetal and early postnatal stage. At the ultrastructural level, however, vascular and non-vascular SMC can divide without substantially losing the myofilamentous organization [256]. Owens and Thompson [80] also demonstrated that the proliferative event takes place irrespective of the SM-type α-actin content of vascular SMC. The peculiar differentiative plasticity of SMC, partially dissociated from the proliferative behavior, might facilitate the reparative process of vascular wall but, at the same time, might be an obstacle in a regenerative process where loss of somatic cell specialization and acquisition of a specialization for cell division occur together [257].

6.2
In vitro

When adult vascular and non-vascular SM tissues are grown *in vitro* in primary cultures the structural and functional properties developed *in vivo* change markedly and exhibit the so-called "synthetic SMC phenotype" [40,41,258]; see also Fig. 13). As mentioned above, with the exception of chicken aortic SMC, there is no evidence that such a "synthetic phenotype" exists *in vivo* before SMC are grown *in vitro*. Thus, due to peculiar differentiative plasticity or more so to the instability in *ex-vivo* environment, all or part of SMC show the tendency to acquire a distinct cell phenotype. It is also known there are a number of endogeneous and exogeneous factors that can dramatically influence the final differentiative and proliferative outcome. For example: the type of tissue (media or intima), the age of donor, the explant *vs.* the enzymatic technique used to prepare the tissue culture, the serum concentration, cellular density, number of passages, duration of cultivation, etc. All of these can indeed affect the morphological and molecular level attained in culture. Thus, we are faced with two linked and, perhaps distinct, problems: (1) the extreme environmental-factor sensitivity of cultured SMC, and (2) the existence of an "intrinsic" structural-functional SMC heterogeneity as revealed, for example, by the unique distribution of some differentiation markers *in vivo* or by cell cloning. In other words, it is of extreme importance to distinguish the basic "genetic" behavior from the environmental ("epige-

Phenotypic changes in primary cultured SMC
from adult smooth muscle tissue

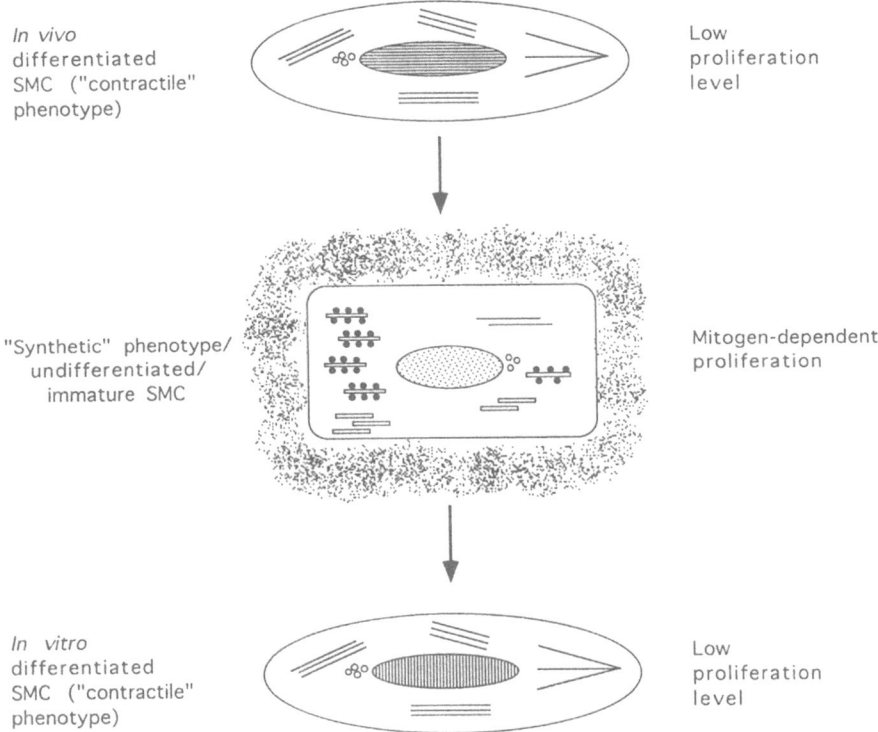

In vivo
differentiated
SMC ("contractile"
phenotype)

Low
proliferation
level

"Synthetic" phenotype/
undifferentiated/
immature SMC

Mitogen-dependent
proliferation

In vitro
differentiated
SMC ("contractile"
phenotype)

Low
proliferation
level

Fig. 13. Changes of some structural and functional characteristics in cultured SMC. Identity between *in vivo* and *in vitro* "contractile" phenotypes is still debated.

netic") factors that drive the growth and differentiation changes in cultured SMC. Before going into details about these two problems, we shall discuss some structural aspects pertaining to the two major SMC phenotypes *in vitro*.

Numerous laboratories have observed that the cell structure as well as cytoskeletal and cytocontractile protein content is markedly different between the two opposite SMC phenotypes. In the "synthetic" phenotype SM2, and to a lesser extent SM1, SM-type α-actin, *h*-caldesmon, calponin and desmin content/distribution are decreased, whereas NM-type MyHC-

A, -B and MyHC$_{pla}$, and NM-type β-actin isoforms are increased [10,94,125,259–263]. The loss or diminution in the expression of SM-specific markers is in part linked to the manipulations used to set up the tissue culture, while the increased expression of NM-type markers is almost constant. A more direct approach to the phenotypic-related expression of some proteins in cultured SMC was undertaken by Shanahan et al. [108]. Using a differential cDNA screening procedure and RNAs from freshly dispersed, differentiated and scarcely differentiated late-passaged cultured rat aortic SMC, they have been able to demonstrate that seven cDNAs are more strongly expressed in the former compared to the latter cultures: SM22α, CHIP28 (a putative membrane channel protein), tropoelastin, SM-type α- and γ-actin, calponin and phospholamban. It must also be taken into account that changes in the level of synthesis and content of some cytocontractile proteins are temporally dissociated [2]. Thus, some caution must be exerted in comparative evaluation of gene expression between cultured SMC.

SMC become competent for proliferation in the presence of exogeneous/endogeneous growth-stimulating factors and adequate substrate for cell attachement [41,264]. It has been hypothesized that the "phenotypic modulation" from the "contractile" to the "synthetic" state is a prerequisite, though not sufficient, for the proliferative event to take place. This hypothesis has been questioned by other groups [265,266] that believe the difficulty in growing SMC *in vitro* is not based on a cause-effect relationship existing between a certain phenotype and the onset of proliferation.

A number of growth factors, cytokines and hormones can modify, to some extent in a reversible manner, the growth and differentiation state of cultured SMC. This has been taken as evidence of "phenotypic modulation" processes [41,264]. The well-defined developmental changes of SMC phenotype in the SMC lineage, ranging from a poorly (fetal-type) to a fully (adult-type) differentiated SMC, and the existence of SMC clones place the problem of the phenotypic modulation in a new context. It might be that in medial aorta two potentially distinguishable SMC populations exist: one that shows a fetal/immature *behavior* (tendency to undergo a "phenotypic modulation" from an apparently "contractile" to a "synthetic" phenotype accompanied by the acquisition of a proliferative capacity), and the other one that displays an adult-type phenotype that can eventually proliferate without changing the cell phenotype. The first cell population goes through a series of transitional steps that can be identified by their dependency on exogeneous or autocrine-acting endogeneous mitogens.

While in this latter circumstance initiation of proliferation is related to changes in the differentiation pattern, in the second cell population the two parameters are not necessarily associated. It is also possible the level of proliferation and motility is comparatively lower than the first one. Clearly this hypothesis awaits adequate experimental support to be fully accepted.

7
Alterations in the stability of the differentiated state and proliferative level in the smooth muscle cells

Compared to cardiac and skeletal muscle, the SM in the adult possesses a wider phenotypic diversification that stems from the lack of antagonism between differentiation and growth. It is likely that all muscle systems share a common rule regarding maintenance of the differentiative state: a continous and precise regulation of gene expression. In skeletal muscle, the expression of muscle genes is triggered by unique transcription factors, i.e., the master regulatory genes such as the MyoD family that includes MyoD, myogenin, myf-5 and myf-6 [78,267]. Other factors such as the protein Id are thought to inhibit differentiation whilst the cells are proliferating [268]. Unfortunately none of the regulatory genes yet found for SMC are able to act in the way commonly attributed to specific cell factors that regulate cell differentiation: by activating coordinating cell type-specific genes (see Refs. in Table 1). Two homeobox genes, HoxB7 and HoxC9, showed a restricted mRNA expression in human fetal SM that might be indicative of some role in establishing the phenotypic pattern at this stage of SMC lineage pathway [271]. Since spatiotemporal expression of some fundamental SMC-linked genes is not the same in the SMC lineage pathway, it is highly probable that distinct regulatory factors control each gene, possible acting in a combinatorial way [78,117] with "default" muscle transcriptional factors (existence of multiple, independent cis-regulatory regions). The picture is even more complicated by the functional coexistence of genes involved in regulating growth and differentiation and by the presence of at least two SMC lineages that might have distinct growth/differentiation regulatory pathways.

Though the adult vascular SM tissue exhibits a heterogeneous distribution of some markers (e.g., the NM-MyHC), it is only when vascular SMC are grown in vitro or when subjected to in vivo conditions involving

Table 1

Genes involved in controlling avian and mammalian SMC differentiation

Gene	Cells/tissues expressing the specific transcript	References
- *MHox*	VSMC, SK, C, NMc	[269]
- *Hox1.11*	VSMC, NMc	[270]
- *HoxA5, HoxA11, HoxB1,*	VSMC, NMc	[271]
HoxB7, HoxC9	fSMC	
- *Gax*	VSMC, C	[272,273]
- *Pax3*	VSMC, C	[274]
- *Th1*	GSMC, C, NMc	[275]
- *GATA-6*	VSMC, BrSMC, BlSMC, C	[276]
- *GATA-5*	BrSMC, BlSMC, C	[277]
- *MEF2s*	aVSMC, SK, C	[278]
- *NKx2-3*	GSMC, NMc	[279]

Key: SMC, smooth muscle cells; VSMC, vascular SMC; aVSMC, adult-type VSMC; fSMC, fetal human SMC grown *in vitro*; SK, skeletal muscle; C, cardiac muscle; NMc, non-muscle cells; GSMC, gastrointestinal SMC; BrSMC, brochial SMC; BlSMC, bladder SMC.

proliferation/migration or hypertrophy/hyperploidy that the potential functional heterogeneity of SMC is revealed and the specific regulatory influence of growth factors, cytokines and hormones are disclosed.

7.1
Regulatory factors of growth and differentiation

In large arterial vessels, interactions of SMC with endothelial cells on the one side, and with adventitial tissue on the other (Fig. 2) are of crucial importance in maintaining SMC growth and differentiation in a stable condition. In addition, SMC may secrete factors such as PDGF, basic fibroblast growth factor (bFGF) and TGFβ, which through autocrine-paracrine mechanisms [253], can contribute to the general homeostasis of arterial wall. One major drawback is that such mechanisms have been documented *in vitro*. Thus, the existence of autocrine growth *in vivo* remains conjectural especially if one considers the large structural homogeneity of adult SMC and their unresponsiveness to growth factors. Functional or structural perturbation of endothelial cell integrity achieved, for example in atherosclerosis, hypertension or in experimentally-induced endothelial denudation, can alter the phenotypic stability and growth properties of

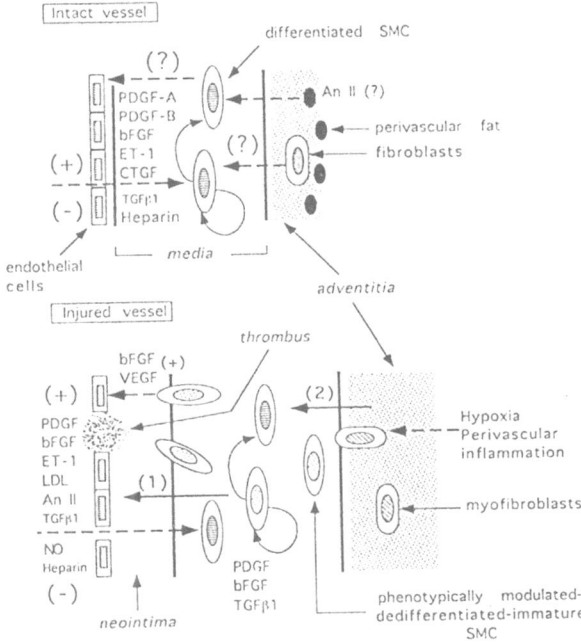

Fig. 14. Factors putatively involved in controlling the differentiation and growth properties of medial SMC in intact and injured vessels. In these vessels, growth, migration and differentiation of medial SMC are controlled by both positive (+) and negative (-) factors: platelet-derived growth factor-A and -B (PDGF-A, PDGF-B); basic fibroblast growth factor (bFGF); endothelin-1 (ET-1); connective tissue growth factor (CTGF); transforming growth factor β1 (TGF β1); low-density lipoprotein (LDL); angiotensin II (AnII); nitric oxide (NO); vascular endothelial growth factor (VEGF). Factors released from endothelial cells, SMC and adventitial cells contribute to the medial SMC homeostasis by autocrine and paracrine mechanims. Perturbation of arterial wall integrity (both in structural and functional terms) induces the release/activation of a number of factors, in part identical to those that control the normal wall homeostasis, and the expansion of medial SMC comparment accompanied by the formation of the neointima. Adventitial cells can variably partecipate in this process *via* : perivascular inflammation; release of inflammatory cytokines; cellular conversion to myofibroblasts; and tissue hypoxia induces by *vasa vasorum* alteration. The numbers in parantheses indicate tthe two major migratory processes occuring in the arterial wall after endothelial injury, namely, from the media to the subendothelial space (1), and from the adventitia to the media (2). Dashed arrows indicate processes mainly observed in vitro

SMC, though, as mentioned above, these two characteristics are not necessarily linked. Since endothelial cells and SMC are quiescent in the adult, it might be that a potential continuous, bidirectional flow of information, at a low level between these cells, is partecipating in the reciprocal regulation of growth and differentiation (see Fig. 14).

In vitro studies have revealed that a number of mitogens are released by endothelial cells: PDGF-A and -B [280,281], bFGF [282,283], endothelin-1 [284], and connective tissue growth factor [285]. On the other hand, endothelial cells can produce factors that antagonize the SMC proliferation such as heparin [286], and TGFβ [287,288], the latter showing either proliferative or inhibitory properties for SMC.

The general approach to investigating the putative effect of endothelial cells on SMC is the use of co-culture systems [289], or of endothelial cell conditioned medium [290]. Endothelial cell-released factors or identical factors added to the culture medium of SMC are able to down-regulate markedly the expression of two SM-specific proteins, such as SM-type α-actin and MyHC, and to increase SMC proliferation [290-293]. However, as mentioned above, high levels of differentiation are not incompatible with proliferation. In contrast with these data, other authors have found that endothelial cells have the ability to induce the in vitro SMC differentiation accompanied by inhibition of their growth [294].

Using concentrations of heparin, that are inhibitory to SMC proliferation, causes SM-type α-actin expression to be up-regulated [295,296]. In plasma-derived serum however, heparin is not able to modify α-actin expression, indicating that its action is related to an antiproliferative action [296]. Barzu et al. [297], have isolated in vitro a subpopulation of rat aortic SMC that is heparin-resistant in terms of growth inhibition, but differentiation responsive (increased SM-type α-actin expression) to heparin treatment. This finding is in line with the theory that heparin acts on SMC differentiation and proliferation through distinct pathways.

TGFβ1 is a bifunctional regulator as regards migration and proliferation in vitro [298]. At relatively low concentration TGFβ1 enhances the migration of SMC, but at higher concentration dose dependently inhibits the PDGF-dependent migration [299]. Similarly, in confluent cultures TGFβ1 acts as mitogen, while in subconfluent cultures it inhibits SMC proliferation [300,301]. TGFβ1, when applied to cultured rat aortic SMC, reduces the serum-dependent proliferation [296], and induces the synthesis of SM-type α-actin in cultured human SMC [302]. An immediate but limited proliferative action of TGFβ1 in vitro relies on the autocrine production of PDGF-AA [303]. By contrast, an enhanced but delayed mitogenic activity is achieved when this cytokine is administered to cultured SMC in combination with EGF, bFGF or PDGF-BB [304].

No studies on cultured cells have been performed regarding the putative effect of adventitial fibroblasts on growth and differentiation of SMC, though a role for angiotensin II is possible in this layer. In fact, angiotensinogen, the ultimate precursor of angiotensin II converted to the biologically active form by angiotensin converting enzyme, can also be produced in the normal vessel wall where it was localized the perivascular fat [305,306]. The AT-1 and and AT-2 angiotensin II receptors are also present in the normal wall and angiotensin converting enzyme is made by medial SMC and endothelial cells [7]. Angiotensin II infusion induces vascular wall hypertrophy and proliferation [307], whereas treatment of cultured aortic SMC with angiotensin II causes an increase in RNA and protein synthesis with little or no immediate increase of DNA synthesis [308]. This agonist not only regulates SMC growth but also promotes SMC differentiation in as much it is able to increase the SM-type α-actin expression [309]. A significant dose-dependent mitogenicity can, however, be demonstrated *in vitro* after a more chronic administration of angiotensin II [310]. Activation of AT-1 receptor by angiotensin II in turn stimulates the production of PDGF-A, TGF β1 and bFGF [311]. It has been hypothesized that angiotensin II induces both a proliferative and anti-proliferative action through two distinct pathways in cultured SMC. Non-mitogenic growth response of SMC to angiotensin II would result from two opposite effects: the anti-proliferative action of autocrine TGF β1 and the proliferative action of autocrine bFGF but not of PDGF-A [312]. This hypothesis might be questioned in light of: (1) different biological assay conditions used by the various investigators, and (2) the existence of distinct SMC lineages in the aorta that might have different angiotensin II or AT receptor distribution. It is interesting in this regards that AT-2 receptors are involved in the trophic effect of angiotensin II, while AT-1 receptors can be associated with changes in SMC phenotype, such as the increased expression of NM-type MyHC$_{pla}$ and "fetal-type" fibronectin isoform [313].

The mRNA induction of PDGF, bFGF and TGF β1 in vascular SMC *in vitro* and *in vivo* as a direct consequence of an endothelial lesion or atherosclerotic plaque formation is preceded by the activation of several immediate-early genes [314,315]. In particular, PDGF and TGF β1 expression, unlike that of bFGF, is dependent on early protein synthesis [315]. There is compelling evidence that the products of these immediate-early genes (c-*myc*, c-*myb*, c-*fos*, c-*jun*, etc.) have a role in controlling SMC growth. For example, deregulated expression of c-*myc* can promote SMC proliferation and induce partial dedifferentiation and apoptosis [151,250-252]. PDGF and bFGF contribute independently to the maintenance of c-*myc* expression and c-Myc down-regulation is a pre-requisite for growth arrest of vascular SMC [251].

7.2
Growth and differentiation response of smooth muscle tissues to pathophysiological stimuli

Figure 15 shows that mechanical, hormonal, pharmacological and neural factors can influence directly or indirectly the phenotypic stability of SM-/NM-type MyHC-based SMC populations in rabbit aorta. It is important to underline that the reduction in the "postnatal"-type SMC population achieved with various interventions is never accompanied by the formation of a neointima, i.e. a newly formed SMC-containing tissue with a variable proportion of "contaminating" inflammatory cells [7,253]. By contrast, expansion in size of such medial SMC population may be eventually associated with the loss of positional control, proliferation and development of a directional movement from the media to the subendothelial space [150,199]. Ultimately, this behavior is dictated, on the one hand, by the local availability of mitogens and chemiotactic substances, and, on the other hand, by a modified responsiveness of medial SMC. This latter property might be the cause or the consequence of the differentiation profile of SMC. Since the phenotypic change, i.e., the expansion of the "postnatal"-SMC phenotype, is not necessarily followed by the formation of a neointimal tissue, it seems reasonable to assume that the differentiation pattern and proliferation/migration properties are dissociable parameters [27,42]. In the next two sections we shall discuss in particular those factors that have a major impact on growth-differentiation properties of SMC.

7.2.1
Mechanical stress

Mechanical factors such as shear stress, longitudinal stretch and wall tension in vascular SM tissue [29,39,263,316-318], and wall stretching in non-

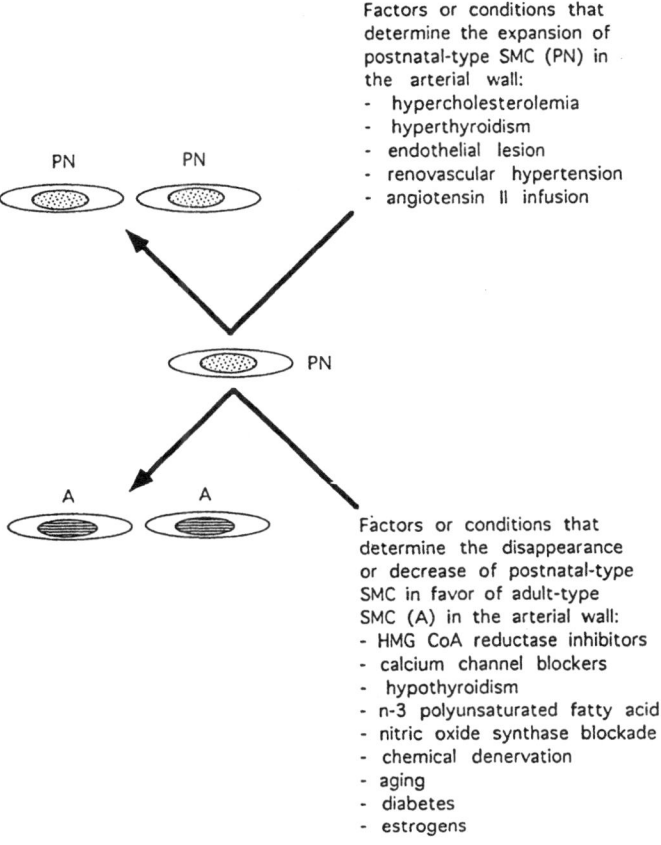

Factors or conditions that
determine the expansion of
postnatal-type SMC (PN) in
the arterial wall:
- hypercholesterolemia
- hyperthyroidism
- endothelial lesion
- renovascular hypertension
- angiotensin II infusion

Factors or conditions that
determine the disappearance
or decrease of postnatal-type
SMC in favor of adult-type
SMC (A) in the arterial wall:
- HMG CoA reductase inhibitors
- calcium channel blockers
- hypothyroidism
- n-3 polyunsaturated fatty acid
- nitric oxide synthase blockade
- chemical denervation
- aging
- diabetes
- estrogens

Fig. 15. Factors and experimental conditions which can expand or reduce the size of postnatal-type SMC (based on SM- and NM-MyHC$_{pla}$ isoform expression in rabbit aorta/carotid artery; [42]).

vascular SM tissues [90,319,320] can modulate growth, differentiation and ECM protein synthesis [321].

In vivo, increased intraluminal pressure can induce stretching in SMC, whereas shear stress affects endothelial cells which eventually can influence the activation of adjacent vascular SMC. Though blood pressure and SMC growth patterns are not always correlated, SMC of large vessels undergo hypertrophy with or without hyperploidy [308]. Depending on the type of model used to induce hypertension, aortic SMC proliferation with [322] or without migration [323] in the subendothelial space occurs. Since hyperploidy develops only in a subset of medial SMC, the question exists whether such an event is associated with a specific subpopulation. Hypertrophy and hyperplasia in SMC are not mutually exclusive as both occur in rabbit pulmonary artery as a direct effect of wall stretching [324].

In organ cultures of rabbit thoracic aorta, Bardy et al. [325] found that pressure and flow exert different effects on protein and DNA synthesis. In particular, in vessels subjected to high pressure and to a constant flow there was an increased protein synthesis without an increment of DNA synthesis in the presence of serum. If the same experiment is performed in the absence of serum, DNA synthesis increases markedly. "Fetal"-type expression of fibronectin was enhanced with high transmural pressure, exactly as happens with spontaneous hypertensive rats [179]. A different picture was seen when cultured SMC were subjected to pressure-induced stretching; in this circumstance DNA synthesis was increased [326]. Recently, Reusch et al. [293], confirming in part data from Birukov et al. [263], discovered that cyclic mechanical forces can up-regulate the expression of SM1- and SM2- MyHC and down-regulate that of MyHC-A and -B. Such effects are amplified if antibodies to PDGF-AB are added to the system, thus indicating that this growth factor, besides acting as inducer of proliferation/migration of SMC, is partially involved in the differentiation response of mechanically activated SMC [327]. To explain the PDGF paradox, Reusch et al. [293] put forward the idea that two distinct SMC populations are selectively involved in the differentiative response: one that might be prone to proliferation and the other one to developing the "differentiated" myosin pattern.

Some links between changes in the expression of SM-specific proteins and growth regulator levels are also evident in non-vascular SM tissues, though their interdependence is not clear. In the bladder, SM can develop a hypertrophic or hyperplastic growth response as consequence of alterations in the bladder wall tension obtained by a partial outflow obstruction [328]. In the rabbit and rat, where SMC hypertrophy-hyperploidy and ECM changes occur, growth is accompanied by: (1) marked increase of SM-type γ-actin mRNA and protein with a decrease of NM-type β-actin and substantial invariance of SM-type α–actin isoform [86], (2) decrease of SM2 mRNA transcript and protein [90,320], and (3) elevation of bFGF, c-*myc*, c-*fos*, Ha and Ki-*ras* protooncogenes [329,330], TGF β2, TGF β3 and TGF α mRNAs and diminution of TGF β1 transcript [331].

7.2.2
Neointima formation

Atherosclerosis, restenosis after angioplasty and vein-graft all share a common theme: the accumulation of *de novo* SMC, partly from locally resident or newly migrated SMC, and partly from the underlying media. The neointima tissue also contains variable amounts of inflammatory cells and ECM, is temporally unstable and potentially able to undergo a necrotic process with serious local and general consequences [7,253]. Experimental atherosclerosis, produced using cholesterol-enriched diet or taking advantage of endogeneously hypercholesterolemic or genetically manipulated animals, along with endothelial denudation are the procedures commonly used to study the growth pattern and differentiation profile of arterial SMC. From a morphological point of view, about 7 days from surgery ballooned rat carotid artery shows the presence of "synthetic" and proliferating SMC in the region facing the lumen. Mitoses are no longer detectable after 14 days and the neointima is now composed of quiescent SMC showing a mixed "synthetic"-"contractile" phenotype, intermingled with ECM components [332]. In the next two sections, modifications of gene expression pertaining to structural proteins and growth-related factors (only protein expression and factors shared by neointimal and developing SMC), will be examined.

7.2.2.1
Down- and up-regulation of structural proteins

As pointed out by several authors [2,7,42], neointimal SMC display phenotypical aspects quite similar if not identical to developing SMC. For example, expression of some SM-specific markers (e.g. SM2, SM-type α-actin, *h*-caldesmon) are down-regulated whereas that of some SM-related NM markers (e.g., MyHC-B, $MyHCA_{pla}$, NM β-actin, keratin 8/18, *l*-caldesmon) are up-regulated in experimental and spontaneous atherosclerosis and endothelial denudation or restenosis [11,56,57,138,157,166,176,181,333-336].

The level of proliferation attained by SMC as part of the wall response to endothelial injury has already been described by Schwartz and coworkers [7] as an appearance of consecutive, time- and space-regulated "waves". Each "wave" is characterized by a precise set of growth factors/hormones/cytokines whose activity imposes a specific pattern of proliferation and/or migration on SMC. Based on a comparative analysis performed on developing and endothelial damaged wall arteries, Belknap et al. [27] came to the conclusion that SMC found in the neointima reverted to a pre-elastogenic cell phenotype rather than to a "fetal" one. This interpretation resulted from the observation that elastogenesis follows SMC replication in the neointima, while both events take place concomitantly during early development. As we reported earlier [42], and confirmed more recently by others [27], expression of "immature"-type phenotypic properties by SMC is not strictly associated with migration into the neointima.

Comparative tissue or cell clonal analysis between cell populations obtained from the thickened intima *vs.* normal adult or newborn media, and expression of growth-promoting genes in the intimal thickening *vs.* newborn vessel have highlighted some aspects of the mechanism of neointima formation [240,244,337]. Neointimal SMC 15 days after endothelial denudation display, both *in vivo* and *in vitro*, a down-regulation of SM-type α-actin and SM-MyHC, accompanied by increased proliferative activity, when compared to the subjacent media. Sixty days after injury, proliferative and differentiative properties of neointimal SMC are almost normalized to that of the underlying media [337]. Bochaton-Piallat and coworkers [243,244] also compared the proliferative/differentiative features of cloned SMC obtained from neointima 15 days after injury with those of the underlying media. It turns out that the majority of neointimal clones were of "cobblestone"-type (showing a high [^{3}H]-thymidine incorporation when growing at low cell density and a relatively high migratory ability), whereas the "spindle-shaped"-type clones were more numerous in the media [244]. More importantly there was no phenotypic modulation between the two types of clones, possibly indicating the two clones derive from distinct precursors [244]. However, the "cobblestone"-type clones appeared phenotypically less stable, being susceptible to modulation by retinoic acid into the "spindle-shaped"-type [338]. Five of the six retinoid receptors were in fact expressed in aortic SMC and all-*trans* retinoic acid can antagonize PDGF-BB-stimulated SMC growth [339]. Interestingly, a retinoid signal has been localized to the neural crest-derived mesenchymal cells surrounding the *ductus arteriosus* [340]. This muscular artery is subjected to a combined contraction and intimal thickening formation that brings about its definitive closure in coincidence with the establishment of normal pulmonary circulation around birth [341]. This process has been thought to mimic the neointimal formation in the adult. In addition, the two types of "intimal thickenings" appear in some aspects to be quite similar. In fact, in the *ductus arteriosus* closure, retinoic acid induces the precocious expression of SM2 isoform (i.e., a SM-type differentiation; Figs. 5 and 6; [340,342]. This expression also occured with adult neointimal clones where retinoic acid is involved in shifting the "cobblestone"-type subset to that of "spindle-shaped"-type [338]. Since the "cobblestone"-type SMC subset expresses the cellular-retinol binding protein-1 and keratin 8 which is a marker of "immature"-type SMC, it is tempting to speculate that a specific SMC population, possibly of neural crest origin, is deeply involved in the formation of neointima at least in the arterial essels proximal to the heart.

7.2.2.2
Up-regulation of growth associated genes

Neointimal and developing SMC also show a strong similarity when examined for the expression profile of a number of growth-related genes. These growth factors are released locally from different cellular sources and can act as inducers of SMC proliferation and/or migration or as chemoattractant agents both alone or in combination with other factors (see Fig. 14). Besides those released by non-SMC, some of the growth factors engaged in promoting proliferation and/or migration are secreted by SMC themselves and act through a paracrine/autocrine mechanism. The PDGF-B and PDGF-α receptor mRNAs content in cultured neointimal SMC from adult carotid artery is quite similar to neonatal ("pup"-type) SMC grown *in vitro* [238]. Work from the same laboratory also confirms the presence of PDGF-β receptor in the neointima and the presence of a SMC subpopulation expressing PDGF-B. This raises the possibility that the difference in PDGF-B distribution within *in vitro* cloning [240] is due to the existence of distinct cell populations *in vivo*, possibly with unique migratory properties [343]. Once available in the intimal thickening, PDGF-BB may disclose its powerful effect on SMC migration possibly via PDGF-dependent expression of α_1 and α_2 type VIII collagen, particularly in SMC that show a dedifferentiated phenotype [344,345]. The presence of this collagen type might decrease cell adherence to the substrate and thus facilitate the migration of SMC [345]. The PDGF-A is a weak mitogen and, contrary to PDGF-BB, its expression does not correlate with proliferation level attained by neointimal SMC [346].

Similar to PDGF, osteopontin represents another example of a growth-related protein that is up-regulated in rat "pup" SMC grown *in vitro* and in neointimal SMC [347]. This protein contains the arginine-glycine-aspartate (RGD) sequence that facilitates cell-matrix and cell-cell attachment by means of recognition of the adhesion-promoting cell surface receptors [348,349]. Its expression is enhanced by TGF β1, bFGF, angiotensin II and

PDGF-B [347,350]. Osteopontin is not associated to proliferating SMC [351], but can affect SMC adhesion and chemiotaxis. Liaw et al. [352] reported that anti-osteopontin antibodies can inhibit neointima formation essentially without interfering in SMC proliferation, but inhibit SMC migration from the media to the subendothelial space. This confirms the important role played by cell-matrix interaction in favoring neointimal growth as also demonstrated by the blocking of intimal formation using integrin antagonists [353].

8
Smooth muscle cells from nonαmuscle cells

Experiments performed in NM cell systems might be of potential interest for understanding the "response-to-injury" process in the arterial wall. The possibility exists that NM cells or cells with a hybrid NM-SM cell phenotype can participate in the reparative process that follows in the vascular lesion by converting to SMC (see Fig. 16; [354]). It is, in fact, in this cell phenotypic conversion process that some aspects of the basic SMC biology might be disclosed in terms of gene expression and mechanisms of SMC differentiation [355].

Non-muscular segments of rat lung microvessels in hyperoxic pulmonary hypertension become muscularized using locally recruited fibroblasts or cells with morphological properties between SMC and pericytes [356]. In addition, the cytocontractile apparatus of myofibroblasts found in hypertrophic scars and in the stromal reaction to mammary carcinoma express SM-MyHC [95,357]. It is also worth noting that stromal cells from human long-term bone marrow cultures can differentiate in a time-dependent manner following a SM differentiation pathway characterized, among the other SM markers, by SM myosin expression [358]. The *in vivo* counterpart of these stromal cells, whose phenotype resembles that of the immature SMC, is believed to be important in maintaining granulomonopoiesis. Colony-derived stroma cell lines, established from normal human marrow and showing high proliferative capacity, exhibit a vascular SMC-like differentiation pattern and express the SM1 isoform, i.e., resemble the cell phenotype found in the atherosclerotic lesion [359].

In the bladder wall, which is also composed of SM-NM tissue interfaces similar to the vascular wall (see Fig. 2), there is an alteration of phenotypic stability of submesothelial mesenchymal cells of the serosa layer. This is caused either by partial outflow obstruction or by local necrotizing injury. As a result, it induces a three-step conversion process whereby locally resident fibroblasts first become myofibroblasts and subsequently SMC (Fig. 16; [159,167,360,361]. Myofibroblasts have been found in abnormal wound healing and stromal reaction to breast cancer [362,363]. In these human settings, it was found that myofibroblasts stem from resident fibroblasts [96,364] and, to a lesser extent, from vascular SMC [96]. In the myofibroblast there is a coexistence of some structural and functional elements common to fibroblasts (abundant rough endoplasmic reticulum) and SMC (cytoplasm filled with myofilaments; [362,363]). In the bladder serosa, this cellular transition was identified essentially on the basis of spatiotemporal-specific expression of some cytoskeletal, cytocontractile and ECM proteins markers, such as vimentin, desmin, SM-type α-actin, procollagen α1(I) mRNA and MyHC, and NM-type MyHC isoforms ([159,167,360,361]; our unpublished results; see also Fig. 17).

There are some interesting analogies between the phenotypic modifications of rabbit bladder serosa and those obtained in the adventitial layer of experimental animals after endothelial denudation by angioplasty. In both models, as part of the "response-to-injury" process there is: (1) a transient inflammation, followed by the respective tissue tickening, (2) the formation of myofibroblasts from resident fibroblasts, and (3) the activating role

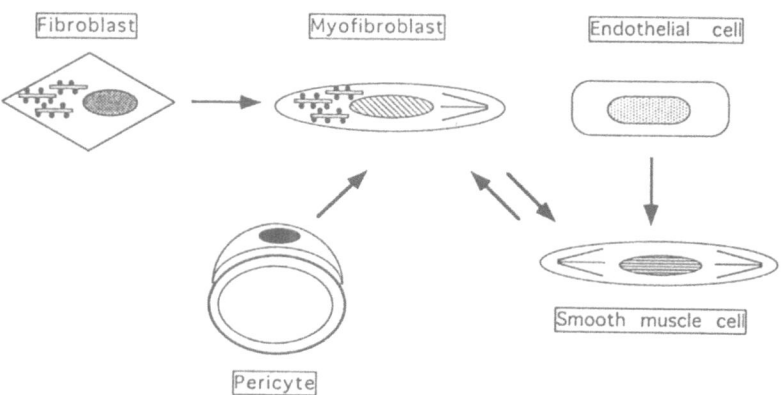

Fig. 16. Myofibroblast-based conversion cell network.

Phenotypic modulation of adventitial fibroblasts

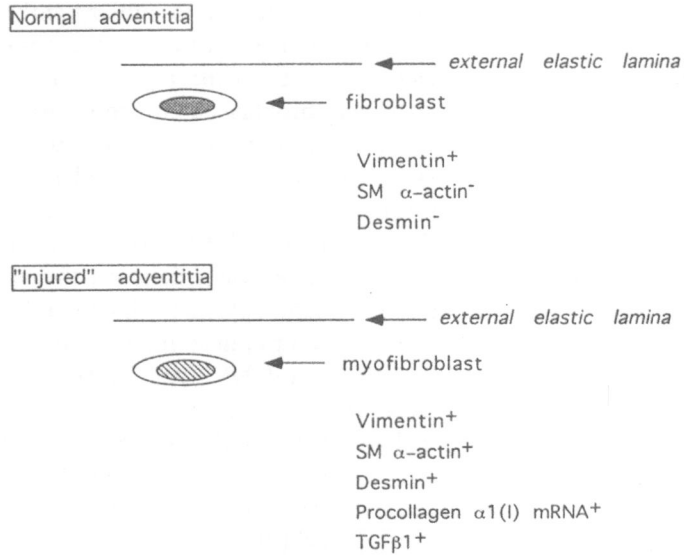

Fig. 17. Phenotypic changes of adventitial cells after experimental wall injury.

of TGFβ1 as inducer of cell proliferation ([7,253,354]; our unplished results). In the serosal thickening there is the "ectopic" formation of SMC that is followed by SMC bundling and the formation of SMC fascicles [15]. Interestingly, both the bladder and the arterial wall are subjected to mechanical stress and it has been hypothesized that this influences the fibroblast to myofibroblast transition [365]. One can argue that cytokines released from the inflammatory and resident cells combined with parietal stress can be responsible for such a phenomenon even in the adventitia [7,366,367].

Recent data obtained with balloon overstretching injury in porcine coronary arteries and vein-to-artery graft using pulse-labeling experiments with bromo-deoxyuridine, an analogue of thymine, have shed some light in this process [367–370]. Some neointimal cells produced after overstretching injury could come from the adventitia [367,369], particularly when the lesion involves the adventitia [369]. A conclusive demonstration of the potential importance of adventitial cells in contributing to the

Remodeling of autologous saphenous vein graft interposed into porcine carotid artery involves fibroblast infiltration from the adventitia to the media

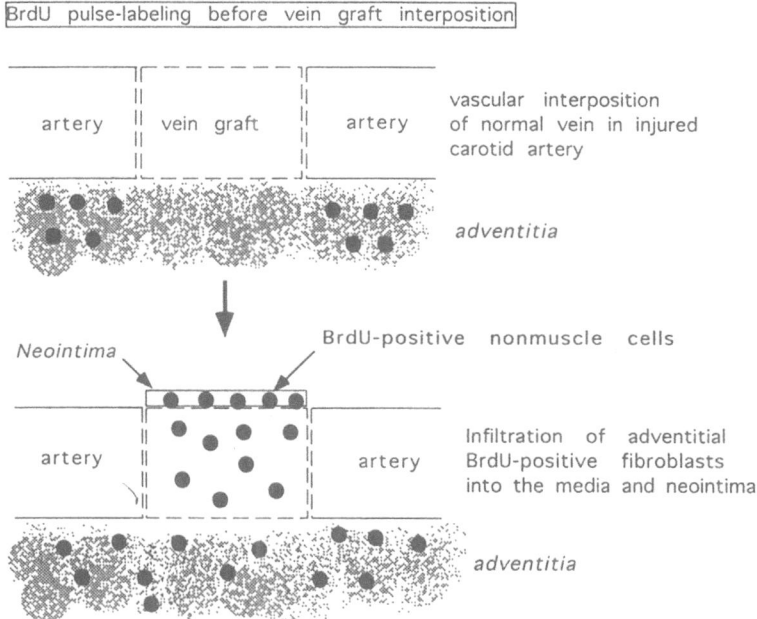

Fig.18. Identification of the role played by converted adventitial cells in the neointima formation. Carotid artery was first pulse-labeled with bromo-deoxyuridine (BrdU) soon after overstretching injury (black dots), then an autologous (not treated) segment of saphenous vein was interposed into the injured/treated artery. Three weeks after surgery, BrdU-positive cells were found both in the "arterialized" venous media and in the neointima (see Ref. [370] for details).

neointima formation has been achived by Shi et al using a model of vein-to-artery interposition ([370]; see Fig. 18). Bromo-deoxyuridine labeled cells from the adventitial layer of artery segments in contact with the vein graft were found both in the "arterialized" media and the neointima.

Depending on the retinoic acid concentration, a continous cellular spectrum ranging from fibroblast/myofibroblast to SMC can be obtained from P19 embryonal cells grown *in vitro* [371]. Since fibroblasts are considered a heterogeneous cell population, possibly with a dual differentiative capacity, it is likely that appropriate availability of specific regulating factors govern can govern the fibroblast to SMC conversion. In any case, adventitial cell plasticity could contribute to the vascular wall remodeling which accompanies angioplasty in humans and endothelial denudation in experimental animals [372], as well as to the narrowing of blood vessels consequent to atherosclerosis and hypertension [7,373,374].

9
Conclusions

In this review we have emphasized that the adult structural-functional alterations involving proliferation/migration of SMC generate a pre-defined molecular and cellular response based on the "immature"-type SMC retrieval. It is likely that the response is driven by the activation of a specific genetic program whose execution may be variably subjected to specific environmental cues (mechanical stress, hormones, growth factors, cytokines, pharmaceutical drugs, etc.). It is clear that deciphering SMC behavior in pathological conditions, such as in the course of neointima formation, relies on the understanding of the mechanisms that govern the early stages of vascular morphogenesis. Obviously, the partial or complete recapitulation of an ontogenetic event in the adult poses a problem for the interactions of newly formed SMC with the environment. Thus, the search for factors that may warrant the acquisition of the maturation profile by SMC is certainly necessary, but not sufficient. In fact, proliferation and differentiation in SMC are not antagonistically linked and SMC can develop a wide range of structural-functional plasticity. This particular aspect of SMC biology, along with the existence of well-defined SMC populations with distinct repertoires of proliferation, differentiation and migration, exemplifies the enormous adaptive potential of this type of muscle tissue with respect to the two sarcomeric muscle tissues. It can be hypothesized that the activation and expansion of specific medial SMC population(s) could imply the acquisition of the competence for migration or migration/proliferation. This cellular response could be (partially) independent from the phenotypical changes that the medial SMC undergo as consequence of the limited environmental perturbations.

10
Some open questions

1. Obviously, the search for SM "master" gene(s), if any, and the delineation of SM gene program [78], have the top priority in SMC biology studies, inasmuch as the identification of endogeneous and exogeneous elements are of crucial importance in the early phase of SMC lineage pathways and in the maintenance of SMC differentiation state. The use of inducible differentiation procedures *in vitro* systems seems to be a suitable approach for evaluating exogeneous factors (for example, retinoic acid) as effective agents able to induce SMC commitment in an undetermined cell line [375].

2. A more complete characterization of the two (or more) SMC lineages existing in the arterial wall is absolutely necessary in order to interpret correctly the pathophysiology of SMC. Spatial- or clonal-type structural and functional differences among vascular SMC have been evidenced early in development (chicken) and both postnatally and in adults (rat). At the moment, no spatiotemporal link has been established between SMC populations that have emerged at different time periods in the course of development. In other words, the delineation of SMC lineage pathways shows a gap in both species around birth/hatching that must be filled with new information. Ideally, this issue could be tackled by the identification of a specific marker that is permanently expressed in developing and adult SM tissue within a given SMC lineage. The existence of SMC lineage(s) should also be tested in veins and in non-vascular SM tissues.

3. If the two SMC lineages have a different tendency to be involved in hypertrophy/hyperploidy or proliferation/migration, as might be inferred from the different experimental models examined so far, it would of interest to ascertain whether, at the level of single clone, apoptosis is differently expressed. While we have found that a precise correlation does not exist between levels of differentiation and development of the apoptotic process in the rabbit [150], others have reported that, at least in closure of *ductus arteriosus* in human, a relationship can be established between these two parameters [376]. It may be that SMC from different lineages display an inherent and specific weakness in cell survival that make these cells more or less vulnerable to cell death.

4. SMC are very plastic cells that are able to change their phenotype *in vitro* and *in vivo* between two SM-type cellular extremes: the "synthetic" and the "contractile" phenotypes. In addition, SMC are developmentally

and experimentally related to other cell types. For example, during development transdifferentiation processes can convert endothelial cells into SMC [230] or SMC into skeletal muscle [377], whereas in some models fibroblasts, myofibroblasts or pericytes can be converted into SMC (see Fig. 16). More recently, putative endothelial cell precursors have been isolated from human peripheral blood [378], raising the intriguing possibility that SMC might ultimately derive also from this source as occurs during vasculogenesis. On this basis, it can be postulated that SMC in the adult, in addition to using their own SMC lineage pathways, might eventually use alternative strategies to afford tissue loss caused by necrosis.

5. Molecular approaches to the inhibition of SMC proliferation are particularly useful developing ways to prevent the development of restenotic process after angioplasty. Data obtained in experimental animals subjected to endothelial denudation are encouraging. As discussed by Schwartz et al. [7] and Post et al. [373], the similarity of the response to angioplasty in man and experimental endothelial lesion remains to be assessed in terms of vascular wall remodeling and/or intimal thickening formation. Gene therapy for vascular diseases might contribute to limiting or preventing restenosis and partially reducing the general impact consequent to neointimal formation [379]. In different models of vascular injury, anti-sense oligonucleotides against PDGF-β receptor [380], c-*myc* and c-*myb* protooncogenes [151,250], positive cell cycle control genes [381], arterial gene transfer of negative cell cycle control genes [382] and growth stimulatory genes [383] can variably affect SMC proliferation. In the light of SMC heterogeneity existing in the blood vessel wall, it would be of interest to evaluate the efficacy, as anti-proliferative agent and/or differentiative modulator, of these potentially useful therapeutic approaches on different SMC clones.

*Acknowledgements.*We are indebted to Prof. Richard A Murphy for his critical reading of this manuscript. We wish to thank Ms. Lisa Galliani for her excellent editorial assistance

References

1. Schwartz SM, Heimark RL, Majesky MW (1990) Developmental mechanisms underlying pathology of arteries. Physiol Rev 70: 1177-1209
2. Owens GK (1995) Regulation of differentiation of vascular smooth muscle cells. Physiol Rev 75:487-517
3. Swynghedauw B (1986) Developmental and functional adaptation of contractile proteins in cardiac and skeletal muscle. Physiol Rev 66:710-770
4. Boheler KR, Schwartz K (1992) Gene expression in cardiac hypertrophy. Trends Cardiovasc Med 2:176-182
5. Schiaffino S, Reggiani C (1996) Molecular diversity of myofibrillar proteins: gene regulation and functional significance. Physiol Rev 76:371-423
6. Pette D, Staron RS (1997) Mammalian skeletal muscle fiber type transitions. Intn Rev Cytol 170:143-223.
7. Schwartz SM, DeBlois D, O'Brien ERM (1995) The intima. Soil for atherosclerosis and restenosis. Circ Res 77:445-465
8. Moss NS, Benditt EP (1970) Spontaneous and experimentally induced arterial lesions.I. An ultrastructural survey of the normal chicken aorta. Lab Invest 22:166-183
9. Lauper NT, Unni KK, Kotke BA, Titus JL (1975) Anatomy and histology of aorta of White Carneau pigeon. Lab Invest 32:536-551
10. Zanellato AMC, Borrione AC, Giuriato L, Tonello M, Scannapieco G, Pauletto P, Sartore S (1990) Myosin isoforms and cell heterogeneity in vascular smooth muscle. I. Developing and adult bovine aorta. Dev Biol 141:431-446.
11. Zanellato AMC, Borrione AC, Tonello M, Scannapieco G, Pauletto P, Sartore S (1990) Myosin isoform expression and smooth muscle cell heterogeneity in normal and atherosclerotic rabbit aorta. Arteriosclerosis 10:996-1009.
12. Frid MG, Moiseeva EP, Stenmark KR (1994) Multiple phenotypically distinct smooth muscle cell populations exist in the adult and developing bovine pulmonary arterial media in vivo. Circ Res 75:669-681
13. Frid MG, Dempsey EC, Durmowicz AG, Stenmark KR (1997) Smooth muscle cell heterogeneity in pulmonary and systemic vessels. Importance in vascular disease. Arterioscler Thromb Vasc Biol 17:1203-1209
14. de Boer WI, Schuller AGP, Vermey M, van der Kwast TH (1994) Expression of growth factors and receptors during specific phases in regenerating urothelium after acute injury in vivo. Am J Pathol 145:1199-1207
15. Faggian L, Pampinella F, Roelofs M, Paulon T, Franch R, Chiavegato A, Sartore S (1997) Phenotypic changes in the regenerating rabbit bladder muscle. Role of interstitial cells and innervation on smooth muscle cell differentiation. Histochem Cell Biol
16. Cunha GR, Battle E, Young P, Brody J, Donjacour A, Hayashi H (1992) Role of epithelial-mesenchymal interactions in the differentiation and spatial organization of visceral smooth muscle. Epithelial Cell Biol 1:76-83

17. Noguchi S, Yura Y, Sherwood ER, Kakinuma H, Kashihara N, Oyasu R (1990) Stimulation of stromal cell growth by normal rat urothelial cell-derived epidermal growth factor. Lab Invest 62:538–544

18. Campbell DJ, Habener JF (1986) Angiotensinogen gene is expressed and differentially regulated in multiple tissues of the rat. J Clin Invest 78:31–39

19. Gabella G, Uvelius B (1990) Urinary bladder of rat: fine structure of normal and hypertrophic musculature. Cell Tissue Res 262:67–79

20. Somlyo AP, Somlyo AV (1968) Vascular smooth muscle: I. Normal structure, pathology, biochemistry and biophysics. Pharmacol Rev 20:197–272.

21. Levy BJ, Wight TN (1995) The role of proteoglycans in bladder structure and function. Adv Exp Med Biol 385:191–203

22. Mironov AA, Rekhter MD, Kolpakov VA, Andreeva ER, Polishchuk RS, Bannykh SI, Filippov SV, Peretjatko LP, Kulida LV, Orekhov AN (1995) Heterogeneity of smooth muscle cells in embryonic human aorta. Tissue & Cell 27:31–38

23. Davis EC (1993) Smooth muscle cell to elastic lamina connections in developing mouse aorta. Role in aortic medial organization. Lab Invest 68:89–99

24. Sosa-Melgarejo JA, Berry CL (1995) Myoendothelial contacts in the human fetal aorta. Arch Med Res 26:431–435

25. Cliff WJ (1967) The aortic tunica media in growing rats studied with the electron microscope. Lab Invest 17:599–615

26. Gerrity RD, Cliff WJ (1975) The aortic tunica media of the developing rat. Lab Invest 32:585–600

27. Belknap JK, Grieshaber NA, Schwartz PE, Orton EC, Reidy MA, Majack RA (1996) Tropoelastin gene expression in individual vascular smooth muscle cells. Circ Res 78:388–394

28. Bendeck MP, Langille BL (1991) Rapid accumulation of elastin and collagen in the aortas of sheep in the immediate perinatal period. Circ Res 69:1165–1169

29. Stenmark KR, Mecham RP (1997) Cellular and molecular mechanisms of pulmonary vascular remodeling. Annu Rev Physiol 59:89–144

30. Olivetti G, Anversa P, Melissari M, Loud AV (1980) Morphometric study of early postnatal development of thoracic aorta in the rat. Circ Res 47:417–424

31. Boudreau N, Turley E, Rabinovitch M (1991) Fibronection, hyaluronan, and hyluronan binding protein contribute to increased ductus arteriosus smooth muscle cell migration. Dev Biol 143:235–247

32. Slomp J, Van Munsteren JC, Poelmann RE, DeReeder EG, Bogers AJJC, Gittenberger-de Groot A (1992) Formation of intimal cushions in the ductus arteriosus as a model for vascular intimal thickening: an immunohistochemical study of changes in the extracellular matrix components. Atherosclerosis 93:25–39

33. Giuriato L, Scatena M, Chiavegato A, Guidolin D, Pauletto P, Sartore S (1993) Rabbit ductus arteriosus during development: anatomical structure and smooth muscle cell composition. Anat Rec 235:95–110

34. Manasek FJ (1971) The ultrastructure of embryonic myocardial blood vessels. Dev Biol 26:42–54

35. Sexton AJ, Turmaine M, Cai WQ, Burnstock G (1996) A study of the ultrastructure of developing human umbilical vessels. J Anat 188:75–85
36. Ward SM, Torihashi S (1995) Morphological changes during ontogeny of the canine proximal colon. Cell Tissue Res 282:93–108
37. Gabella G (1989) Development of smooth muscle: ultrastructural study of the chick embryo gizzard. Anat Embryol 180:213–226
38. Chou R-GR, Stromer MH, Robson RM, Huiatt TW (1992) Assembly of contractile and cytoskeletal elements in developing smooth muscle cells. Dev Biol 149:339–348
39. Carey DJ (1991) Control of growth and differentiation of vascular cells by extracellular matrix proteins. Ann Rev Physiol 53:161–177
40. Chamley-Campbell JH, Campbell GR, Ross R (1979) The smooth muscle cell in culture. Physiol Rev 59:1–61
41. Thyberg J (1996) Differentiation properties and proliferation of arterial smooth muscle cells in culture. Intn Rev Cytol 169:183–265
42. Sartore S, Scatena M, Chiavegato A, Faggin E, Giuriato L, Pauletto P (1994) Myosin isoform expression in smooth muscle cells during physiological and pathological vascular remodeling. J Vasc Res 31:61–81
43. Horowitz A, Menice CB, Laporte R, Morgan KG (1996) Mechanisms of smooth muscle contraction. Physiol Rev 76:967–996
44. Somlyo AP (1993) Myosin isoforms in smooth muscle: how may they affect function and structure? J Muscle Res Cell Motil 14:557–563
45. Kelley CA, Adelstein RS (1994) Characterization of isoform diversity in smooth muscle myosin heavy chains. Can J Physiol Pharmacol 72:1351–1360
46. Miano JM, Cserjesi P, Ligon KL, Periasamy M, Olson EN (1994) Smooth muscle myosin heavy chain exclusively marks the smooth muscle lineage during mouse embryogenesis. Circ Res 75:803–812
47. Babij P, Periasamy M (1989) Myosin heavy chain isoform diversity in smooth muscle is produced by differential RNA processing. J Mol Biol 210:673–679
48. Nagai R, Kuro-o M, Babij P, Periasamy M (1989) Identification of two types of smooth muscle myosin heavy chain isoforms by cDNA cloning and immunoblot analysis. J Biol Chem 264:9734–9737
49. Hamada Y, Yanagisawa M, Katsuragawa Y, Coleman JR, Nagata S, Matsuda G, Masaki T (1990) Distinct vascular and intestinal smooth muscle myosin heavy chain mRNAs are encoded by a single-copy gene in the chicken. Biochem Biophys Res Commun 170:53–58
50. Babij P (1993) Tissue specific and developmentally regulated alternative splicing of a visceral isoform of smooth muscle myosin heavy chain. Nucleic Acids Res 21:1467–1471
51. Kelley CA, Takahashi M, Yu JH, Adesltein RS (1993) An insert of seven aminoacids confers functional differences between smooth muscle myosin from the intestines and vasculature. J Biol Chem 268:12848–12854

52. White S, Martin AF, Periasamy M (1993) Identification of a novel smooth muscle myosin heavy chain cDNA: Isoform diversity in the S1 head region. Am J Physiol 264 (Cell Phsysiol 33):C1252–C1258

53. Matsuoka R, Yoshida MC, Furutani Y, Imamura S-i, Kanda N, Yanagisawa M, Masaki T, Takao A (1993) Human smooth muscle myosin heavy chain gene mapped to chromosomal region 16q12. Am J Med Gen 46:61–67

54. Deng Z, Liu P, Marlton P, Claxton DF, Lane S, Callen DF, Collins FS, Siciliano MJ (1993) Smooth muscle myosin heavy chain locus (MYH11) maps to 16p13.13–p13.12 and establishes a new region of conserved synteny between human 16p and mouse 16. Genomics 18:156–159

55. Kuro-o M, Nagai R, Tsuchimochi H, Katoh H, Yazaki Y, Ohkubo A, Takaku F (1989) Developmentally regulated expression of vascular smooth muscle myosin heavy chain isoforms. J Biol Chem 264:18272–18275

56. Kuro-o M, Nagai R, Nakahara K, Katoh H, Tsai RC, Tsuchimochi H, Yazaki YH, Ohkubo H, Takaku F (1991) cDNA cloning of a myosin heavy chain isoform in embryonic smooth muscle and its expression during vascular development and in arteriosclerosis. J Biol Chem 266:3768–3773

57. Frid MG, Printseva OY, Chiavegato A, Faggin E, Koteliansky VE, Pauletto P, Glukhova MA, Sartore S (1993) Myosin heavy-chain composition in developing and adult human aortic smooth muscle. J Vasc Res 30:279–292

58. Eddinger TJ, Wolf JA (1993) Expression of four myosin heavy chain isoforms with development in mouse uterus. Cell Motil Cytoskel 25:358–368

59. Cavaillé F, Fournier T, Dallot E, Dhellemes C, Ferr F (1995) Myosin heavy chain isoform expression in human myometrium. Cell Motil Cytoskel 30:183–193

60. Capriani A, Chiavegato A, Franch R, Azzarello G, Vinante O, Sartore S (1997) Oestrogen-dependent expression of SM2 smooth muscle type myosin isoform in rabbit myometrium. J Muscle Res Cell Motil 18:413–427

61. Chiavegato A, Scatena M, Roelofs M, Ferrarese P, Pauletto P, Passerini-Glazel G, Pagano F, Sartore S (1993) Cytoskeletal and cytocontractile protein composition of smooth muscle cells in developing and obstructed rabbit bladder. Exp Cell Res 207:310–320

62. Morano I, Erb G, Sogl B: Expression of myosin heavy and light chains changes during pregnancy in the rat uterus (1993) Eur J Physiol 1993;423:434–441

63. Mohammad MA, Sparrow MP (1988) Changes in myosin heavy chain stochiometry in pig tracheal smooth muscle during development. FEBS lett 228:109–112

64. Mohammad MA, Sparrow MP (1989) Distribution of heavy-chain isoforms of myosin in airways smooth muscle from adult and neonate humans. Biochem J 260:421–426

65. Woodcock-Mitchell J, White S, Stirewalt W, Periasamy M, Mitchell J, Low R (1993) Myosin isoform expression in developing and remodeling rat lung. Am J Resp Cell Mol Biol 8:617–625

66. Chiavegato A, Pauletto P, Sartore S (1996) Smooth muscle-type myosin heavy chain isoforms in bovine smooth muscle and non-muscle tissues. Biol Cell 86:27–38

67. Rovner AS, Freyzon Y, Trybus KM (1997) An insert in the motor domain determines the functional properties of expressed smooth muscle myosin isoforms. J Muscle Res Cell Motil 18:103–110

68. Fisher SA, Ikebe M, Brozovich F (1997) Endothelin-1 alters the contractile phenotype of cultured embryonic smooth muscle cells. Circ Res 80:885–893

69. Katoh Y, Loukianov E, Kopras E, Zilberman A, Periasamy M (1994) Identification of functional promoter elements in the rabbit smooth muscle myosin heavy chain gene. J Biol Chem 269:30538–30545

70. Kallmeier RC, Somasundaram C, Babij P (1995) A novel smooth muscle-specific enhancer regulates transcription of the smooth muscle myosin heavy chain gene in vascular smooth muscle cells. J Biol Chem 270:30949–30957

71. White SL, Low RB (1996) Identification of promoter elements involved in cell-specific regulation of rat smooth muscle myosin heavy chain gene transcription. J Biol Chem 271:15008–15017

72. Watanabe M, Sakomura Y, Kurabayashi M, Manabe I, Aikawa M, Kuro-o M, Suzuki Y, Nagai R (1996) Structure and characterization of the 5'-flanking region of the mouse smooth muscle myosin heavy chain (SM1/SM2) gene. Circ Res 78:978–989

73. Madsen CS, Hershey JC, Hautmann MB, White SL, Owens GK (1997) Expression of the smooth muscle myosin heavy chain gene is regulated by a negative acting GC-rich element located between two positive-acting serum response factor-binding elements. J Biol Chem 272:6332–6340

74. van der Loop FTL, Schaart G, Timmer ED, Ramaekers FC, van Eys GJ (1996) Smoothelin, a novel cytoskeletal protein specific for smooth muscle cells. J Cell Biol 134:401–411

75. van der Loop FTL, Gabbiani G, Kohnen G, Ramaekers FCS, van Eys GJJM (1997) Differentiation of smooth muscle cells in human blood vessels as defined by smoothelin, a novel marker for the contractile phenotype. Arterioscler Thromb Vasc Biol 17:665–671

76. Wehrens XHT, Mies B, Gimona M, Ramaekers FCS, van Eys GJJM, Small JV (1997) Localization of smoothelin in avian smooth muscle and identification of a vascular-specific isoform. FEBS lett 405:315–320

77. Jain MK, Fujita KP, Hsieh C-M, Endege WO, Sibinga NE, Yet S-F, Kashiki S, Lee W-S, Perrella MA, Haber E, Lee M-E (1996) Molecular cloning and characterization of SmLIM, a developmentally regulated LIM protein preferentially expressed in aortic smooth muscle cells. J Biol Chem 271:10194–10199

78. Firulli AB, Olson EN (1997) Modular regulation of muscle gene transcription: a mechanism for cell diversity. Trends in Genetics 13:364–369

79. Vandekerckove J, Weber K (1978) At least six different actins are expressed in a higher mammal: an analysis based on the amino acid sequence of the amino-terminal peptide. J Mol Biol 126:783–802

80. Owens GK, Thompson MM (1986) Developmental changes in isoactin expression in rat aortic smooth muscle cells in vivo. Relationship between growth and cytodifferentiation. J Biol Chem 261:13373–13380

81. Eddinger TJ, Murphy RA (1991) Developmental changes in actin and myosin heavy chain isoform expression in smooth muscle. Arch Biochem Biophys 284:232–237

82. McHugh KM, Crawford K, Lessard JL (1991) A comprehensive analysis of the developmental and tissue-specific expression of the isoactin multigene family. Dev Biol 148:442–458

83. McHugh KM (1995) Molecular analysis of smooth muscle development in the mouse. Dev Dyn 204:278–290

84. Gabbiani G, Schmid E, Winter S, Chaponnier C, De Chastonay C, Vandekerckhove J, Weber K, Franke WW (1981) Vascular smooth muscle cells differ from other smooth muscle cells: predominance of vimentin filaments and a specific α-type actin. Proc Natl Acad Sci USA 78:298–302

85. Fatigati V, Murphy RA (1984) Actin and tropomyosin variants in smooth muscles. J Biol Chem 259:14384–14388

86. Kim YS, Wang Z, Levin RM, Chacko S (1994) Alterations in the expression of the β-cytoplasmic and γ-smooth muscle actins in hypertrophied urinary bladder smooth muscle. Mol Cell Biochem 131:115–124

87. Yamamoto Y, Kubota T, Atoji Y, Suzuki Y (1996) Distribution of α-vascular smooth muscle actin in the smooth muscle cells of the gastrointestinal tract of the chicken. J Anat 189:623–630

88. Sawtell NM, Lessard JL (1989) Cellular distribution of smooth muscle actins during mammalian embyogenesis: expression of the α-vascular but not the γ-enteric isoform in differentiating striated muscle. J Cell Biol 109:2929–2937

89. Kim E, Waters S, Hake L, Hecht N (1989) Identification and developmental expression of a smooth muscle γ-actin in postmeiotic male germ cells of mice. Mol Cell Biol 9:1875–1881

90. Malmqvist U, Arner A, Uvelius B (1991) Contractile and cytoskeletal proteins in smooth muscle during hypertrophy and its reversal. Am J Physiol 260 (Cell Physiol 29):C1085–C1093

91. Schafer BW, Perriard JC (1988) Intracellular targeting of isoproteins in muscle cytoarchitecture. J Cell Biol 106:1161–1170

92. Drew JS, Moos C, Murphy RA (1991) Localization of isoactins in isolated smooth muscle thin filaments by double gold immunolabeling. Am J Physiol 260 (Cell Physiol 29):C1332–C1340

93. North AJ, Gimona M, Lando Z, Small JV (1994) Actin isoform compartments in chicken gizzard smooth muscle cells. J Cell Sci 107:445–455

94. Gabbiani G, Kocher O, Bloom WS, Vandekerckhove J, Weber K (1984) Actin expression in smooth muscle cells of rat aortic intimal thickening, human atheromatous plaque, and cultured rat aortic media. J Clin Invest 73:148–152

95. Lazard D, Sastre X, Frid MG, Glukhova MA, Thiery J-P, Koteliansky VE (1993) Expression of smooth muscle-specific proteins in myoepithelium and stromal myofibroblasts of normal and malignant human breast tissue. Proc Natl Acad Sci USA 90:999–1003

96. Rønnov-Jessen L, Petersen OW, Koteliansky VE, Bissell MJ (1995) The origin of the myofibroblasts in breast cancer. J Clin Invest 95:859–873

97. Jahoda CAB, Reynolds AJ, Chaponnier C, Forester JC, Gabbiani G (1991) Smooth muscle alpha-actin is a marker for hair follicle dermis in vivo and in vitro. J Cell Sci 99:627–636

98. Peled A, Zipori D, Abramsky O, Ovadia H, Shezen E (1991) Expression of smooth alpha-actin in murine bone marrow stromal cells. Blood 78:304–309

99. Gabbiani G (1996) The cellular derivation and life span of the myofibroblast. Path Res Pract 192:708–711

100. Ruzicka DL, Schwartz RJ (1988) Sequential activation of alpha actin genes during avian cardiogenesis: vascular smooth muscle alpha-actin gene transcripts mark the onset of cardiomyocyte differentiation. J Cell Biol 107:2575–2586

101. Woodcock-Mitchell J, Mitchell JJ, Low RB, Kieny M, Sengel P, Rubbia L, Skalli O, Jackson B, Gabbiani G (1988) Alpha-smooth muscle actin is transiently expressed in embryonic rat cardiac and skeletal muscles. Differentiation 39:161–166

102. Hungerford JE, Owens GK, Argraves WS, Little CD (1996) Development of the aortic vessel wall as defined by vascular smooth muscle and extracellular matrix markers. Dev Biol 178:375–392

103. Mitchell JJ, Reynolds SE, Leslie KO, Low RB, Woodcock-Mitchell J (1990) Smooth muscle cell markers in developing rat lung. Am J Respir Cell Mol Biol 3:515–523

104 Blank RS, McQuinn TC, Yin KC, Thompson MM, Takeyasu K, Schwartz RJ, Owens GK (1992) Elements of the smooth muscle α-actin promoter required in cis for transcriptional activation in smooth muscle. J Biol Chem 267:984–989

105. Shimizu RT, Blank RS, Jervis R, Lawrenz-Smith SC, Owens GK (1995) The smooth muscle α-actin gene promoter is differentially regulated in smooth muscle versus non- muscle cells. J Biol Chem 270:7631–7643

106. Li L, Miano JM, Mercer B, Olson EN (1996) Expression of the SM22α promoter in transgenic mice provides evidence for distinct transcriptional regulatory programs in vascular and visceral smooth muscle cells. J Cell Biol 132:849–859

107. Lees-Miller JP, Heeley DH, Smillie LB, Kay CM (1987) Isolation and characterization of an abundant and novel 22-kDa protein (SM22α) from chicken gizzard smooth muscle. J Biol Chem 262:2988–2993

108. Shanahan CM, Weisseberg PL, Metcalf JC (1993) Isolation of gene markers of differentiated and proliferating vascular smooth muscle cells. Circ Res 73:193–204

109. Solway J, Seltzer J, Samaha FF, Kim S, Alger LE, Niu Q, Morrisey EE, Ip HS, Parmacek MS (1995) Structure and expression of a smooth muscle cell-specific gene, SM22α. J Biol Chem 270:13460–13469

110. Ayme-Southgate A, Lasko P, French C, Pardue ML (1989) Characterization of the gene for mp20: a Drosophila muscle protein that is not found in asynchronous oscillatory flight muscle. J Cell Biol 108:521–531

111. Ren W-Z, Ng GYK, Wang R-x, Wu PH, O'Dowd BF, Osmond DH, George SR, Liew C-C (1994) The identification of NP25: a novel protein that is differentially expressed by neuronal subpopulations. Mol Brain Res 22:173–185

112. Gimona M, Sparrow MP, Strasser P, Herzog M, Small JV (1992) Calponin and SM22 isoforms in avian and mammalian smooth muscle. Eur J Biochem 205:1067–1075

113. Duband J-L, Gimona M, Scatena M, Sartore S, Small JV (1993) Calponin and SM22 as differentiation markers of smooth muscle: spatiotemporal distribution during avian embryonic development. Differentiation 55:1–11

114. Li L, Miano P, Cserjesi P, Olson EN (1996) SM22α, a marker of adult smooth muscle, is expressed in multiple myogenic lineages during embryogenesis. Circ Res 78:188–195

115. Nishida W, Kitami Y, Hiwada K (1993) cDNA cloning and mRNA expression of calponin and SM22 in rat aorta smooth muscle cells. Gene 130:297–302

116. Moessler H, Mericskay M, Li Z, Nagl S, Paulin D, Small JV (1996) The SM22 promoter directs tissue-specific expression in arterial but not in venous or visceral smooth muscle cells in transgenic mice. Development 122:2415–2425

117. Kim S, Ip HS, Lu MM, Clendenin C, Parmacek MS (1997) A serum response factor-dependent transcriptional regulatory program identifies distinct smooth muscle cell lineages. Mol Cell Biol 17:2266–2278

118. Li L, Liu Z-c, Mercer B, Overbeek P, Olson EN (1997) Evidence for serum response regulatory networks governing SM22α transcription in smooth, skeletal, and cardiac muscle cells. Dev Biol 187:311–321

119. El-Mezgueldi M (1996) Calponin. Int J Biochem 28:1185–1189

120. Sobue K, Sellers JR (1991) Caldesmon, a novel regulatory protein in smooth muscle and nonmuscle actomyosin systems. J Biol Chem 266:12115–12118

121. Winder SJ, Walsh MP (1993) Calponin: thin-filament linked regulation of smooth muscle contraction. Cell Signal 5:677–686

122. Takahashi K, Nadal-Ginard B (1991) Molecular cloning and sequence analysis of smooth muscle calponin. J Biol Chem 266:13284–13288

123. North AJ, Gimona M, Cross RA, Small JV (1994) Calponin is localised in both the contractile apparatus and the cytoskeleton of smooth muscle cells. J Cell Sci 107:437–444

124. Hayashi K, Fujio Y, Kato I, Sobue K (1991) Structural and functional relationships between h- and l-caldesmons. J Biol Chem 266:355–361

125. Applegate D, Feng W, Gree RS, Taubman MB (1994) Cloning and expression of a novel acidic calponin isoform from rat aortic vascular smooth muscle. J Biol Chem 269:10683–10690

126. Sakurai H, Matuoka R, Furutani Y, Imamura S-i, Takao A, Momma K (1996) Expression of four myosin heavy chain genes in developing blood vessels and other smooth muscle organs in rabbits. Eur J Cell Biol 69:166–172

127. Miano JM, Olson EN (1996) Expression of the smooth muscle cell calponin gene marks the early cardiac and smooth muscle cell lineages during mouse embryogenesis. J Biol Chem 271:7095–7103

128. Strasser P, Gimona M, Moessler H, Herzog M, Small JV (1993) Mammalian calponins. Identification and expression of genetic variants. FEBS lett 330:13–18

129. Hayashi K, Yano K, Hashida T, Takeuchi R, Takeda O, Asada K, Takahashi E, Kato I, Sobue K (1992) Genomic structure of human caldesmon gene. Proc Natl Acad Sci USA 89:12122–12126

130. Yano H, Hayashi K, Haruna M, Sobue K (1994) Identification of two distinct promoters in the chicken caldesmon gene. Biochem Biophys Res Commun 201:618–626

131. Yano H, Hayashi K, Momiyama T, Saga H, Haruna M, Sobue K (1995) Transcriptional regulation of the chicken caldesmon gene. J Biol Chem 270:23661–23666

132. Frid MG, Shekhonin BV, Koteliansky VE, Glukhova MA (1992) Phenotypic changes of human smooth muscle cells during development: late expression of heavy caldesmon and calponin. Dev Biol 153:185–193

133. Glukhova MA, Frid MG, Koteliansky VE (1991) Phenotypic changes of human aortic smooth muscle cells during development and in the adult vessel. Am J Physiol Suppl 261:78–80

134. Borrione AC, Zanellato AMC, Scannapieco G, Pauletto P, Sartore S (1989) Myosin heavy-chain isoforms in adult and developing rabbit vascular smooth muscle. Eur J Biochem 183:413–417

135. Gaylinn BD, Eddinger TJ, Martino PA, Monical PL, Hunt DF, Murphy RA (1989) Expression of nonmuscle myosin heavy and light chains in smooth muscle. Am J Physiol 257 (Cell Physiol 26):C997–C1004

136. Kawamoto S, Adelstein RS (1991) Chicken nonmuscle myosin heavy chains: differential expression of two mRNAs and evidence for two different polypeptides. J Cell Biol 112:915–924

137. Murakami N, Elzinga M (1992) Immunohistochemical studies on the distribution of cellular myosin II isoforms in brain and aorta. Cell Motil Cytoskel 22:281–295

138. Aikawa M, Nalla Sivam P, Kuro-o M, Kimura K, Nakahara K, Takewaki S, Ueda M, Yamaguchi H, Yazaki Y, Periasamy M, Nagai R (1993) Human smooth muscle myosin heavy chain isoforms as molecular markers for vascular development and atherosclerosis. Circ Res 73:1000–1012

139. Sun W, Chantler PD (1992) Cloning of the cDNA encoding a neuronal myosin heavy chain from mammalian brain and its differential expression within the central system. J Mol Biol 224:1185–1193

140. Takahashi M, Kawamoto S, Adelstein RS (1992) Evidence for inserted sequences in the head region of nonmuscle myosin specific to the nervous system. J Biol Chem 267:17864–17871

141. Murakami N, Trenkner E, Elzinga M (1993) Changes in expression of nonmuscle myosin heavy chain isoforms during muscle and nonmuscle tissue development. Dev Biol 157:19–27

142. Simons M, Wang M, McBride OW, Kawamoto S, Yamakawa K, Gdula D, Adelstein RS, Weir L (1991) Human nonmuscle myosin heavy chains are encoded by two genes located on different chromosomes. Circ Res 69:530–539

143. Phillips CL, Yamakawa K, Adelstein RS (1995) Cloning of the cDNA encoding human nonmuscle myosin heavy chain-B and analysis of human tissues with isoform-specific antibodies. J Muscle Res Cell Motil 16:379–389

144. Choi OH, Park C-S, Itoh K, Adelstein RS, Beaven MA (1996) Cloning of the cDNA encoding rat myosin heavy chain-A and evidence for the absence of myosin heavy chain-B in cultured rat mast (RBL-2H3) cells. J Muscle Res Cell Motil 17:69–77

145. Kelley CA, Sellers JR, Gard DL, Bui D, Adelstein RS (1996) Xenopus nonmuscle myosin heavy chain isoforms have different subcellular localization and enzymatic activities. J Cell Biol 134:675–687

146. Itoh K, Adelstein RS (1995) Neuronal cell expression of inserted isoforms of vertebrate nonmuscle myosin heavy chain-IIB. J Biol Chem 270:14533–14540

147. Pato MD, Sellers JR, Preston YA, Harvey EV, Adelstein RS (1996) Baculovirus expression of chicken nonmuscle heavy meromyosin II-B. J Biol Chem 271:2689–2695

148. Murakami N, Mehta P, Elzinga M (1991) Studies on the distribution of cellular myosin with antibodies to isoform-specific synthetic peptides. FEBS lett 278:23–25

149. Giuriato L, Scatena M, Chiavegato A, Tonello M, Scannapieco G, Pauletto P, Sartore S (1992) Non-muscle myosin isoforms and cell heterogeneity in developing rabbit vascular smooth muscle. J Cell Sci 101:233–246

150. Sartore S, Chiavegato A, Franch R, Faggin E, Pauletto P (1997) Myosin gene expression and cell phenotypes in vascular smooth muscle during development, in experimental models, and in vascular disease. Arterioscler Thromb Vasc Biol 17:1210-1215

151. Simons M, Rosenberg RD (1992) Antisense nonmuscle myosin heavy chain and c-myb oligonucleotides suppress smooth muscle cell proliferation in vitro. Circ Res 70:835–843

152. Kelley CA, Oberman F, Yisraeli JK, Adelstein RS (1995) A Xenopus nonmuscle myosin heavy chain isoform is phosphorylated by cyclin-p34^{cdc2} kinase during meiosis. J Biol Chem 270:1395–1401

153. Elledge SJ (1996) Cell cycle checkpoints: preventing an identity crisis. Science 274:1664–1672

154. Kriajevska MV, Cardenas MN, Grigorian MS, Ambartsumian NS, Georgiev GP, Lukanidin EM (1994) Non-muscle myosin heavy chain as a possible target for protein encoded by metastasis-related mts-1 gene. J Biol Chem 269:19679–19682

155. Maupin P, Phillips CL, Adelstein RS, Pollard TS (1994) Differential localization of myosin-II isozymes in human cultured cells and blood cells. J Cell Sci 107:3077–3090

156. Amore B, Chiavegato A, Paulon T, Pauletto P, Sartore S (1996) Atherosclerosis resistance in rats correlates with lacking of expansion of an immature smooth muscle cell population. J Vasc Res 33:442–453

157. De Leon H, Scott NA, Martin F, Simonet L, Bernstein KE, Wilcox JN (1997) Expression of nonmuscle myosin heavy chain-B isoform in the vessel wall of porcine coronary arteries after ballon angioplasty. Circ Res 80:514–519

158. Giuriato L, Chiavegato A, Pauletto P, Sartore S (1995) Correlation between the presence of an immature smooth muscle cell population in tunica media and the development of atherosclerotic lesion. A study on different-sized rabbit arteries from cholesterol-fed and Watanabe heritable hyperlipemic rabbits. Atherosclerosis116:77–92

159. Buoro S, Ferrarese P, Chiavegato A, Roelofs M, Scatena M, Pauletto P, Passerini-Glazel G, Pagano F, Sartore S (1993) Myofibroblast-derived smooth muscle cells during remodeling of rabbit urinary bladder wall induced by partial outflow obstruction. Lab Invest 69:589–602

160. Chiavegato A, Capriani A, Azzarello G, Vinante, Pauletto P, Sartore S (1996) Expression of non-muscle myosin isoforms in rabbit myometrium is estrogen-dependent. Cell Tissue Res 283:7–18

161. Turley H, Pulford KAF, Gatter KC, Mason DY (1988) Biochemical evidence that cytokeratins are present in smooth muscle. Br J Exp Pathol 69:433–440

162. Jahn L, Fouquet B, Rohe K, Franke WW (1987) Cytokeratins in certain endothelial and smooth muscle cells of two taxonomically distant vertebrate species, Xenopus laevis and man. Differentiation 36:234–254

163. Bader B, Jahn L, Franke WW (1988) Low levels of cytokeratin 8, 18 and 19 in vascular SMCs of human umbilical cord and cultured cells derived therefrom, and analyis of the locus containing the cytokerain 19 gene. Eur J Cell Biol 47:300–319

164. Johansson B, Eriksson A, Virtanen I, Thornell L-S (1997) Intermediate filament proteins in adult human arteries. Anat Rec 247:439–448

165. Frank ED, Warren, L (1981) Aortic smooth muscle cells contain vimentin instead of desmin. Proc Natl Acad Sci USA 78:3020–3024

166. Jahn L, Kreuzer J, von Hodenberg E, Kubler W, Franke WW, Allenberg J, Izumo S (1993) Cytokeratin 8 and 18 in smooth muscle cells. Arterioscler Thromb 13:1631–1639

167. Pampinella F, Roelofs M, Castellucci E, Chiavegato A, Guidolin D, Passerini-Glazel G, Pagano F, Sartore S (1996) Proliferation of submesothelial mesenchymal cells during early phase of serosal thickening in the rabbit bladder is accompanied by transient keratin 18 expression. Exp Cell Res 223:327–339

168. Glukhova M, Koteliansky VE, Fondacci C, Marotte F, Rappaport L (1993) Laminin variants and integrin laminin receptors in developing and adult human smooth muscle cells. Dev Biol 157:437–447

169. Dupl^a C, Couffinhal T, Dufourcq P, Llanas B, Moreau C, Bonnet J (1997) The integrin very late antigen-4 is expressed in human smooth muscle cell. Circ Res 80:159–169

170. Sheppard AM, Onken MD, Rosen GD, Noakes PG, Dean DC (1994) Expanding roles for α4 integrin and its ligands in development. Cell Adhesion Commun 2:27–43

171. Ruoslahti E (1988) Fibronectin and its receptors. Annu Rev Biochem 57:375–413

172. Schwartzbauer JE (1991) Alternative splicing of fibronectin: three variants, three functions. Bioassay 13:527–533

173. fFrench-Constant C, Hynes RO (1989) Alternative splicing of fibronectin is temporally and spatially regulated in the chicken embryo. Development 106:375–388

174. Glukhova MA, Frid MG, Shekhonin BV, Balabanov YV, Koteliansky VE (1990) Expression of fibronectin variants in vascular and visceral smooth muscle cells in development. Dev Biol 141:193–202

175. Pagani F, Zagato L., Vergani C, Casari G, Sidoli A, Bartalle FE (1991) Tissue-specific splicing pattern of fibronectin messanger RNA precursor during development and aging in rat. J Cell Biol 113:1223–1229

176. Dubin D, Peters JH, Brwon LF, Logan B, Kent KC, Berse B, Berven S, Cercek B, Sharifi BG, Pratt RE, Dzau VJ, Van de Water L (1995) Balloon catheterization induces arterial expression of embryonic fibronectins. Arterioscler Thromb Vasc Biol 15:1958–1967

177. Hosi M, Takahashi I, Pavlova-Rezakova A, Himeno H, Chobanian AV, Brecher P (1993) Selective induction of an embryonic fibronectin isoform in the rat aorta in vitro. Circ Res 73:689–695

178. Glukhova MA, Frid MG, Shekhonin BV, Vasilevskaya TD, Grunwald J, Saginati M, Koteliansky VE (1989) Expression of extradomain A fibronectin sequence is phenotypic dependent. J Cell Biol 109:357–366

179. Contard F, Sabri A, Glukhova M, Sartore S, Marotte F, Pomies JP, Schiavi P, Guez D, Samuel J-L, Rappaport L (1993) Arterial smooth muscle cell phenotype in stroke-prone spontaneously hypertensive rats. Hypertension 22:665–676

180. Himeno H, Crawford DC, Hosoi M, Chobanian AV, Brecher P (1994) Angiotensin II alters aortic fibronectin independently of hypertension. Hypertension 23[part2]:823–826

181. Bauters C, Marotte F, Hamon M, Oliviero P, Farhadian F, Robert V, Samuel J-L, Rappaport L (1995) Accumulation of fetal fibronectin mRNAs after balloon denudation of rabbit arteries. Circulation 92:904–911

182. Mecham RP, Stenmark KR, Parks WC (1991) Connective tissue production by vascular smooth muscle in development and disease. Chest 99:43S–47S

183. Noguchi A, Samaha H (1991) Developmental changes in tropoelastin gene expression in the rat lung studied by in situ hybridization. Am J Respir Cell Mol Biol 5:571–578

184. Noguchi A, Samaha H, DeMello DE (1992) Tropoelastin gene expression in the rat pulmonary vasculature: a developmental study. Pediatric Res 31:280–285

185. Durmowicz AG, Frid MG, Worley JD, Stenmark KR (1996) Expression and localization of the tropoelastin mRNA in the developing bovine pulmonary artery is dependent on vascular phenotype. Am J Respir Cell Mol Biol 14:569–576

186. Lowell Langille B (1993) Remodeling of developing and mature arteries: endothelium, smooth muscle, and matrix. J Cardiovasc Pharmacol 21 (Suppl 1):S11–S17

187. Osborn M, Caselitz J, Weber K (1981) Heterogeneity of intermediate filament expression in vascular smooth muscle: a gradient of desmin positive cells from the rat aortic arch to the level of the arteria iliaca communis. Differentiation 20:196–202

188. Kacem K, Seylaz J, Aubineau P (1996) Differential processes of vascular smooth muscle cell differentiation within elastic and muscular arteries of rats and rabbits: an immunofluorescence study of desmin and vimentin distribution. Histochem J 28:53–61

189. Skalli O, Ropraz P, Trzeciak A, Benzonana G, Gillessen D, Gabbiani G (1986) A monoclonal antibody against α-smooth muscle actin: a new probe for smooth muscle differentiation. J Cell Biol 103:2787–2796

190. Kocher O, Gabbiani G (1986) Expression of actin mRNAs in rat aortic smooth muscle cells during development, experimental intimal thickening, and culture. Differentiation 32:245–251

191. Holifield B, Helgason T, Jemelka S, Taylor A, Navran S, Allen J, Seidel C (1996) Differentiated vascular myocytes: are they involved in neointimal formation? J Clin Invest 97:814–825

192. Meer DP, Eddinger TJ (1996) Heterogeneity of smooth muscle myosin heavy chain expression at the single cell level. Am J Physiol 270 (Cell Physiol 39):C1819–CC1824

193. Seidel CL, Helgason T, Allen JC, Wilson C (1997) Migratory abilities of different vascular cells from the tunica media of canine vessels. Am J Physiol 272 (Cell Physiol 41):C847–C852

194. Wohrley JD, Frid MG, Moiseeva EP, Orton EC, Belknap JK, Stenmark KR (1995) Hypoxia selectively induces proliferation in a specific subpopulation of smooth muscle cells in the bovine neonatal pulmonary arterial media. J Clin Invest 96:273–281

195. Yablonka-Reuveni Z, Schwartz SM, Christ B (1995) Development of chicken aortic smooth muscle: expression of cytoskeletal and basement membrane proteins defines two distinct cell phenotypes emerging from a common lineage. Cell Mol Biol Res 41:241–249

196. Babaev VR, Bobryshev YV, Stenina OV, Tararak EM, Gabbiani G (1990) Heterogeneity of smooth muscle cells in atheromatous plaque of human aorta. Am J Pathol 136:1031–1042

197. Pauletto P, Da Ros S, Capriani A, Chiavegato A, Pessina AC, Sartore S (1995) Smooth muscle cell types at different aortic levels and in microvasculature of rabbits with renovascular hypertension. J Hypertens 13:1679–1685

198. Price RJ, Owens GK, Skalak TC (1994) Immunohistochemical identification of arteriolar development using markers of smooth muscle differentiation. Evidence that capillary arterialization proceeds from terminal arterioles. Circ Res 75:520–527

199. Schwartz SM, Majesky MW, Murry CE (1995) The intima: development and monoclonal response to injury. Atherosclerosis 118 (Suppl):S125–S140

200. Gilbert SF (1988) Developmental biology. Sinauer Associates, Inc. Publishers, Sunderland, Massachusetts

201. Noden DM (1989) Embryonic origins and assembly of blood vessels. Am Rev Respir Dis 140:1097–1103

202. Poole TJ, Coffin JD (1989) Vasculogenesis and angiogenesis: two distinct morphogenetic mechanisms establish embryonic vascular pattern. J Exp Zool 251:224–231

203. Risau W, Flamme I (1995) Vasculogenesis. Annu Rev Cell Dev Biol 11:73–91
204. Schaper W, Wulf I (1996) Molecular mechanisms of coronary collateral vessel growth. Circ Res 79:911–919
205. Pardanaud L, Yassine F, Dieterlen-Lievre (1989) Relationship between vasculogenesis, angiogenesis and haemopoiesis during avian ontogeny. Development 105:473–485
206. Cossu G, Tajbakhsh S, Buckingham M (1996) How is myogenesis intiated in the embryo? Trends Genetics 12:218–222
207. Topouzis S, Majesky MW (1996) Smooth muscle lineage diversity in the chick embryo. Two types of aortic smooth muscle cells differ in growth and receptor-mediated transcriptional responses to transforming growth factor-β. Dev Biol 178:430–445
208. Le Lievre C, Le Douarin N (1975) Mesenchymal derivatives of the neural crest: analysis of chimeric quail and chick embryos. J Embryol Exp Morphol 34:125–154
209. Kirby ML, Waldo KL (1995) Neural crest and cardiovascular patterning. Circ Res 77:211–215
210. Ito K, Sieber-Blum M (1993) Pluripotent and developmentally restricted neural-crest-derived cells in posterior visceral arches. Dev Biol 156:191–200
211. Brody JR, Cunha GR (1989) Histologic, morphometric, and immunocytochemical analysis of myometrial development in rats and mice. I. Normal development. Am J Anat 186:1–20
212. Simons-Assmann P, Kedinger M (1993) Heterotypic cellular cooperation in gut morphogenesis and differentiation. Semin Cell Biol 4:221–230
213. Minoo P, King RJ (1994) Epithelial-mesenchymal interactions in lung development. Annu Rev Physiol 56:13–45
214. Baskin LS, Hayward SW, Young PF, Cunha GR (1996) Ontogeny of the rat bladder: smooth muscle and epithelial differentiation. Acta Anat 155:163–171
215. Mikawa T, Gourdie RG (1996) Pericardial mesoderm generates a population of coronary smooth muscle cells migrating into the heart along with ingrowth of the epicardial organ. Dev Biol 174:221–232
216. Poelmann RE, Gittenberger-de Groot AC, Mentink MMT, Bökenkamp R, Hogers B (1993) Development of the cardiac coronary vascular endothelium, studied with antiendothelial antibodies, in chicken-quail chimeras. Circ Res 73:559–568
217. Vrancken Peeters MPFM, Gittenberger-de Groot AC, Mentink MMT, Hungerford JE, Little CD, Poelmann RE (1997) The development of the coronary vessels and their differentiation into arteries and veins in the embryonic quail heart. Dev Dyn 208:338–348
218. Hood LC, Rosenquist TH (1992) Coronary artery development in the chick: origin and deployment of smooth muscle cells, and the effects of neural crest ablation. Anat Rec 234:291–300
219. Waldo KL, Kumiski DH, Kirby ML (1994) Association of the cardiac neural crest with development of the coronary arteries in the chick embryo. Anat Rec 239:315–331

220. Flamme I, Risau W (1992) Induction of vasculogenesis and hematopoiesis in vitro. Development 116:435–439

221. Davis S, Aldrich TH, Jones PF, Acheson A, Compton DL, Jain V, Ryan TE, Bruno J, Radziejewski C, Maisonpierre PC, Yancopoulos GD (1996) Isolation of angiopoietin-1, a ligand for the TIE2 receptor, by secretion-trap expression cloning. Cell 87:1161–1169

222. Folkman J, D'Amore P (1996) Blood vessel formation: what is its molecular basis? Cell 87:1153–1155

223. Flamme I, Breier G, Risau W (1995) Expression of vascular endothelial growth factor (VFGF) and VFGF-receptor2 (flk-1) during induction of hemangioblastic precursurs and vascular differentiation in the quail embryo. Dev Biol 169:699–712

224. Cleaver O, Tonissen KF, Saha MS, Krieg PA (1997) Neovascularization of the *Xenopus* embryo. Dev Dyn 210:66–77

225. Venuti JM, Cserjesi P (1996) Molecular embryology of skeletal myogenesis. Curr Top Dev Biol 34:169–206

226. Saint-Jeannet J-P, Levi G, Girault J-LM, Koteliansky V, Thierry J-P (1992) Ventro-lateral regionalization of *Xenopus laevis* mesoderm is characterized by the expression of α-smooth muscle actin. Development 115:1165–1173

227. Takahashi Y, Imanaka T, Takano T (1996) Spatial and temporal pattern of smooth muscle cell differentiation during develoment of the vascular system in the mouse embryo. Anat Embryol Berl 194:515–526

228. Lee SH, Hungerford JE, Little CD, Iruela-Arispe ML (1997) Proliferation and differentiation of smooth muscle cell precursors occurs simultaneously during the development of the vessel wall. Dev Dyn 209:342–352

229. Hungerford JE, Hoeffer JP, Bowers CW, Dahm LM, Falchetto R, Shabanowitz J, Hunt DF, Little LD (1997) Identification of a novel marker for primordial smooth muscle and its differential expression in contractile versus non-contractile cells. J Cell Biol 137:925–937

230. DeRuiter MC, Poelmann RE, VanMunsteren JC, Mironov V, Markwald RR, Gittenberger-de Groot A (1997) Embryonic endothelial cells transdifferentiate into mesenchymal cells expressing smooth muscle actins in vivo and in vitro. Circ Res 80:444–451

231. Markwald RR, Fitzharris TP, Adams-Smith WN (1975) Structural analysis of endocardial cytodifferentiation. Dev Biol 42:160–180

232. Wrenn RW, Raeuber CL, Herman LE, Walton WJ, Rosenquist TH (1993) Transforming growth factor-beta: signal transduction via protein kinase C in cultured embryonic vascular smooth muscle cells. In vitro Cell Dev Biol 29A:73–76

233. Gadson PF, Dalton ML, Patterson E, Svoboda DD, Hutchinson L, Schram D, Rosenquist TH (1997) Differential response of mesoderm- and neural crest-derived smooth muscle to TGFβ1: regulation of c-myb and α1(I) procollagen genes. Exp Cell Res 230:169–180

234. Topouzis S, Catravas JD, Ryan JW, Rosenquist TH (1992) Influence of vascular smooth muscle heterogeneity on angiotensin converting enzyme activity in chicken embryonic aorta and in endothelial cells in culture. Circ Res 71:923–931

235. Dreher KL, Cowan K (1991) Expression of antisense transcripts encoding an extracellular matrix protein by stably transfected vascular smooth muscle cells. Eur J Cell Biol 54:1–9

236. Blaes N, Bourdillon M-C, Lamaziere JMD, Michaille J-J, Andujar M, Covacho C (1991) Isolation of two morphologically distinct cell lines from rat arterial smooth muscle expressing high tumorigenic potential. In Vitro Cell Dev Biol 27A:725–734

237. Ehler E, Jat PS, Nobme MD, Citi S, Draeger A (1995) Vascular smooth muscle cells of H-2K^b-tsA58 transgenic mice. Circulation 92:3289–3296

238. Majesky MW, Giachelli CM, Reidy MA, Schwartz SM (1992) Rat carotid neointimal smooth muscle cells reexpress a developmentally regulated mRNA phenotype during repair of arterial injury. Circ Res 71:759–768

239. Majesky MW, Benditt EP, Schwartz SM (1988) Expression and developmental control of platelet-derived growth factor A-chain and B-chain/Sis genes in rat aortic smooth muscle cells. Proc Natl Acad Sci USA 85:1524–1528

240. Lemire JM, Covin CW, White S, Giachelli CM, Schwartz SM (1994) Characterization of cloned aortic smooth muscle cells from young rats. Am J Pathol 1441068–1081

241. Schwartz SM, Foy L, Bowen-Pope DF, Ross R (1990) Derivation and properties of platelet-derived growth factor-independent rat smooth muscle cells. Am J Pathol 136:1417–1428

242. Bochaton-Piallat M-L, Gabbiani F, Ropraz P, Gabbiani G (1992) Cultured aortic smooth muscle cells from newborn and adult rats show distinct cytoskeletal features. Differentiation 49:175–185

243. Bochaton-Piallat M-L, Gabbiani F, Ropraz P, Gabbiani G (1993) Age influences the replicative activity and the differentiation features of cultured rat aortic smooth muscle cell populations and clones. Arterioscler Thromb 13:1449–1455

244. Bochaton-Piallat M-L, Gabbiani F, Ropraz P, Gabbiani G (1996) Phenotypic heterogeneity of rat arterial smooth muscle cell clones. Arterioscler Thromb Vasc Biol 16:815–820

245. Clowes AW, Reidey MA, Clowes MM (1983) Kinetics of cellular proliferation after arterial injury. Lab Invest 49:327-333

246. Lombardi DM, Reidy MA, Schwartz SM (1991) Methodological considerations important in the accurate quantitation of aortic smooth muscle cell replication in the normal rat. Am J Pathol 138:441–446

247. Cook CL, Weiser MCM, Schwartz PE, Jones CL, Majack RA (1994) Developmentally timed expression of an embryonic growth phenotype in vascular smooth muscle cells. Circ Res 74:189–196

248. Wyllie AH (1992) Apoptosis and the regulation of cell numbers in normal and neoplastic tissues: an overview. Cancer Metastasis Rev 11:95–103

249. Yeh ETH (1997) Life and death in the cardiovascular system. Circulation 95:782–786

250. Bennett MR, Angelini S, McEwan JR, Jagoe R, Newby AC, Evan GI (1994) Inhibition of vascular smooth muscle cell proliferation in vitro and in vivo by c-*myc* antisense oligodeoxynucleotides. J Clin Invest 93:820–828

251. Bennett MR, Littlewood TD, Hancock DC, Evan GI, Newby AC (1994) Down-regulation of the c-*myc* proto-oncogene inhibition of vascular smooth muscle cell proliferation: a signal for growth arrest? Biochem J 302:701–708

252. Bennett MR, Evan GI, Newby AC (1994) Deregulated c-*myc* oncogene expression blocks vascular smooth muscle cell inhibition mediated by heparin, interferon-γ, mitogen depletion and cyclic nucletide analogues and induces apoptotic cell death. Circ Res 74:525–536

253. Ross R (1993) The pathogenesis of atherosclerosis: a perspective for the 1990s. Nature 362:801–809

254. Bennett MR, Evan GI, Schwartz SM (1995) Apoptosis of human vascular smooth muscle cells derived from normal vessels and coronary atherosclerotic plaques. J Clin Invest 95:2266–2274

255. Maione R, Amati P (1997) Interdependence between muscle differentiation and cell-cycle control. Biochim Biophys Acta 1332:M19–M30

256. Imai H, Lee KJ, Lee SK, Lee KT, O'Neal RM, Thomas WA (1979) Ultrastructural features of aortic cell in mitosis in control and cholesterol-fed swine. Lab Invest 23:401–415

257. Hay ED (1968) Dedifferentiation and metaplasia in vertebrate and invertebrate regeneration. In Ursprung H ed. The stability of differentiated state. Springer-Verlag, Heidelberg, pp. 85–108

258. Chamley-Campbell JH, Campbell GR, Ross R (1981) Phenotype-dependent response of cultured aortic smooth muscle to serum mitogens. J Cell Biol 89:379–383

259. Owens GK, Loeb A, Gordon D, Thompson MM (1986) Expression of smooth muscle- specific α-isoactin in cultured vascular smooth muscle cells: relationship between growth and cytodifferentiation. J Cell Biol 102:343–352

260. Rovner AS, Murphy RA, Owens GK (1986) Expression of smooth muscle and nonmuscle myosin heavy chains in cultured vascular smooth muscle cells. J Biol Chem 261:14740–14745.

261. Kawamoto S, Adelstein RS (1987) Characterization of myosin heavy chains in cultured aorta smooth muscle cells. J Biol Chem 262:7282–7288

262. Babij P, Kawamoto S, White S, Adelstein RS, Periasamy M (1992) Differential expression of SM1 and SM2 myosin isoforms in cultured vascular smooth muscle. Am J Physiol 262 (Cell Physiol 31): C607–C613

263. Birukov KG, Shirinsky VP, Stepanova OV, Thachuk VA, Hahn AWA, Resink TJ, Smirnov VN (1995) Stretch affects phenotype and proliferation of vascular smooth muscle cells. Mol Cell Biochem 144:131–139

264. Bobik A, Campbell JH (1993) Vascular derived growth factors: cell biology, pathophysiology, and pharmacology. Pharmacol Rev 45:1–42

265. De Mey JG, Uitendaal MP, Boonen HC, Vrijdag MJ, Daemen MJ, Struyker-Boudier HA (1989) Acute and long-term effects of tissue culture on contractile reactivity in renal arteries of the rat. Circ Res 65:1125–1135

266. Holycross BJ, Peach MJ, Owens GK (1993) Angiotensin II stimulates increased protein synthesis, not increased DNA synthesis, in intact rat aortic segments in vitro. J Vasc Res 30:80–86

267. Olson EN (1993) Regulation of muscle transcription by the MyoD family: the heart of the matter. Circ Res 72:1–6

268. Kemp PR, Metcalf JC, Grainger DJ (1995) Id - A dominant negative regulator of skeletal muscle differentiation - is not involved in maturation or differentiation of vascular smooth muscle cells. FEBS lett 368:81–86

269. Cserjesi PB, Lilly L, Bryson Y, Wang Y, Sassoon DA, Olson EN (1992) MHOX: a mesodermal restricted homeodomain protein that binds an essential site in the muscle creatine kinase enhancer. Development 115:1087–1101

270. Patel CV, Gorski DH, Lepage DF, Lincecum J, Walsh K (1992) Molecular cloning of a homeobox transcription factor from adult aortic smooth muscle. J Biol Chem 267:26085–26090

271. Miano JM, Firulli AB, Olson EN, Hara P, Giachelli CM, Schwartz SM (1996) Restricted expression of homeobox genes distinguishes fetal from adult human smooth muscle cells. Proc Natl Acad Sci USA 93:900–905

272. Gorski DH, Lepage DF, Patel CV, Copemand NG, Jenkins NA, Walsh K (1993) Molecular cloning of a diverged homeobox gene that is rapidly downregulated during the Go/G1 transition in vascular smooth muscle cells. Mol Cell Biol 13:3722–3733

273. Skopicki HA, Lyons GE, Schatteman G, Smith RC, Andrés V, Schirm S, Isner J, Walsh K (1997) Embryonic expression of the Gax homeodomain protein in cardiac, smooth and skeletal muscle. Circ Res 80:452–462

274. Epstein JA (1996) Pax3, neural crest and cardiovascular development. Trends Cardiovasc Med 6:255–261

275. Hollenberg SM, Sternglanz R, Cheng PF, Weintraub H (1995) Identification of a new family of tissue-specific basic helix-loop-helix proteins with a two-hybrid system. Mol Cell Biol 15:3813–3822

276. Morrisey EE, Ip HS, Lu MM, Parmacek MS (1996) GATA-6: a zinc finger transcription factor that is expressed in multiple cell lineages derived from lateral mesoderm. Dev Biol 177:309–322

277. Morrisey EE, Ip HS, Tang Z, Lu MM, Parmacek MS (1997) GATA-5: a transcriptional activator expressed in a novel temporally and spatially-restricted pattern during embryonic development. Dev Biol 183:21–36

278. Firulli AB, Miano JM, Bi W, Johnson AD, Casscells W, Olson EN, Schwartz JJ (1996) Myocyte enhancer binding factor-2 expression and activity in vascular smooth muscle cells. Circ Res 78:196–204

279. Pabst O, Schneider A, Brand T, Arrold A-H (1997) The mouse NKX2-3 homeodomain gene is expressed in gut mesenchyme during pre- and postnatal mouse development. Dev Dyn 209:29–35

280. Collins T, Ginsburg D, Boss JM, Orkin SH, Pober J (1985) Cultured human endothelial cells express platelet-derived growth factor B chain: cDNA cloning and structural analysis. Nature 316:748–750

281. Collins T, Pober JS, Gimbrone MA, Hammach B, Betsholtz B, Westermark B, Heldin C-H (1987) Cultured human endothelial cells express platelet- growth factor A chain. Am J Pathol 126:7–12

282. Hannan RL, Kourembanas S, Flanders KC, Roberts AB, Faller DV, Klagsbrun M (1988) Endothelial cells synthesize basic fibroblast growth factor and transforming growth factor beta. Growth Factors 1:7–17

283. D'Amore PA, Smith SR (1993) Growth factor effects on cells of the vascular wall: a survey. Growth Factors 8:61–75

284. Komuro I, Kurihara H, Sugiyama T, Takaku F, Yazaki Y (1988) Endothelin stimulates c-*fos* and c-*myc* expression and proliferation of vascular smooth muscle cells. FEBS lett 238:249–252

285. Bradham DM, Igarashi A, Poter RL, Grotendorst GR (1991) Connective tissue growth factor: a cysteine-rich mitogen secreted by human vascular endothelial cells is related to the SRC-induced immediate early gene product CEF-10. J Cell Biol 114:1285–1294

286. Castellot JJjr, Addonizo MJ, Rosenberg R, Karnovsky MJ (1981) Cultured endothelial cells produce a heparin-like inhibitor of smooth muscle growth. J Cell Biol 90:372:379

287. Moses HL, Yang EY, Pietenpol JA (1990) TGF-beta stimulation and inhibition of cell proliferation: new mechanistic insights. Cell 63:245–247

288. Owens GK, Geisterfer AA, Yang YW, Komoriya A (1988) Transforming growth factor- beta-induced growth inhibition and cellular hypertrophy in cultured vascular smooth muscle cells. J Cell Biol 107:771–780

289. Kunzelman U, Dartsch PC (1992) Expression of smooth muscle alpha-actin and the proliferative activity of human smooth muscle cells in culture is influenced by endothelial- cell and fibroblast-conditioned medium. Cell Physiol Biochem 2:49–56

290. Vernon SM, Campos MJ, Haystead T, Thompson MM, DiCorleto PE, Owens GK (1997) Endothelial cell-conditioned medium downregulates smooth muscle contractile protein expression. Am J Physiol 272(Cell Physiol 41):C582–C591

291. Blank RS, Owens GK (1990) Platelet-derived growth factor regulates actin isoform expression and growth state in cultured rat aortic smooth muscle cells. J Cell Physiol 142:635–642

292. Holycross BJ, Blank RS, Thompson MM, Peach MJ, Owens GK (1992) Platelet-derived growth factor-BB-induced suppression of smooth muscle cell differentiation. Circ Res 71:1525–1532

293. Reusch P, Wagdy H, Reusch R, Wilson E, Ives HE (1996) Mechanical strain increases smooth muscle and decreases nonmuscle myosin expression in rat vascular smooth muscle cells. Circ Res 79:1046–1053

294. Campbell GR, Campbell JH (1986) Endothelial cells influences on vascular smooth muscle phenotype. Annu Rev Physiol 48:295–306

295. Desmouliére A, Rubbia-Brandt L, Gabbiani G (1991) Modulation of actin isoform expression in cultured arterial smooth muscle cells by heparin and culture conditions. Arterioscler Thromb 11:244–253

296. Orlandi A, Ropraz P, Gabbiani G (1994) Proliferative activity and α-smooth muscle actin expression in cultured rat aortic smooth muscle cells are differently modulated by transforming growth factor-β1 and heparin. Exp Cell Res 214:528–536

297. Barzu T, Hereber I-M, Desmouliére A, Carayon P, Pascal M (1994) Characterization of rat aortic smooth muscle cells resistant to the antiproliferative activity of heparin following long-term heparin treatment. J Cell Physiol 160:239–248

298. Majack RA (1987) Beta-type transforming growth factor specifies organizational behaviour in vascular smooth muscle cell cultures. J Cell Biol 105:465–471

299. Koyama N, Koshikawa T, Morisaki N, Saito Y, Yoshida S (1990) Bifunctional effects of transforming growth factor-β on migration of cultured rat aortic smooth muscle cells. Biochem Biophys Res Commun 169:725–729

300. Goodman LV, Majack RA (1989) Vascular smooth muscle cells express distinct transforming growth factor-β receptor phenotypes as a function of cell density in culture. J Biol Chem 264:5241–5244

301. Davidson JM, Zonia O, Liu J-M (1993) Modulation of transforming growth factor-beta 1 stimulated elastin and collagen production and proliferation in porcine vascular smooth muscle cells and skin fibroblasts by basic fibroblast growth factor, transforming growth factor-β, and insulin-like growth factor-1. J Cell Physiol 155:149-156

302. Björkerud S (1991) Effects of transforming growth factor-beta 1 on human arterial smooth muscle cells in vitro. Arterioscler Thromb 11:892–902

303. Majack RA, Majesky MW, Goodman LV (1990) Role of PDGF-A expression in the control of the vascular smooth muscle cell growth by transforming growth factor-beta. J Cell Biol 111:239–247

304. Stoufer GA, Owens GK (1994) TGF-β promotes proliferation of cultured SMC via both PDGF-AA-dependent and PDGF-AA-independent mechanisms. J Clin Invest 93:2048–2055

305. Campbell GR, Campbell JH, Manderson JA, Horrigan S, Rennick RE (1988) Arterial smooth muscle: a multifunctional mesenchymal cell. Arch Pathol Lab Med 112:977–986

306. Cassis LA, Lynch KR, Peach MJ (1988) Localization of angiotensinogen messanger RNA in the rat aorta. Circ Res 62:1259–1262

307. Daemen MJAP, Lombardi DM, Bosman FT, Schwartz SM (1991) Angiotensin II induces smooth muscle cell proliferation in the normal and injured rat arterial wall. Circ Res 68:450–456

308. Owens GK (1989) Control of hypertrophic versus hyperplastic growth of vascular smooth muscle cells. Am J Physiol (Heart Circ Physiol 26):H1755-H1765

309. Turla MB, Thompson MM, Corjay MH, Owens GK (1991) Mechanisms of angiotensin II- and arginine vasopressin-induced increases in protein synthesis and content in cultured rat aortic smooth muscle cells. Circ Res 68:288-299

310. Weber H, Taylor DS, Molloy CJ (1994) Angiotensin II induces delayed mitogenesis and cellular proliferation in rat aortic smooth muscle cells. J Clin Invest 93:788-798

311. Pauletto P, Sarzani R, Rappelli A, Chiavegato A, Pessina AC, Sartore S (1994) Differentiation and growth of vascular smooth muscle cells in experimental hypertension. Am J Hypertens 7:661-674

312. Itoh H, Mukoyama M, Pratt RE, Gibbons GH, Dzau VJ (1993) Multiple autocrine growth factors modulate vascular smooth muscle cell growth response to angiotensin II. J Clin Invest 91:2268-2274

313. Sabri A, Levy B, Poitevin P, Caputo L, Faggin E, Marotte F, Rappaport L, Samuel J-L (1997) Differential role of AT1 and AT2 receptor subtypes in vascular trophic and phenotypic changes in response to stimulation with angiotensin II. Arterioscler Thromb Vasc Biol 17:257-264

314. Miano JM, Vlasic N, Tota RR, Stemerman MB (1993) Localization of fos and jun proteins in rat aortic smooth muscle cells following vascular injury. Am J Pathol 142:715-724

315. Miano JM, Vlasic N, Tota RR, Stemerman MB (1993) Smooth muscle cell immediate-early gene and growth factor activation follows vascular injury. Arterioscler Thromb 13:211-219

316. Bondjers G, Glukhova M, Hansson GK, Postnov YV, Reidy MA, Schwartz SM (1991) Hypertension and atherosclerosis. Cause and effect, or two effects with one unknown cause? Circulation 84(supplVI):VI-2-VI-16

317. Hsieh HJ, Li NH, Frangos JA (1991) Shear stress increases endothelial platelet-derived growth factor mRNA levels. Am J Physiol 260:H642-H646

318. Rosati C, Garay G (1991) Flow-dependent stimulation of sodium and cholesterol uptake and cell growth in cultured vascular smooth muscle. J Hypertens. 9:1029-1033

319. Karim OMA, Pienta K, Seki N, Mostwin JL (1992) Stretch-mediated visceral smooth muscle growth in vitro. Am J Physiol 262(Regulatory Integrative Comp Physiol):R895-R900

320. Wang Z, Gopalakurup SK, Levin RM, Chacko S (1995) Expression of smooth muscle myosin isoforms in urinary bladder smooth muscle during hypertrophy and regression. Lab Invest 73:244-251

321. Owens GK, Vernon SM, Madsen CS (1996) Molecular regulation of smooth muscle cell differentiation. J Hypertens 14(suppl 5):S55-S64

322. Pauletto P, Scatena M, Chiavegato A, Giuriato L, Faggin E, Sarzani R, Rappelli A, Grisenti A, Pessina AC, Sartore S (1994) Hyperplastic growth of aortic smooth muscle cells in renovascular hypertensive rabbits is accompanied by the expansion of an immature cell phenotype. Circ Res 74/774-788

323. Owens GK, Reidy M (1985) Hyperplastic growth response of vascular smooth muscle cells following induction of acute hypertension in rats by aortic coarctation. Circ Res 57:695–705

324. Kolpakov V, Rekhter MD, Gordon D, Wang WH, Kulik TJ (1995) Effect of mechanical forces on growth and matrix protein synthesis in the in vitro pulmonary artery. Circ Res 77:823–831

325. Bardy N, Karillon GJ, Merval R, Samuel J-L, Tedgui A (1995) Differential effects of pressure and flow on DNA and protein synthesis and on fibronectin expressin by arteries in a novel organ culture system. Circ Res 77:684–694

326. Hishikawa K, Nakaki T, Marumo T, Hayashi M, Suzuki H, Kato R, Saruta T (1994) Pressure promotes DNA synthesis in rat cultured vascular smooth muscle cells. J Clin Invest 93:1975–1980

327. Owens GK (1996) Role of mechanical strain in regulation of differentiation of vascular smooth muscle cells. Circ Res 79:1054–1055

328. Levin RM, Monson FC, Haugaard N, Buttyan R, Hudson A, Roelofs M, Sartore S, Wein AJ (1995) Genetic and cellular characteristics of bladder outlet obstruction. Urol Clin North Am 22:263–283

329. Chen MW, Krasnapolsky L, Levin RM, Buttyan R (1994) An early molecular response induced by acute overdistension of the rabbit urinary bladder. Mol Cell Biochem 132:39–44

330. Levin RM, Wein AJ, Buttyan R, Monson FC, Longhurst PA (1994) Update on bladder smooth-muscle physiology. World J Urol 12:226–232

331. Baskin LS, Sutherland RS, Thomson AA, Hayward SW, Cunha GR (1996) Growth factors and receptors in bladder development and obstruction. Lab Invest 75:157–166

332. Thyberg J, Blomgren K, Hedin U, Dryjski M (1995) Phenotypic modulation of smooth muscle cells during the formation of neointimal thickenings in the rat carotid artery after balloon injury: an electron-microscopic and stereological study. Cell Tissue Res 281:421–433

333. Kocher O, Gabbiani F, Gabbiani G, Reidy MA, Cokay MS, Peters H, Hüttner I (1991) Phenotypic features of smooth muscle cells during the evolution of experimental carotid artery intimal thickening. Lab Invest 65:459–470

334. Simons M, Leclerc G, Safian RD, Isner JM, Weir L, Baim DS (1993) Relation between activated smooth-muscle cells in coronary artery lesions and restenosis after atheroctomy. N Engl J Med 328:608–613

335. Reckless J, Fleetwood G, Tilling L, Huber PAJ, Marston SB, Pritchard K (1994) Changes in the caldesmon isoform content and intimal thickening in the rabbit carotid artery induced by a silicone elastomer collar. Arterioscler Thromb 14:1837–1845

336. Okamoto E, Suzuki T, Aikawa M, Imataka K, Fujii J, Kuro-o M, Nakahara K, Hasegawa A, Yazaki Y, Nagai R (1996) Diversity of the synthetic state smooth-muscle cells proliferating in mechanically and hemodynamically injured rabbit arteries. Lab Invest 74:120–128

337. Orlandi A, Ehrlich P, Ropraz P, Spagnoli LG, Gabbiani G (1994) Rat aortic smooth muscle cells isolated from different layers and at different times after endothelial denudation show distinct biological features in vitro. Arterioscler Thromb 14:982–989

338. Neuville P, Geinoz A, Benzonana G, Redard M, Gabbiani F, Ropraz P, Gabbiani G (1997) Cellular retinol-binding protein-1 is expressed by distinct subsets of rat arterial smooth muscle cells in vitro and in vivo. Am J Pathol 150:509–521

339. Miano JM, Topouzis S, Majesky MW, Olson EN (1996) Retinoic receptor expression and all-trans retinoic acid-mediated growth inhibition in vascular smooth muscle cells. Circulation 93:1886–1895

340. Colbert MC, Kirby ML, Robbins J (1996) Endogeneous retinoic acid signaling colocalizes with advanced expression of the adult smooth muscle myosin heavy chain isoform during development of the ductus arteriosus. Circ Res 78:790–798

341. Gittenberger-de Groot A (1979) Morphology of the normal human ductus arteriosus. In Heyman MA, Rudolph AM, eds. The ductus arteriosus. Ross Conferences on Pediatric Research. Columbus, Ohio, Ross Laboratories:3–9

342. Kim HS, Aikawa M, Kimura K, Kuro-o M, Nakahara K, Suzuki T, Katoh H, Okamoto E, Yazaki Y, Nagai R (1993) Ductus arteriosus advanced differentiation of smooth muscle cells demonstrated by myosin heavy chain isoform expression in rabbits. Circulation 88:1804–1810

343. Lindner V, Giachelli CM, Schwartz SM, Reidy MA (1995) A subpopulation of smooth muscle cells in injured rat arteries expresses platelet-derived growth factor-B chain mRNA. Circ Res 76:951–957

344. Bendeck MP, Regenass S, Tom WD, Giachelli CM, Schwartz SM, Hart C, Reidy MA (1996) Differential expression of α_1 type VIII collagen in injured platelet-derived growth factor-BB-stimulated rat carotid arteries. Circ Res 79:524–531

345. Sibinga NES, Foster LC, Hsieh C-M, Perella MA, Lee W-S, Endege WO, Sage EH, Lee M-E, Haber E (1997) Collagen VIII is expressed by vascular smooth muscle cells in response to vascular injury. Circ Res 80:532–541

346. Murry CE, Bartosek T, Giachelli CM, Alpers CE, Schwartz SM (1996) Platelet-derived growth factor-A mRNA expression in fetal, normal adult, and atherosclerotic human aortas. Circulation 93:1095–1106

347. Giachelli CM, Schwartz SM, Liaw L (1995) Molecular and cellular biology of osteopontin. Trends Cardiovasc Med 5:88–95

348. Liaw L, Almeida M, Hart CE, Schwartz SM, Giachelli CM (1994) Osteopontin promotes vascular cell adhesion and spreading and is chemotactic for smooth muscle cells in vitro. Circ Res 74:214–224

349. Weintraub AS, Giachelli CM, Krauss RS, Almeida M, Taubman MB (1996) Autocrine secretion of osteopontin by vascular smooth muscle cells regulates their adhesion to collagen gels. Am J Pathol 149:259–272

350. Wang X, Louden C, Ohlstein E, Stadel JM, Gu J-L (1996) Osteopontin expression in platelet-derived growth factor-stimulated vascular smooth muscle cells and

carotid artery after balloon angioplasty. Arterioscler Thromb Vasc Biol 16:1365–1372

351. Newman CM, Bruun BC, Porter KE, Mistry PK, Shanahan CM, Weissberg PL (1995) Osteopontin is not a marker for proliferating human vascular smooth muscle cells. Arterioscler Thromb Vasc Biol 15:2010–2018

352. Liaw L, Lombardi DM, Almeida MM, Schwartz SM, deBlois D, Giachelli CM (1997) Neutralizing antibodies directed against osteopontin inhibit rat carotid neointimal thickening after endothelial denudation. Arterioscler Thromb Vasc Biol 17:188–193

353. Choi ET, Engel L, Callow AD, Sun S, Trachtenberg J, Santoro S, Ryan US (1994) Inhibition of neointimal hyperplasia by blocking $\alpha_v\beta_3$ integrin with a small peptide antagonist GpenGRGDSPCA. J Vasc Surg 19:125–134

354. Zalewski A, Shi Y (1997) Vascular myofibroblasts. Arterioscler Thromb Vasc Biol 17:417–422

355. Okada TS (1986) Can specialized cells change their phenotype? Curr Topics Dev Biol. 20:XXV–XXXI

356. Jones R (1992) Ultrastructural analysis of contractile cell development in lung microvessels in hyperoxic pulmonary hypertension. Am J Pathol 141:1491–1505

357. Chiavegato A, Bochaton-Piallat M-L, D'Amore E, Sartore S, Gabbiani G (1995) Expression of myosin heavy chain isoforms in mammary epithelial cells and in myofibroblasts from different fibrotic settings during neoplasia. Virchows Archiv B 426: 77–86

358. Galmiche MC, Koteliansky VE, Brère J, Hervé P, Charbord P (1993) Stromal cells from human long-term marrow cultures are mesenchymal cells that differentiate following a vascular smooth muscle differentiation pathway. Blood 82:66–76

359. Li, J, Senseb L, Hervé P, Charbord (1995) Nontransformed colony-derived stromal cell lines from normal human marrows. II. Phenotypic characterization and differentiation pathway. Exp Hematol 23:133–141

360. Roelofs M, Wein AJ, Monson FC, Passerini-Glazel G, Koteliansky V, Sartore S, Levin RM (1995) Contractility and phenotype transitions in serosal thickening of obstructed rabbit bladder. J Applied Physiol 78:1432–1441

361. Pampinella F, Roelofs M, Castellucci E, Passerini-Glazel G, Pagano F, Sartore S (1997) Time dependent remodeling of the bladder wall in growing rabbits after partial outlet obstruction. J Urol 157:677–682

362. Sappino AP, Schürch W, Gabbiani G (1990) Differentiation repertoire of fibroblastic cells: expression of cytoskeletal proteins as marker of phenotypic modulations. Lab Invest 63:144–161

363. Schmitt-Gräff A, Desmouliére A, Gabbiani G (1994) Heterogeneity of myofibroblast phenotypic features: an example of fibroblastic plasticity. Virchows Archiv 425:3–24

364. Darby I, Skalli O, Gabbiani G (1990) α-Smooth muscle actin is transiently expressed by myofibroblasts during experimental wound healing. Lab Invest 63:21–29

365. Vracko R, Thorning D (1991) Contractile cells in rat myocardial scar tissue. Lab Invest 65:214–227

366. Andersen HR, Maeng M, Thorwest M, Falk E (1996) Remodeling rather than neointimal formation explains luminal narrowing after deep vessel wall injury. Circulation 93:1716–1724

367. Scott NA, Cipolla GD, Ross CE, Dunn B, Martin FH, Simonet L, Wilcox JN (1996) Identification of a potential role for the adventitia in vascular lesion formation after balloon overstretch injury of porcine coronary arteries. Circulation 93:2178–2187

368. Shi Y, Pieniek M, Frad A, O'Brien J, Mannion JD, Zalewski (1996) Adventitial remodeling after coronary arterial injury. Circulation 93:340–348

369. Shi Y, O'Brien JE, Fard A, Mannion JD, Wang D, Zalewski A (1996) Adventitial myofibroblasts contribute to neointimal formation in injured porcine arteries. Circulation 94:1655–1664

370. Shi Y, O'Brien JE, Mannion JD, Morrison RC, Chung W, Fard A, Zalewski A (1997) Remodeling of autologous saphenous vein grafts. The role of perivascular myofibroblasts. Circulation 95:2684–2693

371. Suzuki T, Kim H-S, Kurabayashi M, Hamada H, Fujii H, Aikawa M, Watanabe M, Watanabe N, Sakomura Y, Yazaki Y, Nagai R (1996) Preferential differentiation of P19 mouse embryonal carcinoma cells into smooth muscle cells. Circ Res 78:595–404

372. Mintz GS, Popma JJ, Pichard AD, Kent KM, Satler LF, Wong C, Hong MK, Kovach JA, Leon MB (1996) Arterial remodeling after coronary angioplasty. Circulation 94:35–43

373. Post MJ, Borst C, Pasterkamp G, Haudenschield (1995) Arterial remodeling and restenosis: a vague concept of a distinct phenomenon. Atherosclerosis 118(suppl):S115–S123

374. Schwartz SM, Reidy MA, deBlois D (1996) Factors important in arterial narrowing J Hypertens 14(suppl5):S71–S81

375. Blank RS, Swartz EA, Thompson MM, Olson EN, Owens GK (1995) A retinoic acid-induced clonal cell line derived from multipotential P19 embryonal carcinoma cells expresses smooth muscle characteristics. Circ Res 76:742–749

376. Slomp J, Gittenberger-de Groot A, Glukhova MA, van Munsteren JC, Kockx MM, Schwartz SM, Koteliansky VE (1997) Differentiation, dedifferentiation, and apoptosis of smooth muscle cells during the development of the human ductus arteriosus. Arterioscler Thromb Vasc Biol 17:1003–1009

377. Patapoutian A, Wold BJ, Wagner RA (1995) Evidence for developmentally programmed transdifferentiation in mouse esophageal muscle. Science 270:1818–1821

378. Asahara T, Murohara T, Sullivan A, Silver M, van der Zee R, Li T, Witzenbichler A, Shatteman G, Isner JM (1997) Isolation of putative progenitor endothelial cells for angiogenesis. Science 275:964–967

379. Nadel EG (1995) Gene therapy for vascular diseases. Atherosclerosis 118(suppl):S51–S56

380. Sirois MG, Simons M, Edelman ER (1997) Antisense oligonucleotide inhibition of PDGFR-β receptor subunit expression directs suppression of intimal thickening. Circulation 95:669–676

381. Morishita R, Gibbons GH, Ellison KE, Nakajima M, von der Leyen H, Zhang L, Kaneda Y, Ogihara T, Dzau VJ (1994) Intimal hyperplasia after vascular injury is inhibited by antisense cdk 2 kinase oligonucleotides. J Clin Invest 93:1458–1464

382. Chang MW, Barr E, Lu MM, Barton K, Leiden JM (1995) Adenovirus-mediated over-expression of the cyclin/cyclin-dependent kinase inhibitor, p21, inhibits vascular smooth muscle cell proliferation and neointima formation in the rat carotid artery model of balloon angioplasty. J Clin Invest 96:2260–2268

383. Nabel EG, Yang Z, Liptay S, San H, Gordon D, Haudenschild CC (1993) Recombinant platelet-derived growth factor B gene expression in porcine arteries induces intimal hyperplasia in vivo. J Clin Invest 91:1822–1829

Springer
and the
environment

Springer